Inspections and Reports on Dwellings

Inspections and Reports on Dwellings is a three-volume series that comprehensively explores the process of independent professional home assessment required for the purchase of residential property. This fully updated second edition of *Inspecting* retains a focus on the needs of the surveyor to recognise and interpret the significance of observations on site whilst also updating the market context within which surveyors and valuers are operating.

Inspecting includes a consideration of the important benchmarking by the Royal Institution of Chartered Surveyors (RICS) of three distinct survey service levels for independent surveyors and a review of the wider choice of survey options professional surveyors can now offer to potential clients in addition to the RICS Home Survey range. There is additional content on preparing for the inspection and on reporting, and there are expanded or completely new sections on a variety of subjects such as conservatories, renewable energy technologies, and innovative techniques and forms of construction. With over 500 colour illustrations and an enhanced structure, the new edition reflects the very latest approach to inspecting and reporting on services, risk and legal matters.

This book is essential reading for all those engaged in inspecting dwellings, whether experienced, newly qualified or studying for appropriate qualifications to become members of professional institutions.

Philip Santo FRICS has over 35 years' experience as a residential surveyor and valuer. As a consultant for RICS he has contributed to numerous important Guidance Notes and Information Papers. Regularly contributing to the RICS *Property Journal*, he writes the acclaimed *Case Notes* series, and his photographs often feature in professional publications.

Inspections and Reports on Dwellings

Inspecting

Second edition

Philip Santo FRICS

*First edition published 2006
by Estates Gazette*

Routledge
Taylor & Francis Group

LONDON AND NEW YORK

First edition published 2006 by Estates Gazette

Second edition published 2016
by Routledge
2 Park Square, Milton Park, Abingdon, Oxon OX14 4RN

and by Routledge
711 Third Avenue, New York, NY 10017

Routledge is an imprint of the Taylor & Francis Group, an informa business

© 2016 new edition material, Philip Santo; previous edition material, Ian A. Melville and Ian A. Gordon

British Library Cataloguing in Publication Data
A catalogue record for this book is available from the British Library

Library of Congress Cataloging in Publication Data
Melville, Ian A.
  Inspections and reports on dwellings. Inspecting / Ian A. Melville, FRICS, and Ian A. Gordon, FRICS . -- Second edition / Philip Santo, FRICS.
  pages cm.
  Includes bibliographical references and index.
  1. Dwellings—Inspection. 2. House construction—Standards. I. Gordon, Ian A. II. Santo, Philip. III. Title.
  TH4817.5.M45 2016
  690'.21--dc23
  2015016991

ISBN: 978-0-080-97131-5 (pbk)
ISBN: 978-0-080-97143-8 (ebk)

Typeset in Bembo and Univers
by Florence Production Ltd, Stoodleigh, Devon, UK

For my children, James and Mary,
not only treasured blessings since birth,
but close friends, good companions,
and a continuing source of parental pride and joy.

Contents

Photographs

Listed by chapter, section and photograph number.

Chapter 17: External joinery

Chapter 18: Roof space and pitched roof structures

Chapter 27: Garages, conservatories and outbuildings

Chapter 28: Gardens and boundaries

Chapter 29: Risks

Chapter 30: Matters for legal advisers

Figures

Listed by chapter, section and figure number.

Boxes and tables

Listed by chapter and section, with box or table number, title and page number.

Preface

This book and its companions, *Inspections and Reports on Dwellings: Assessing Age* and *Reporting for Buyers*, are successors to a series written by Ian Melville and Ian Gordon in the expectation of a fundamental change in the home-buying process: the requirement by law for vendors to commission and provide for prospective buyers a Home Information Pack which would include a Home Condition Report in England and Wales, and a Home Report in Scotland, to include a Single Survey with valuation. In the event, only in Scotland were the changes implemented.

The need to remove references to the ill-fated Home Information Packs (HIPs) during the preparation of this edition was self-evident, but what quickly became apparent was the need to reflect how far the residential surveying profession itself has moved since the abandonment of HIPs in 2010. The most significant changes have been a consequence of RICS (Royal Institution of Chartered Surveyors) focussing particular attention on the housing sector and the important role residential surveyors and valuers have always played in this major part of the national economy. The only RICS-branded residential survey product previously available, for the 30 years prior to 2010, has been completely overhauled and transformed into a suite of residential survey products, with further enhancements anticipated in the future. These offer differing levels of surveys and reports to meet varying client needs and have a much clearer focus on communication. Equally significantly, the third edition of the RICS Guidance Note *Surveys of Residential Property*, published in 2013, sets out defined levels for residential survey inspections across the profession. These important changes have established benchmarks which will apply to residential surveyors for the foreseeable future, and it has been necessary to reflect them with a new opening chapter.

After the necessary updating, a key objective for this edition has been to make the content very much more accessible. It is recognised that many will use the book for reference purposes so the contents have been reorganised into separate chapters and clearly signposted. Although some first edition photographs are now showing their age, it has been possible to retain the majority, but this edition has nearly 200 new photographs. Another major improvement has been to list the 500 plus images by chapter to assist identification and location.

It may once have been the case that the bricks and mortar of buildings were a largely unchanging background to the residential surveying profession. Now, however, new technologies, in building services as well as in building construction, are imposing ever-greater demands on practitioners while the regulatory environment is similarly subject to rapid change. As the areas requiring investigation and consideration by the surveyor continue to develop, the surveyor's approach to the inspection equally needs to be refined. In the light of these challenges, the chapters dealing with services have been largely rewritten and new chapters have been added on subjects including risk and innovative construction while other parts of the original text – for example, on sustainable energy generation and on garages and outbuildings – have been greatly expanded.

Those familiar with the first edition may notice some omissions. The ease with which the internet now allows access to information sources, such as the Approved Documents to the Building Regulations, has reduced the need to simply reproduce technical data, which previously occupied significant space. Some of the content of the first edition, such as the historical background of brick making or the national sources of natural stone, while of interest, relate less to the day-to-day work of most practitioners than other topics for which they have made way. Certain areas, such as the identification of trees, are probably better served either with a pocket book on site, an internet search or a reference book in the office. Nevertheless, it would be regrettable to lose much fascinating material so selected sections of the original text remain available as downloads from the companion website to this book (www.routledge.com/cw/santo).

It is a measure of Melville and Gordon's longevity that I encountered their book *Structural Surveys of Dwelling Houses* when I was first employed in the mid 1970s. It became a constant and irreplaceable point of reference in my work as a residential surveyor and valuer. Indeed, the same much-thumbed copy still sits within reach of my desk. To follow in the footsteps of Melville and Gordon was a daunting prospect and it was with some trepidation that I embarked on the challenge of producing this second edition. Despite the many changes, the majority of this book derives directly from the first edition and I have strived to retain its essence. It is my hope that it will continue to provide the same level of support and assistance to practising professionals as have, for many decades, the many books of Melville and Gordon, to whom I am pleased to acknowledge my personal gratitude.

PHILIP SANTO

Acknowledgements

The author is grateful to the following for permission to reproduce the copyright material set out below.

The Building Research Establishment

Photographs numbered:

145, 261.

Copyright IHS, reproduced by permission

Photographs numbered:

146, 252, 254, 260, 278, 280, 296, 300, 302, 303.

MIB Facades Ltd

Photograph numbered:

309.

The National Assembly for Wales

Photographs numbered:

294, 295, 369.

The Northern Consortium of Housing Authorities

Photographs numbered:

251, 253, 281, 285, 291, 292, 297, 299, 301, 393.

Paul Daniel

Photograph numbered:

259.

Peter Cox Ltd

Photographs numbered:

198, 427, 428.

RICS

Extracts from Guidance Note *Surveys of Residential Property*, 3rd edition, 2013, and Information Paper *Flat Roof Coverings*, 2nd edition, 2011.

Stewart Milne Timber Systems

Photographs numbered:

305, 306.

The Timber Research and Development Association

Figures 33–43.

Tom Littler

Photographs numbered:

15, 92, 413, 473–475, 482.

Unite Group plc

Photographs numbered:

311 and 312.

The following photographs are copyright Philip Santo:

2–10, 12, 16–18, 20, 26, 27, 30, 37, 39–41, 45, 47, 49–52, 54–57, 59–66, 68–77, 79, 80, 87–89, 93–102, 105, 108, 110–112, 115, 121, 122, 124–128, 132, 133, 135–137, 139, 141–144, 149–152, 155, 157, 159, 160, 162, 167, 169–173, 175, 178–181, 183, 187–189, 192–195, 197, 200, 203, 205, 207, 208, 210, 211, 214–217, 219–231, 234, 236–238, 242, 245–250, 256, 257, 262–266, 268, 269, 272, 273, 279, 282, 283, 286, 287, 289, 293, 298, 304, 307, 308, 310, 313–317, 319, 322, 323, 326, 327, 330, 332, 335, 336, 338, 339, 342, 345, 354–358, 360, 363, 364, 367, 368, 370–372, 374, 377, 378, 380, 381, 385, 388–390, 395, 398, 402, 404, 412, 416, 421, 423–425, 430, 432–438, 440, 442, 443, 446, 447, 449–451, 453–462, 464, 466, 471, 476–481, 483–506.

All remaining photographs are copyright Ian A. Melville and Ian A. Gordon.

Introduction

This is a book about inspecting residential properties. The context anticipated most is an inspection upon the sale of a dwelling to a purchaser who requires information about the condition of the property being purchased. While surveys of dwellings may be required for many reasons, these frequently involve technical investigations into specific aspects of a building that is already known to the client. By contrast, when undertaking a pre-purchase inspection, the surveyor must report on the whole of a building without knowing in advance what may be encountered. This means the surveyor must identify the form of construction, recognise and interpret often quite subtle clues pointing to the presence of defects, analyse the causes of those defects and make recommendations to remedy them. Additionally, there is an increasing expectation that the surveyor will also report on a range of other matters not directly related to the construction of the main building but which might affect its value or which may have a bearing on its future occupation by the new owner.

Traditionally, pre-purchase surveys provided the bread and butter of many urban practitioners in broad-based local general practices, perhaps operating from two or three offices. Since the 1970s, however, sweeping changes in the professional landscape, the demands of increasing specialisation, and opportunities provided by information technology have led to a divergence. In the main, residential practitioners now work either for large national surveying brands or as very small, often one-person operations, frequently operating from home offices. There are relatively few of the old-style medium-sized general practices now remaining.

In the same way, the traditional method of training new entrants to this sector of the profession was somewhat akin to an apprenticeship with the novice surveyor acting as an assistant to an experienced practitioner, absorbing much learning by observing and being mentored while on the job. Since the 1960s, the requirement for entrants to take academic routes into the profession has militated against this process, and one of the consequences has been a reduction in the numbers choosing this practice specialisation. Consequently, as many in the current generation of experienced residential surveyors reach retirement over the next decade, there is an increasing likelihood that there will be a shortage of suitably experienced younger surveyors ready to take their place. Fortunately the pendulum is now starting to swing back as the benefits of practical learning while working, especially in the residential sector, are being recognised. Likewise the abilities of those in associated occupations are being recognised as providing a possible route into the profession.

This book is not intended to replace or duplicate the many excellent technical publications that deal with particular aspects of dwelling construction or defects. Rather, it provides an overview of the whole process of inspecting a dwelling, especially when advising a client contemplating a purchase. It is written for a broad readership but, mindful of the changes in the profile of the profession, it will hopefully be of particular assistance to those who do not have the benefit of an experienced mentor immediately to hand.

PART 1

Inspections and inspectors

1

Inspections

Report types

The surveying landscape envisioned when the first edition of this series was published has been transformed in ways that could not have been anticipated. The last-minute decision in 2007 by the government of the day to cancel the much-heralded requirement for vendors to commission a Home Condition Report before marketing their property effectively holed their Home Information Pack (HIP) policy below the waterline. The sole remaining mandatory provision of an Energy Performance Certificate currently has little or no bearing on most people's purchase decisions, though this may change in time.

There is no prospect of the current government attempting to reintroduce any similar legislation for mandatory pre-purchase surveys in the foreseeable future, but it does share the widespread recognition across a range of influential national bodies that purchasers should seek professional advice from qualified surveyors before they commit themselves to buying a home.

In 2008 the consumer affairs magazine *Which?* reported that the cost of unexpected works found to be necessary by purchasers who had not commissioned a private survey averaged £2,500, with one buyer spending more than £10,000. Where purchasers had commissioned a private survey, 44 per cent managed to negotiate reductions in price and a further 10 per cent ensured that the problems were rectified before completing their purchases. Research by RICS in 2010 and 2013 produced similar findings. Consumer organisations such as the Consumers' Association, professional bodies such as the Law Society, the Council of Mortgage Lenders (CML) and the Building Societies Association (BSA) representing lenders and, unsurprisingly, RICS have all attempted to promote the take-up of pre-purchase surveys; but, with rare exceptions, market research consistently demonstrates that the proportion commissioning a private survey prior to making a purchase remains stubbornly at around one-fifth of all purchasers. Briefly, the majority of the house-buying public persist in the sadly misguided belief that instructing a private survey on their prospective purchase is not worthwhile, especially if they will be receiving a courtesy copy of a mortgage valuation from their lender.

After the failure of the Home Condition Report proposals, RICS developed a combination of updated and completely new report formats. Within the space of a few years, RICS established the RICS HomeSurvey licensed product range, intended to enable chartered surveyors to offer a variety of survey options with consistent branding. First, it redesigned the long-established mid-range HomeBuyer format to incorporate 'traffic light' style Condition Ratings. It then added a basic Condition Report and a high-end Building Survey with similar formats. Creation of a dedicated valuation module is likely at some stage to facilitate removal of the valuation element from the HomeBuyer Report. This would renew access to the HomeBuyer product for the many surveyors who are not also Registered Valuers and have been unable to use the format since RICS introduced Valuer Registration in 2011.

Needless to say, the newly developed RICS product range did not impress all within the profession. Some groupings such as the Independent Surveyors and Valuers Association and the Residential Property Surveyors Association have their own mid-range equivalent to the HomeBuyer Report and many practitioners continue to offer Building Surveys of their own design, saving on license fees and in most cases resisting the inclusion of colour-coded Condition Ratings. Whatever the shortcomings of the report formats themselves, however, the RICS range arguably offers a national benchmark against which all other reports and formats can be measured. Whether or not this was considered a good thing came down to a matter of opinion, but it became difficult to imagine a scenario where the courts would not be mindful of the RICS HomeSurvey range and the supporting Practice Notes and documentation when considering the competence or otherwise of a practitioner, whichever level of report or individual format had been adopted in the circumstances under the court's scrutiny.

Any remaining doubts about benchmarking were removed at the end of 2013 when RICS published a Guidance Note entitled *Surveys of Residential Property*, which defines and contrasts differing levels of survey services. This publication is distinct from the separate Practice Notes for each of the branded RICS products and also replaced the 2004 Guidance Note on *Residential Building Surveys*. It provides definitive guidance for practitioners who offer independent services and gives clear points of comparison between the different levels of surveys for members of the public and their advisers. Most importantly, for the first time it facilitates direct comparisons between competing survey products which may be available, and any previous uncertainties around differing levels of inspection and reporting have been removed. The principles outlined in this book, namely how the surveyor can ascertain forms of construction and identify defects, apply to all types of survey inspection; but the surveyor now also needs to ensure that the detailed level of the inspection and, especially, the content of the final report are appropriate for the level agreed in advance with the client. This new reality is reflected in the remainder of this chapter.

From its title, it should be clear that this book deals with the inspection part of this process and in the main it does not differentiate between the varying levels of report which an inspection might be addressing. Now that definitive guidance has been made available, however, the surveyor must understand how it applies and be clear about the differing expectations which may be appropriate in varying circumstances. Consequently it is necessary to briefly discuss the inspection and reporting requirements that the residential surveyor must fulfil and how these vary, depending on which of the respective levels of report ultimately needs to be completed. Even before that, however, it is helpful to understand how the profession has reached this point; and there is no better place to start than to consider the basic and crucial, but often misunderstood, distinction between a valuation and a survey.

Valuation or survey?

At one end of the spectrum of residential property inspection requirements is the need for a basic valuation. This might be for any one of many purposes including presale advice from an estate agent, a valuation for civil purposes such as a matrimonial dispute, or taxation reasons such as a probate valuation. The inspection for such a valuation report will usually be completed in well under an hour and will involve a valuer walking around the dwelling to review its accommodation and note its overall appearance and superficial condition insofar as these might affect the value, but there will be no investigation or analysis of the building structure itself. The report will be short and often follow a prescribed format. At the opposite end of the spectrum is a building survey, involving a detailed technical examination of the structure, often taking many hours. This will usually be undertaken on behalf of a prospective purchaser who will receive a lengthy and detailed written report and possibly a preliminary verbal report of the main findings.

Most valuations will fall within the requirements of the RICS *Red Book*, which defines the matters that need to be taken into account and those which need to be reported. Although they also need to comply with the *Red Book*, valuations undertaken for mortgage purposes are regarded as a special case. The standards for mortgage valuations are set out in the current global and UK editions of the *Red Book*. In the 2014 edition, the relevant sections were UK Valuation Standard 3, 'The Valuations of Residential Property', and UK Appendix 10, 'RICS Residential Mortgage Valuation Specification'. Similar types of valuations, such as valuing for buy-to-let mortgages, were covered by UK Appendix 11, 'Application of the RICS Residential Mortgage Valuation Specification to Related Purposes'.

The RICS 'Residential Mortgage Valuation Specification' prescribes a series of assumptions that the valuer can make without further investigation and without reference to the client when valuing for a lender. This eliminates the need for formally confirming instructions and simplifies and speeds up the process of inspecting and reporting. There is a misconception that the 'Residential Mortgage Valuation Specification' has been negotiated and agreed with lenders, if not actually imposed by them. The reality is that the final decisions about the Specification are made by RICS alone, although naturally it informs and consults lenders through their representative organisations, the CML and the BSA, and indeed also consults its own RICS members prior to implementing any changes.

The most significant change to the Residential Mortgage Valuation Specification in recent years was the removal in 2011 of the mandatory require-ment for a 'head and shoulders' inspection of the roof space. Controversial at the time of its introduction in the early 1980s, this removal was greeted with mixed reactions from lenders as well as valuers, but one of the main rationales for removing the requirement was to emphasise even more clearly the difference between a valuation and a survey.

The terms 'valuer' and 'surveyor' tend to be used without differentiation by those within the profession as well as outside, but a clear distinction ought to be made. A *valuation* is not the same thing as a *survey* of a property. In the same way, the activities and outputs required of a *valuer* are not the same as those of a *surveyor*. Admittedly, one individual, during one visit to a property, may complete both a survey and a valuation, but the two activities are distinct and the unfortunate blurring of the two roles has been compounded during recent decades, not least by the profession itself.

As already noted, a valuation is essentially based on a walk-through type of inspection, whether for sale, probate, matrimonial or other purposes. Any serious matters impacting on the value will be reflected in the valuation and, at times, this may include significant structural defects noted in the course of the valuation inspection. In general terms, a mortgage valuation is also essentially a walk-through inspection, hence the exclusion of the 'head and shoulders' roof space inspection from the 'Residential Mortgage Valuation Specification'. A valuer will conduct a valuation inspection to produce only a valuation. If a report on the condition of the building's structure is required then a survey should be commissioned. This clear and unambiguous message should be understood and emphasised by all who advise purchasers, not only property professionals but also legal and financial advisers. Purchasers cannot rely on the valuation inspection and report for information on the condition of a building. If information on condition is considered important by a purchaser – and why would it not be? – the purchaser must commission a survey.

Why has it become so confused?

Most people require a mortgage to complete a house purchase, but until 1980 the mortgage valuation report commissioned by the prospective lender was considered confidential to the lender and the contents and recommendations were not disclosed to mortgage applicants. The only part of the report released to borrowers was details of any repair or improvement works reported by the valuer to the lender as essential to preserve the property and subsequently imposed by the lender on the borrowers as a condition of granting the mortgage loan. Those buying a property had the simple choice of completing their purchase without the benefit of any guidance other than that which might be gleaned from friends and acquaintances or commissioning a detailed and relatively expensive building survey, known at the time as a 'structural survey'.

After 1980, recognising the public relations issues around collecting valuation fees from their mortgage applicants and not sharing the results with them, lenders started to release courtesy copies of their mortgage valuation reports. Additionally, recognising the marketing opportunity of offering additional services to their customers, some lenders also began to offer a more detailed inspection and report than the basic mortgage valuation, but one which came in a prescribed format that was both cheaper and less detailed than a bespoke structural survey. This development was rapidly followed by RICS issuing their own intermediate level of survey report, the RICS House Buyers Report and Valuation, and other alternative mid-range report formats also followed. Over the ensuing quarter of a century, developments in surveying practice, decisions in the courts and changes in public expectations resulted in much confusion about the respective roles of the valuer and the surveyor in the house-buying process.

As soon as mortgage valuation reports started being released to applicants, and especially after the decision in *Smith* v *Bush* 1987, mortgage valuers came under conflicting pressures when completing their reports. Lenders still needed to know of any defects significant enough to affect their lending decisions, but purchasers increasingly expected to be told about anything that might affect their purchase decision. Valuers became increasingly uncertain about their legal liabilities and tended to include more and more detailed comments about defects that weren't strictly related to the mortgage application and lending process. For an extended period, mortgage valuation reports received by some house purchasers virtually amounted to a mini survey, but within the format and for the price of a mortgage valuation. This contradictory and plainly unsustainable position seriously blurred the distinction between a mortgage valuation and a survey, not least in the eyes of an increasingly litigious public.

The dangers of this trend were eventually recognised, and increasingly as the new century approached, lenders tried to educate their valuers to differentiate between defects that a lender needed to know about and those which, while possibly of interest to a purchaser, were irrelevant to the lending decision and should not be reported with the mortgage valuation. At the same time, advances in information technology meant that lenders were increasingly streamlining internal processes. Standard paragraphs were introduced to improve consistency in valuation reports and, often, much of the mortgage valuer's descriptive role in reporting became limited to choices between preset options and ticking boxes. While undoubtedly improving lenders' internal processes, these changes meant that scope for individual valuer comment in lender report forms was progressively reduced.

While expectations around mortgage valuations fluctuated during the 1980s and 1990s, the world for surveyors carrying out building surveys or intermediate-level reports for private purchasers was rather more stable. A building survey report had no standard format and would typically follow a layout established by a particular surveyor or firm over many years; however, it would probably be markedly different from one completed by an equally competent surveyor from a different firm in the same locality. Some reports would deal with a property room by room and others, element by element. Some would have an executive summary at the front, some would include a valuation, and some would estimate the likely cost of remedying defects; but many would have none of these items. There was no standardisation and no national standard to suggest what a report might contain or what it might look like. Indeed there was no guidance at all on the subject from RICS until 1981. Even after that, surveyors continued to rely to a large extent on their understanding and interpretation of established practice combined with advice from a gradually increasing number of independent books from experienced and respected practitioners sharing their own thoughts on the subject. This followed the lead originally set in 1964 by Ian Melville and Ian Gordon with their seminal *Structural Surveys of Dwelling Houses*.

Unlike building surveys, the various types of intermediate-level reports were designed to be completed in a particular way and on purpose-designed forms. In the days before word processors, the text relating to specific sections needed to be fitted into fixed spaces on preprinted sheets. When extra text was required, it had to be carried over on to continuation sheets, making the reports awkward to produce and unhelpful for the client reading the report. RICS had a standard format for this level of report, but different lenders also had their own designs which they linked to their mortgage valuation requirements. At the time, this was seen as adding value to the mortgage product offering, saving money for the mortgage applicant by killing two inspection birds with one surveyor stone. Unfortunately, it also compounded the wider lack of understanding among the house-buying public about the difference between a private survey and a lender valuation.

Towards the end of the 1990s and in the early years of the new century, the need to make a clearer distinction became more widely recognised and most lenders separated the mortgage valuation element from anything required privately by the purchaser. At the same time, the political desire for a vendor survey began to gather momentum, and gradually, ways of bringing this into effect were explored. Despite serious reservations in many quarters, legislation was enacted and the matter pursued until a last-minute realisation in 2007 that the Home Condition Report proposals were unworkable and the HIP scheme suffered an ignominious collapse.

Freed from the imposition of a politically imposed process across the residential property market, but conscious of the need to be more consumer focussed, RICS gradually started to develop its single in-house survey product into a broader portfolio that could be offered as a range of options to home purchasers. The intermediate-level survey was relaunched as the RICS HomeBuyer Report, and this was followed by the lower-level RICS Condition Report. When the higher-level RICS Building Survey was launched in 2012, the final part of the portfolio was put into place, and for the first time, the RICS HomeSurvey products provided members with a full range of RICS-branded residential surveys to offer their clients.

In the longer term, RICS intend that their branded HomeSurvey range will be extended by the addition of dedicated modules to meet specific client needs. This will enable purchasers to opt for additional modules, beyond the basic survey, that relate to their particular areas of concern or interest, perhaps on topics such as flooding, contaminated land or energy sustainability. The resulting report will then be capable of meeting the needs of individual clients rather than requiring the surveyor to cover all areas in the same level of detail in every report, whether the client wants this or not.

The difference between surveys and mortgage valuations was made more distinct when RICS undertook a fundamental revision of the Residential Mortgage Valuation Specification and excluded the mandatory roof space inspection in 2011. Finally, in 2013 the Guidance Note *Surveys of Residential Property* was published, which for the first time defined three levels of survey service, setting out the professional requirements for each. At last, formal documentation intended to overcome the long-standing lack of clarity had been put in place, although it would take something more than an optimist to believe that the problem, especially that of public perception, would no longer exist.

Scope and method of inspection

The latest RICS guidance is intended to be a flexible framework in which practitioners can develop and offer their own survey services at all levels while providing clarity so that consistent quality standards in residential property surveys offered by RICS members can be protected and maintained. Surveyors using the RICS licensed product range have always had detailed guidance about the inspection process which must be complied with, but previously, there was no definitive comparison between different survey types for practitioners producing non-branded reports. One of the most important elements of the RICS guidance is that it specifies in some detail the activities which a surveyor is expected to undertake on site and aligns those activities with respective levels of reporting. It is now necessary for every surveyor offering residential property surveys to be familiar with this guidance and to ensure that the services being offered are fully aligned with its requirements or, where there are any variations, that these are clear and, where appropriate, fully explained to the client.

The guidance is not prescriptive and there is no need to repeat its detail here, but one fundamental principle is worth emphasising: the importance of being systematic during the inspection. In time, each surveyor adopts an inspection regime to suit his or her individual preferences, but one of the most essential disciplines for all residential surveyors is to adopt and maintain a consistently methodical approach to each inspection. It is good practice to walk through the property briefly before starting the detailed inspection in order to note the arrangement of the accommodation and gain an initial impression of its condition. After that, some surveyors prefer to inspect the exterior before interior; others, the reverse. When outside, some will go around the property several times, inspecting a different structural element each time: the roof, the walls, the windows, and so on. Others will stand in front of each elevation and, starting with the roof and working downwards, make notes about all of the visible elements before moving around to the next elevation and repeating the process.

When inside the property, some like to start in the roof space and work down; others start at the bottom and work their way upwards. Especially at larger properties, a methodical approach is required to ensure that all rooms are inspected. Working up or down, floor by floor and always moving to the next room on, say, the left usually achieves the objective in all but the very largest houses. The need for a methodical approach extends to inspecting each room to ensure that the appropriate level of inspection has been applied to the ceiling, all walls, windows, doors, the floor, and any fittings or services. In a modest dwelling, this may not be very difficult, but in larger properties, it is not unusual for rooms to have several windows, intercommunicating doors, multiple built-in units and a great deal of furnishing to obstruct the surveyor's view. Having a methodical personal inspection routine, refined on numerous smaller properties, will reduce the risk of missing something when faced with this challenge.

When it comes to note-taking, some advocate recording on a room by room basis while others separate their observations between the respective elements of the building. Each method has its devoted adherents and there are advantages and disadvantages to both. The most important thing is to stick to the preferred inspection and note-taking routine and whenever that is not possible, as will inevitably happen from time to time, to make appropriate adjustments to ensure that nothing is missed.

Many surveyors like to sketch floor plans showing the relative arrangements of the rooms, on which notes can be made about, for example, where dampness has been found or where furnishings restricted access. A sketch plan showing the external plan of the building and the site as a whole is most strongly recommended. This should include the locations of garages and outbuildings, drain inspection chamber covers, confirmed or supposed drain runs, the relative positions of soil, waste and rainwater pipes and gullies, the types and proximity of trees, and a variety of other information observed during the inspection.

It is worth emphasising that the surveyor's site notes are not merely an *aide-memoire* for use in the preparation of the report but are also contemporaneous evidence of the actions and observations made during the inspection. They are the first port of call if any query about the report should arise at any time in the future, and it is essential that the site notes are complete and accurate. Ultimately, they should be capable of unambiguous interpretation by a non-surveyor because if the worst comes to the worst and a complaint or legal action passes beyond the immediate control of the surveyor who carried out the inspection, it is likely that the assessment or judgement regarding the surveyor's inspection, analysis or report on the property will be made by someone who is not a surveyor.

The surveyor should ensure not only that every appropriate part of the property is seen during the course of the inspection but also that the relationship of each part to every other part is considered so that, ultimately, the building has been appraised and assessed as a whole. The order in which the inspection is carried out is less important than ensuring that every part of the property has been adequately inspected to meet the requirements of the service being provided for the client (1).

The writer will not attempt to convince readers of the undoubted merits of his own particular procedure but will share one hard-learned lesson about the inspection process – while still physically on site, the surveyor should attempt as far as possible to understand everything about the building and to decide, at least in principle, what the report will say to the client. That may sound rather obvious, but it is very easy to finish the inspection and leave the site while still uncertain about the cause of a defect. It is even easier to leave while unsure what recommendation will be made to the client about some particular aspect. Experience has unequivocally shown that it is far more beneficial to spend an extra few minutes on site looking around the actual building and attempting to get to the bottom of a particular problem, or at least determining exactly where the uncertainties lie, than it is to try to reach a resolution some hours or days later using personal recollections of the inspection, site notes and photographs.

By the same token, when identifying and recording a defect in the site notes, it usually takes only a moment or two to jot a further note about an appropriate recommendation to include in the report. If the site notes do not include any mention of recommendations along with the defects then it is inevitable that, when it comes to compiling the report, the surveyor will periodically struggle to recall the precise point being made in the original note, with the risk that the report will not contain the right comment.

The RICS guidance rightly states the importance of leaving sufficient time for reflective thought between inspecting the property and completing the report, cautioning against dictating the report during the inspection as this does not allow time for the reflective process.

During this period of sometimes deliberate but often unconscious reflection, pieces of the inspection jigsaw will slip into place and conclusions will crystallise in the surveyor's mind. Occasionally, a warning bell will sound and the thought process will take an unexpected turn. These post-site reflections, however, should be based upon a thorough inspection which has left as little uncertainty as possible. This does not mean that ideas and thoughts about recommendations will not change; it is simply that conclusions are usually easier to reach on site. Thoughts and decisions made during the inspection can be changed on reflection, but a lack of analysis and an absence of decisions during the site inspection increases the risk that the final report will not be soundly based.

The recent guidance also includes some more subtle changes in focus compared with earlier versions which, while relating more specifically to the final report, need to be borne in mind by the surveyor throughout the inspection. It is a requirement for the surveyor, first, to make a clearer distinction between those defects which are of greater and lesser significance and, second, to identify risks that the client should be aware of; that is, matters which, while not necessarily defects, might have an impact on the life of someone living at the property. The rationale for introducing these changes is clear enough: it is a response to the question, what would anyone contemplating a purchase reasonably wish to know about? The primary need is, naturally, a clear pointer to the most serious defects rather than a full but undifferentiated list of all defects. One way of achieving this would be by introducing an objective means

1 There may not be time for a preliminary drive past look at the dwelling before formulating instructions but there are occasions when the surveyor would wish the opportunity had been taken. It also seems unbelievable that a client would not have mentioned the vegetation, but this is perfectly possible if it is considered merely as a decorative adjunct. The surveyor is left with no option but to give warnings about the problems likely to occur and provide comments on the little which can be seen of the roof and walls.

of categorising defects. The use of Condition Ratings (discussed in the next section) is one method, but other ways of making the necessary distinctions are just as valid. The second requirement is to report matters of 'risk' which might not be apparent to a prospective purchaser when viewing a property but which might be significant when in occupation. These risks could include external matters such as noise or smell nuisance from nearby or matters within the property such as potential dangers, perhaps large gaps in staircase balustrades or doors glazed without safety glass. In the past, reporting of some such risks has usually been only within the relevant sections, and commenting on others may not have been considered part of the surveyor's remit. The recent guidance places more of an onus on surveyors to recognise that such matters are likely to be significant to their clients, and rather than potentially having such important information lost in the detail of reports, the guidance suggests that there is a case for these to contain a dedicated 'Risks' section that can be used to summarise the most important areas of risk identified during the inspection.

Reporting and Condition Ratings

While the RICS Guidance Note *Surveys of Residential Property* clarified the distinction between the three levels of survey services, the earlier development of the RICS HomeSurvey range introduced arguably the most significant change to reporting practice since pre-formatted survey reports were launched in 1980: the use of Condition Ratings. Before comparing the three different levels of inspection and reporting which surveyors might be required to undertake, it is worth reviewing how Condition Ratings fit into the process.

Users of the RICS HomeBuyer Report have adapted to the various revisions in its format over the years. The 2010 version introduced the concept of Condition Ratings, which requires the surveyor to allocate a rating to define the condition of each of the main construction elements of the dwelling being inspected. RICS report forms use a traffic light style, with a green-coloured numeral '1' for elements where nothing more than normal maintenance is required and an amber-coloured '2' where repairs are required but the defect is not urgent or serious. Where a defect is serious or requires urgent attention, it warrants a red '3'. A fourth rating, 'NI' (meaning Not Inspected), is given to parts of the dwelling usually included but where an inspection was not possible. The supporting Practice Notes set out in some detail how the ratings should be decided.

This innovation was greeted with mixed feelings by practitioners. Some recognised the advantage of focussing the client's attention on the major areas of concern but others were concerned about the risks of 'dumbing down' the report content, potentially reducing the value of both the report and the surveyor's professional expertise in the eyes of the client. The cynic might suggest that this was more to do with some surveyors' high regard for their own work and their reluctance to change their established practices than any belief that this might be an inappropriate or ineffective way to improve communications with the client. In fact, feedback from clients who received reports with Condition Ratings was consistently positive, so when the Condition Report and Building Survey were subsequently introduced the concept was extended across the whole RICS HomeSurvey suite.

The assessment of Condition Ratings does place additional demands on the surveyor, both on site and when completing the report, but it is a discipline which should be viewed positively. Not only does it aid clarity in communication but it also challenges woolly thinking and any tendency to be indecisive when reporting to the client. It is not unusual for the surveyor to encounter situations where the underlying cause of some problem or particular set of defects is not clear. It may well be appropriate to set out the various options when reporting to the client, but simply leaving matters there is not enough. The client needs guidance on what, if anything, ought to be done to resolve any uncertainty. Likewise, it is all too often that survey reports with accurate observations and incisive technical analysis stop short of providing the client with clear guidance on the significance of what has been described and what steps should be taken as a result. How serious is the matter, how urgent is it and what are the risks of doing nothing? The discipline imposed by the Condition Rating format requires the surveyor to make a decision on these points and thereby help the client to understand their significance, even if the client does not grasp the technical subtleties of the matter described in the report.

The consequence for many surveyors is that defects now tend to be mentally categorised according to where they come on the scale of Condition Ratings, almost irrespective of the type of survey being undertaken. The ratings themselves tend to be independent of the type of report; the difference for the client is the way in which the different levels of report are framed around the results of the inspection and the manner in which the construction details giving rise to those Ratings are reported and explained.

Each of the three levels of survey service will meet differing client needs and, although there are clearly many overlaps, each does require a different approach by the surveyor. Perhaps the easiest way to appreciate the difference between the service levels is to consider the requirements for the most popular version, the intermediate Survey Level Two service which is exemplified by the RICS HomeBuyer Report, and then see how that differs from the more basic Survey Level One, such as the RICS Condition Report, and the top tier Survey Level Three, exemplified by the RICS Building Survey.

The Survey Level Two service – e.g. the RICS HomeBuyer Report

The Survey Level Two service is intended for a broad range of conventionally built properties. Without being too rigid, the Practice Note for the RICS

HomeBuyer Report describes the types of property that are considered most suitable for this style of report. Essentially it is intended for houses, bungalows and flats, including conversions, which are of Victorian to present-day construction, 'conventional in type and construction' and in apparently reasonable condition (**2**). While this range encompasses more than 95 per cent of the national housing stock, the qualifier 'conventional in type' can be understood to exclude properties which are especially large or unusual in some way; for example, where substantial alterations have been carried out to the original structure or where multiple extensions have been added at different times. The requirement to inspect, analyse and report appropriately on such buildings would necessarily deviate from the intended concise nature of the service.

It is also acknowledged that to some extent the limitations of suitability will be defined by the knowledge and experience of the individual surveyor offering the service. Some will feel more able to report, within this format, on certain ages or types of properties than others. Surveyors with particular areas of expertise or experience in less common construction types – such as local vernacular forms, post-Second World War prefabricated properties, or construction using new and developing technologies and materials – are permitted to use the format as long as the inspection and report still conform to the specific requirements set out in the *Description of the RICS HomeBuyer Service*.

The Survey Level Two service is intended for lay clients who want a professional opinion at an economic price, and it is deliberately less comprehensive than the Survey Level Three service or Building Survey. It focusses on an assessment of the general condition of the main elements of a property, identifying and evaluating the particular features that affect its present value and may affect its future resale. The inspection is not exhaustive and the surveyor does not carry out any tests. Areas such as the roof space or underfloor areas, which occupiers might be expected to open up and use, should normally be inspected; but there remains a risk, which the client must accept, that certain defects may not be found that would have been revealed by more thorough inspection. Where there is a 'trail of suspicion', however, as made clear in the judgement in *Roberts v J. Hampson & Co.* 1988, the surveyor 'must take reasonable steps to follow the trail'. This may include either extending the inspection or recommending further investigation, or doing both.

In practical terms, this level of inspection involves entering accessible roof spaces and visually inspecting the roof structure, although access panels are not opened if they are secured. Insulation inside the roof space is not lifted and stored goods and other contents are not moved. Areas vulnerable to deterioration and damage must be given particular attention, including appropriate checks with a moisture meter. In the living areas, the surveyor should open a representative sample of windows. Exposed floor surfaces should be closely inspected and floors assessed with a 'heel drop' test and using an appropriately sized spirit level, but floor coverings and furniture are not moved and fixed floorboards are not lifted; although where loose hatches, panels or floorboards are present, the surveyor should conduct an inverted 'head and shoulder' inspection of the subfloor areas without entering them. No tests are carried out to any of the services, installations or appliances, but accessible drain inspection covers should be lifted, where possible, to allow a visual inspection of the drain chamber. More detailed observations on inspecting and reporting on the services are made in Part 4 of this book.

The report on the Level Two inspection should provide an objective description of the various elements of the property and an assessment of the relative importance of defects or other matters found during the inspection along with brief advice on repairs and any maintenance issues. There is no requirement under the guidance to include a valuation at any level, but as with previous versions, the RICS HomeBuyer Report included a valuation when the 2010 version was released. It is likely that this will eventually be removed and made available as a separate optional module, but until this is done, the RICS HomeBuyer Report can only be carried out by those who are RICS Registered Valuers.

The meat of the RICS HomeBuyer Report is in the parts dealing with the different construction elements of the property. When the client reads the report, the Condition Rating forms the headline at the top of each elemental section of the report. For the surveyor, however, the allocation of the rating comes at the end of the investigation and analysis process. During this process, if there are a number of sub-elements under consideration within one section – for example, a series of different roofs to report on – the condition of each should be considered and rated separately. The worst rating allocated to any of the sub-elements is then used as the overall rating for that element in the report.

For each element of the property, after the headline Condition Ratings, the report includes a descriptive section. Here, the surveyor must, in turn, briefly describe the element, briefly summarise its condition and, if appropriate, briefly describe what must be done to remedy any significant defects in that element. At times, it will also be appropriate to describe the likely consequences of not carrying out the recommended remedial works. This sequence has been memorably described as providing the answers to the 'Four Whats'. These are the questions that a client is entitled to see answered in a survey report: What is an element made of? What is its condition? What should be done about it? What will happen if I don't?

Each section of a Survey Level Two service should seek to answer these questions as accurately and concisely as possible. The RICS *HomeBuyer Report* Practice Note suggests that this should be completed in four or five sentences for each element or sub-element. There seems no reason to think that a competent and experienced surveyor should not to be able to do this for the majority of elements and sub-elements for the majority of the time. On one hand, for anything other than a very straightforward dwelling, it is difficult to see the Four Whats being answered in fewer than four or five sentences.

On the other hand, while it is inevitable that some elements may occasionally require more text, if a report contains sections which are consistently longer then it is likely that the surveyor is not providing the concise summary-style report intended for the Level Two service.

Regrettably, it is all too often that examples of overlong HomeBuyer-style reports are encountered. It has been said that it is much more difficult to write a short speech or article than a long one, but the discipline of cultivating brevity is one which will ultimately reap rewards in improved productivity for the surveyor and in gratitude from the client that the report is not over-technical and can be readily digested.

If a surveyor is finding it difficult to compress the necessary content into this report format, it is likely to be for one of three reasons. Perhaps the property is too big, too old, too unusual or, for some other reason, too complex to allow an adequate report to be completed within the constraints of this format. The RICS guidance describes the normal limits of the Survey Level Two service, and if a property falls outside these limits then a Survey Level Three ought to have been commissioned in the first place. Hopefully, problems like these will be identified during the initial discussions with the client, but there will be occasions when the surveyor will be misled, usually inadvertently, by the client's description or simply caught out by something completely un-expected once on site. Sometimes it is possible to contact the client and seek instructions before proceeding further; but at other times, the surveyor will need to decide whether to bite the bullet and proceed regardless – assuming professional competence to do so – or alternatively to decline the instructions and, regrettably but inevitably, write off the time spent in preparation and travelling into the bargain.

A second reason why reports are insufficiently concise occurs when the surveyor is actually including details and commentary more appropriate to a Survey Level Three report despite ostensibly providing the Survey Level Two service. Surveyors guilty of this are doing themselves no favours; they are probably underselling their services and they are certainly providing more than their clients are entitled to receive under the, hopefully previously agreed, Survey Level Two terms and conditions. They are probably also taking much longer than necessary to produce their reports. Almost worst of all, they are doing a disservice to their fellow professionals because those surveyors who actually comply with the requirements of the Level Two service might seem, by comparison, to be providing an inferior product.

A third cause of overlong reports is simply an inability to report succinctly. Surveyors are not alone in falling prey to the supposition that clients equate report length with value. This is a fallacy. What the client values more than anything in a report is clarity, and in too many reports, clarity is obscured by unnecessary detail, lengthy description, needless technical information and superfluous general observations. This is one reason why the inclusion of Condition Ratings has received such a positive response from the market: no

2 A typical 1930s detached house – the sort of property many will regard as ideal for a Level Two inspection and report where the surveyor expects to examine an essentially straightforward structure but there is scope to analyse, report and recommend where considered necessary.

matter how unclear the surveyor's comments may be, the overall message is simply encapsulated in a coloured number which immediately indicates whether or not a particular element is a cause for concern.

Wanton verbiage can certainly be a problem, but all too frequently, what may seem to be vague and imprecise language is actually completely accurate use of language reflecting ideas that are vague and imprecise. A surveyor who remains uncertain about the cause of a defect or unsure about remedial action is unlikely to provide an incisive report. This uncertainty often manifests itself as a rambling discourse across a series of options without reaching any firm conclusions. The client is likely to interpret such vagueness as coming from a dithering surveyor engaged in back-covering rather than from a competent professional wrestling with a legitimate area of uncertainty.

There will be occasions when, even after post-inspection research and reflective thought, the cause of a particular problem or the most suitable remedy will remain unclear to the surveyor. There is nothing wrong with stating as much in the report. At the same time, though, the report should include guidance on the likely risks, and pointers should be given for the next steps to reduce the uncertainty. It is the surveyor's responsibility to report this clearly so that the client: understands what the surveyor has been able to conclude and where and why the uncertainty exists; knows what options for further action are available; and is in no doubt what course of action the surveyor recommends.

The Survey Level One service – e.g. the RICS Condition Report

The surveyor undertaking a Survey Level One service is required to make an objective assessment of the general condition of the property. The visual inspection for a Level One service is less extensive than for Levels Two or Three. There are three main differences between the practical steps the surveyor takes for the Level One inspection and those described previously for Level Two. The first is that the Level One inspection only requires a 'head and shoulders' inspection of the roof space from the access hatch, meaning that the roof space is not entered and the inspection is limited only to those parts of the roof structure and features which can be seen by the surveyor from the access hatch. Second, only the exposed surfaces of floors are inspected and there is no attempt to lift any loose hatches or boards to inspect any part of the subfloor area. Third, drain inspection covers are not lifted to inspect the chambers beneath. More subtle distinctions include opening only a limited sample of the windows whereas the Level Two inspection requires 'a representative sample'. When appropriate, however, there is still a requirement for the surveyor to take reasonable steps to 'follow the trail', in the same way as with the other levels of inspection.

Just as the Level One inspection is less extensive, the RICS Condition Report might simply be regarded as a cut-down version of the Survey Level Two RICS HomeBuyer Report. It has no valuation but it has a similar appearance, using essentially the same headings and including Condition Ratings. The reports at both levels include objective descriptions of each element and any defects that have been identified, together with an indication of their relative importance. The fundamental difference, which has a major bearing on how a client might be able to use a Level One report, is that the surveyor then makes no further comment. As the name of the RICS Condition Report implies, this solely describes the actual condition of each element whereas the Level Two report also goes on to outline what remedial measures are recommended for any defects and possibly adds warnings about the implications of not doing so. In short, only the first two of the Four Whats are answered by a Level One report.

It is not difficult to anticipate that most people, especially amongst prospective purchasers, would consider themselves seriously short-changed to receive a report without such important information. So who would require such a report and where is the market which this report format is intended to serve? Put another way, for which group of clients will information only about condition be sufficient for their purposes?

The first group is those who are buying a dwelling which has been only relatively recently constructed. This does not relate solely to newly built properties being sold to the first occupier, but also includes those built within the last few years that are being resold. There is unlikely to be very much wrong with such a property and usually purchasers are happy to proceed to completion having assumed that to be the case (**3**). This is not necessarily sound reasoning but, in truth, there is normally little to include by way of recommended remedial works in a Survey Level Two report on a recently built property. This often means the content of such a report can look rather skimpy, in turn risking the client feeling it was an expensive waste of time and money. The more concise Level One report provides an economical way for a client to check that a recently built property really is sound while ensuring that if any significant defects are present, they will be identified. As for any remedial works that are needed, the client will need to decide how to follow this up, but the astute surveyor will doubtless be alert to the potential for further business without compromising the basic principles of the report.

Increasingly over the coming years, there are likely to be more innovations in the construction and service provision of new homes as builders wrestle with the demands of the various codes that require sustainable construction. Moreover, some of these innovations will impose maintenance requirements on homeowners which, if neglected, could result in defects arising. Consequently, it is conceivable that the inspection and appraisal of nearly new dwellings will become an increasingly productive market for the services of residential surveyors, and the Survey Level One service could be an ideal vehicle for making inroads into this sector.

The second group likely to find the Level One service of benefit is not the purchasers but the vendors of residential properties. The rationale behind the ill-fated and politically driven HIP with its compulsory Home Condition Report was that the purchaser would be able to find out about the condition of a property from an independent report prepared for the vendor prior to the property going on the market. The argument was that prospective buyers could then make offers in full knowledge of the condition of properties, resulting in fewer sales falling through due to defects coming to light at a later stage during inspections carried out for a mortgage lender or in a private purchaser survey.

This is not the place to explore the numerous issues around the failure of the HIP proposal, but the principle behind the advantages of full knowledge is sound. The case for the vendor commissioning an independent survey prior to marketing has much merit, not least because the vendor is then able to pitch the price realistically to reflect actual condition. The vendor might even decide to undertake certain remedial works to eliminate potential obstacles to a sale. Either way, prior knowledge is preferable to the situation of problems arising unexpectedly at a later date, possibly necessitating renegotiation of the price or even threatening a sale that has been agreed.

A vendor contemplating a vendor survey will want something quick and economically priced and is unlikely to require all of the recommendations for remedial action and maintenance included in a Level Two report. The Level One report is ideal for their purpose and this is a largely untapped market offering great opportunities for residential surveyors.

A third group for whom the Survey Level One service is potentially useful is not even directly related to buying and selling. Many residential landlords, especially those with modest portfolios – perhaps acquired almost by accident during the collapse in the property market after 2007 – frequently organise the day-to-day management of their properties and tenancies without resorting to professional managing agents, but they rarely have the expertise necessary to deal with the longer-term maintenance needs of their properties. An economical way to manage this would be to have Survey Level One inspections and reports carried out periodically. This would quickly highlight serious matters requiring attention, identify areas of concern needing to be monitored or allow works to be planned to avoid future problems, all without the unwanted detail contained in a HomeBuyer Report.

Arguably the most powerful application of the Level One service, however – and one still not fully appreciated by many in the profession – is the marketing opportunity it can create during the preliminary discussion with the client. It is a fact that many residential surveyors are not noted for their intuitive application of sales techniques, probably because the necessary skills to be a good residential surveyor do not overlap closely with those of a good salesperson. However, the small independent practitioner in particular needs to convert enquiries into sales to ensure commercial survival. Experience has shown that during the conversation about survey options available to a potential client, there is great merit in having the Survey Level One service as the starting point in the range. The economical price is attractive, but once the limitations of the report are explained, the benefits of the more detailed Level Two service are much clearer and the cost is less of a barrier to commitment than when the Level Two service is the starting point and lowest-priced option. The natural inclination of most purchasers of any commodity to opt for a mid-price option also reinforces this process.

The advantage of providing this choice to clients was demonstrated when the Survey Level One RICS Condition Report was launched. Researching demand for the new survey product, and with the consent of their lender client, a national surveying organisation contacted prospective purchasers who had indicated on their application form that they were not planning to have a private survey and, during a short telephone conversation, simply described the three survey options available. After having the options explained, 25 per cent of those contacted changed their minds and decided to have a private survey, despite their previously stated intention. (As an interesting aside, not only is this a higher proportion than usually commission a survey but, by definition, those contacted excluded any who had already decided to have a private survey.) Only 10 per cent of those who changed their minds chose a Condition Report whereas 81 per cent opted for a HomeBuyer Report and a further 9 per cent decided to have a Building Survey.

The relatively low proportion of those opting for the Condition Report might initially be thought disappointing but in fact this outcome has a greater

3 This modern house – on a large estate with no known problems and completed only a few years previously – is unlikely to have any significant defects, so it is well suited for a Level One inspection and report. If anything of significance should be identified, the surveyor must avoid making recommendations within the report format, but there is no reason not to offer further advice as an additional service. The client has the reassurance of a professional inspection without the unnecessary expense of a more detailed inspection format that in all likelihood will have very little to report.

significance. All of the new instructions were generated from a single conversation and the mere presence of the economically priced Condition Report as the starting point contributed to the proportion of higher-value instructions. Marketing experience suggests that if the lowest-priced option had not been available, the total take-up from the two remaining products would have been lower. The higher-priced HomeBuyer Report would have been the entry-level product, which would have deterred a significant proportion of the purchasers who actually went on to select this as the middle option in a range of three. Enabling clients to choose from a range of products produces better results, and the Level One service completes the product range even if the actual take-up for this basic report is likely to be relatively limited.

The Survey Level Three service – e.g. the RICS Building Survey

While the Survey Level One service is ideal for the most straightforward properties and those where the client requires no advice on remedial work, the Level Three service is suited for exactly the opposite circumstances:

complex, old, unusual or large buildings and clients who want detailed analysis and reporting (**4**). The RICS guidance *Surveys of Residential Property* makes it clear that the inspection for the Survey Level Three service is longer, more detailed and more complex than for the Level One or Level Two services, and the detail and extent of the report is expected to be substantially greater.

The guidance makes an interesting distinction between the levels of expertise necessary for surveyors offering the various services. For all levels of report, the surveyor must have sufficient knowledge of the construction type and the area in which the subject property is situated, but those offering Level Three services are expected to have a broader and deeper technical knowledge in order to inspect and report on the more complex buildings, historic buildings or those in poorer condition. The respective Practice Notes make the same point, describing the RICS HomeBuyer as requiring the surveyor to have 'an adequate level of competence' while the RICS Building Survey requires 'a high level of competency'.

During the Level Three inspection, the surveyor opens and closes all windows rather than a sample, lifts the corners of loose and unfitted floor coverings and, as far as is practicable, removes lightly fixed floorboards to allow an inverted 'head and shoulders' inspection. If practicable, within defined limits, the surveyor also enters the subfloor area to carry out a more thorough inspection. The roof space is entered even if, unlike for Level One or Level Two inspections, this means removing lightly secured access panels. Another significant difference from the lower levels of inspection is that the services and appliances are observed in 'normal operation'. This typically includes operating lights, turning on water taps, flushing toilets and observing water flow within drain chambers, although formal tests are still not undertaken.

The surveyor is advised not to limit the time allocated for the Level Three inspection but to devote however long is required to carry out a careful and thorough inspection, recording the construction and any defects that may be evident. The guidance emphasises the need to consider the interdependence and interactions of building elements, such as the roof, walls and floors, so that parts are not merely considered in isolation.

In the report on the RICS Building Survey, as with the other report formats in the RICS portfolio, a Condition Rating is given at the start of each element section. The surveyor is then required to describe the individual elements of the property in sufficient detail to identify their construction, condition and location in order to give a balanced view of the property and to support the surveyor's judgement of the Condition Rating. The guidance requires the surveyor to describe the worst-rated element first and the best, last. If more than one element has the same worst rating then the one which presents the greatest problem – whether considered in terms of cost, access difficulty or future risk to the property – is described first.

In contrast to a Level Two report, which expects the condition of each element to be summarised in the matter of a few sentences, the *RICS Building Survey* Practice Note and the reporting requirements for Level Three Surveys described in *Surveys of Residential Property* each set out in some detail a list of the information the surveyor should include in respect of each element of the construction and services, adding that there is no prescribed length for the completed report. The respective lists are not identical but the variations are minor. The required information is summarised below. The surveyor will find it well worthwhile to absorb the following points and to bear them in mind throughout the whole of the preparation, inspection and reporting process for Level Three surveys. This will not only ensure that the report ultimately contains all necessary information but that the surveyor will be guided during all activities on site, both mentally and in note-taking, as well as during the later reflection prior to compiling the report:

- Describe the form of construction and materials used for each element.
- Outline the performance characteristics of the material or construction.
- Describe obvious defects.
- State the identifiable risk of potential or hidden defects.
- Outline remedial options.
- If considered to be significant, explain the likely consequences of non-repair.
- Make general recommendations in respect of the likely timescale for necessary work.
- Include, where appropriate, recommendations for further investigation prior to commitment to purchase.
- Discuss future maintenance of the property and identify those elements where more frequent or costly maintenance and repairs than normal might be required.
- Identify the nature of risks in areas that have not been inspected.

Indeed, this list sets out a structure which could usefully be adopted as a reliable guideline for the surveyor's activities on site, irrespective of the type of survey report ultimately being produced. Each one of the Four Whats mentioned in the discussion on the Level Two report – those questions which the client is entitled to see answered in the finished report – will be fully addressed when following these guidelines. Although the level of inspection and the depth of the finished report will vary depending on which format the client has commissioned, the surveyor who uses this as an underlying foundation to every inspection will not go far wrong.

Cost guidelines can be included if agreed with the client, but the reservations and limitations of such advice must be made very clear and the client should be advised to obtain formal quotations for any necessary work prior to making a legal commitment to purchase.

Sometimes the surveyor will not be able to reach firm conclusions with a reasonable level of confidence, perhaps in relation to structural movement or regarding the condition of the services; and at times it may not be possible to determine the full extent of a defect or there may be suspicions that a visible

4 Some might be prepared to undertake a Level Two inspection on this cottage-style dwelling; but it has been built in at least three stages with at least three different wall constructions and two different roof coverings, so it is unlikely that the limitations of this format will allow the surveyor to do the building justice. A Level Three inspection is much more appropriate, quite apart from the need for the surveyor to be competent in reporting on thatched roofs and cob wall construction.

defect may be affecting concealed areas. In such cases, it will be necessary to recommend further investigations. However, the guidance cautions against merely recommending further investigation simply because an element is not visible or accessible during a normal inspection. There needs to be evidence to support the suspicion of a defect; such recommendations should not be made simply because the surveyor is unwilling or unable to exercise professional judgement and is seeking protection against possible future liabilities. As far as possible, a Survey Level Three report should provide the client with all the information required to make a purchase decision, and referrals for further investigation should be the exception rather than the rule.

2

Inspectors

Attributes

It almost goes without saying that the surveyor should be mentally alert and physically capable of undertaking the inspection. The physical demands of residential surveying are not insignificant. Gaining access to parts of some properties, while not requiring undue athleticism, may necessitate a degree of agility and confidence. At a time when there is justified sensitivity about undue discrimination, it seems appropriate to question whether a surveyor should still be offering the service if their increased girth, dodgy hip or advancing pregnancy makes it awkward to climb a ladder, enter a roof hatch and clamber safely around a roof space. For the younger in limb, this may not usually present a particular challenge; but it is a concern when more, how shall we say, seasoned professionals – a significant proportion of practising residential surveyors and valuers – fail to acknowledge the limitations imposed by increasing years.

The necessary physical attributes of the inspector relate neither to age nor sex but to health and the capacity to do the work. Tramping around dwellings, going up and down stairs, crawling around in confined spaces and craning the neck to peer intently at some peculiarity of construction can be exhausting. Few are blessed with good health at all times. Working with a sniffle is one thing but aching limbs and a high temperature are liable to cloud the judgement. Surveyors differ in height and circumference – and there is reason to suppose that earlier generations were smaller than those of today, judging by the restricted size of trapdoors to the lofts of older dwellings – so some parts of the inspection may be easier for some and more difficult for others. Good eyesight is an important attribute since it is the inspector's eyes that take the responsibility for the outcome of the inspection. Even for the most clear-sighted, the need for regular eye checks should not be overlooked. It is one thing to defy the passing of the years and struggle with small print but quite another to be placed in the position of the surveyor who had to be shown that when he wasn't wearing his glasses, he could no longer see woodworm holes around the edge of a roof access trap, literally beneath his nose as he entered the roof space.

One of the attractions of a surveyor's life is in being able to objectively apply hard-learned professional skills to buildings. Since it is residential buildings that are being examined, a knowledge of their construction must be a basic attribute. This can be acquired through study, attending lectures and examining drawings and specifications, but it is absolutely necessary to look at them, both in their totality and at the details. The study should not only be inclined towards the construction of new properties but should also embrace the study of older dwellings, their forms of construction and their likely defects. An assiduous reading of books, journals, pamphlets and the like relating to buildings and their defects, so as to keep up to date with ideas and research, forms part of this attribute. It is only by maintaining a keen interest in buildings and their defects throughout a career that experience is gained and consolidated. This experience should be reflected in recommendations from satisfied clients, which is undoubtedly the best way to obtain a continual stream of work.

Knowledge should be extended to a study of the particular characteristics of the residential properties in the area. From what period do they mainly derive, and what bearing does the local topography and history have on their durability and condition? A keen interest in the area, not only awareness of what is going on at the current time but also an understanding of its historical development from old maps and documents, is a highly desirable attribute. Time should be found to look at local newspapers as well as the nationals so that policies of local as well as central government are kept in mind.

It is easy to think that the cost of building repairs is a matter to be left entirely to contractors, but a necessary attribute for the inspector is to be able to put an approximate figure to many of the common repairs that may be recommended. Costs need not be quoted − unless specifically requested and then only if the full extent of the repair is ascertainable and with appropriate caveats − but they need to be kept in mind for their possible effect on the valuation if one is being provided with the report. Being a member of a firm that manages residential property or one that specifies and supervises repair work is helpful; but for those without this assistance, there are publications for this purpose, which are invaluable when used with care. Much knowledge will be gained from involvement in organising building or repair work or managing further investigations on behalf of clients, but not all practitioners wish to or have the opportunity to undertake such work.

One important feature of working as a residential surveyor, whether within a larger organisation or as a sole practitioner, is the necessity to take sole responsibility for one's actions and decisions. In some circumstances, an inspection may be carried out jointly with a colleague or an assistant; but for the majority of surveyors, the inspection is almost always a solo activity and this requires a degree of confidence in one's own abilities at all stages of the inspection and reporting process, which will not suit everyone. This is an area of the profession that does not readily allow sharing or delegation of responsibility. Those who prefer to operate as part of a larger team might not welcome the exposure this focusses on the practitioner. In some practice specialisations, especially with the increasing use of homeworking and for those who are working for themselves, residential surveying can at times be a relatively isolated professional existence with limited face-to-face interaction with colleagues. Surveying-related internet forums and social media can provide access to professional peers, but a strong, fully justified belief in one's professional capabilities and a high degree of self-motivation are certainly desirable attributes.

Tenacity, though at some cost in time, is sometimes an essential requirement during the inspection itself when it is necessary to follow up some minor indication of a defect – 'following the trail' as one judge put it (see p. 15) – so that not only a description of a defect but also, whenever possible, a diagnosis of its cause can be given, together with an indication of its importance and urgency. Suggesting a remedy and putting an approximate cost to it is possible with growing experience and can, on occasion, be done without needing to recommend that the client brings in a 'specialist' to do what amounts to the surveyor's job.

A knowledge of all the law relating to residential property is another essential attribute. While the legal profession can deal with finer points of the law, its practitioners are very unlikely to see the property, so the inspector is probably the only person viewing the building and its site who will appreciate the range of possible legal problems. Knowledge should cover environmental, planning, conservation and energy matters as well as easements and rights in relation to party walls and boundaries.

Writing skills come more naturally to some than others, but the end product of the work under consideration is a report which – however it may appear, whether on paper or screen – has to be put together in words. The most thorough and perceptive of site inspections is of little value if the ensuing report does not encapsulate the findings in a manner that is helpful to the client. Some are better at stringing words together into sentences and paragraphs to provide clear non-ambiguous advice, but all need to hone their skills in this regard. This is especially important given that many reports now require information to be presented within the constraints of particular formats, some with space restrictions. The art of being able to turn out a concise meaningful précis from a collection of notes or a number of rambling disjointed sentences has become even more essential.

The verbal communication skills necessary to exercise tact and understanding when dealing with clients, as well as vendors and occupiers of the home being inspected, and the patience to listen to clients' problems and concerns also deserve continual development. By listening and talking, it is sometimes possible, without seeming to be nosey, to find out the client's occupation. It is not good to be caught out in a report labouring a point about a legal matter or the electrical installation when the client is either a lawyer or an electrical engineer.

Ultimately, however, whether dealing with estate agents, vendors, lawyers or clients, the ability to communicate effectively is a critical attribute for all property inspectors. Whether reporting verbally or in writing, the surveyor is rightly expected to provide insight, guidance and advice about the property in question, and it is important to develop the necessary skills to do this properly. Whether the client is a young family with a limited budget or a national figure in some field, discharging the responsibility of advising private clients on what is usually one of the biggest and most important purchase decisions of their lifetime can be immensely rewarding.

Qualifications

It is understandable that RICS should state in its most recent Guidance Note on the subject, the third edition of *Surveys of Residential Property* published in 2013, that the surveyor must be professionally qualified to provide a satis-factory service at all three levels of inspection. The Note then goes on to list the practical experience for the work, which naturally develops as individuals continually enhance their knowledge and competence through a mixture of initial training, on-the-job learning, instruction, self-managed learning and assessment. Three levels of professional qualification are specifically mentioned. Two of these levels, FRICS (Fellow of the Royal Institution of Chartered Surveyors) and MRICS (Member of the Royal Institution of Chartered Surveyors) are long established and well known, even outside the profession. The third, AssocRICS (Royal Institution of Chartered Surveyors Associate) has been introduced more recently to provide a route to RICS membership for people without a degree. It recognises relevant work experience and voca-tional qualifications and offers a potential stepping stone to chartered status. It is worth spending a moment to understand how this has come about and why it can be considered a suitable level of qualification.

Historically, qualification as a chartered surveyor used to be achieved through apprenticeship within an existing practice while taking examinations set by the Institution. After the Second World War, however, increasing num-bers obtained qualification by studying for a recognised degree and only then taking up employment. The Institution continued to offer its own examin-ations to those favouring the route of working while qualifying, but gradually, it moved the emphasis of surveying education away from the practical to the more academic and, eventually, completely abandoned its own examination system in favour of degree qualification through the universities. Recognising that graduates had no practical experience, a formal process to assess professional competence was introduced during the 1970s to ensure that candidates' post-graduate work experience was adequate prior to granting them chartered status.

An unexpected consequence of these decisions about qualification routes was to severely restrict opportunities for experienced property professionals in other disciplines to migrate towards chartered status and, equally, to prevent anyone who was able but non-academically qualified from becoming a member. The need to bridge this gap was eventually recognised, initially with a Technical grade and now with the Associate qualification. Because AssocRICS has only been available for a relatively short period, some still have uncertainty about the level of expertise that it offers compared with the MRICS level and may question whether it is an adequate level of qualification for the highly skilled work of surveying and valuing residential properties. The MRICS qualification is described as being broad, benefitting from training across a whole range of professional areas – most of which will not be followed once the practitioner starts to specialise in a particular field of work. Meanwhile, AssocRICS has a narrower focus but one which is just as detailed in each specific area of practice as MRICS. Both AssocRICS and MRICS are qualifications, not licences to practise. They define only the level of competence the holder has demonstrated at the time the person was assessed. Once qualified, however, any individual may develop his or her skills through CPD and practical experience, and there is no reason why both should not have equivalent levels of expertise in their practice areas. Some with the AssocRICS qualification will wish to take up the opportunity offered by the process to continue their development by embarking on the wider training needed to qualify at MRICS. Others, meanwhile, will only wish to hold the AssocRICS qualification while working as expert practitioners in their chosen field. The fact that an AssocRICS holds a lower level of qualification does not prevent progression to a higher level of competence and responsibility.

RICS qualification is a prerequisite for using RICS-branded products. There are also alternative providers of survey services, such as members of the Residential Property Surveyors Association (RPSA), who hold qualifications from organisations such as the Awarding Body of the Built Environment (ABBE), the Institution of Civil Engineers, the Institution of Structural Engineers and the Chartered Institute of Building who may have the necessary expertise to inspect and report on residential properties for prospective purchasers. Every one of these professional organisations requires their members to hold professional indemnity insurance, not least to protect the reputation of their respective professions. Institutional self-interest apart, however, proper insurance cover is essential when carrying out the work of inspecting and reporting on residential property, protecting both the individual surveyor and their clients. Such cover, however, is not available to all and sundry.

Insurance

RICS requires its members to take out professional indemnity insurance that meets specific requirements. To ensure that the necessary cover is provided, RICS only authorises a specified range of policies, each of which offers the requisite minimum terms. Beyond the RICS requirements, details of the policy will depend on the type of work being undertaken, the experience and background of the individual or practice seeking cover, and their claims history. The cost of the policy will be a reflection of these factors.

Insurers, in their turn, will only offer cover to those who are considered acceptable risks for insurance. The definition of 'acceptable' varies from time to time, depending on the conditions in the insurance market and the appetite of insurers for business. In the difficult conditions after the 2007 property crash, some practices found it impossible to obtain cover on suitable terms because many insurers withdrew from the market and others imposed unacceptably stringent conditions or quoted premiums too high to be commercially viable. As a consequence, a significant number of practices had to cease trading.

It is understood that certain surveying disciplines present a much higher risk for insurers than others, and residential survey work is one of the areas with the highest proportion of claims in relation to numbers of jobs completed. Fortunately, however, both the average settlement costs and the costs of the most expensive settlements are far below those experienced in practice areas like commercial valuation work. This is easy to explain. The nature of survey inspections means that significant parts of properties cannot be inspected. Once an owner takes possession and is living in a dwelling, they may become aware of matters that had not been reported. Because it is their home, they will tend to notice problems and will be more inclined to seek redress than, say, a commercial occupier of premises who can lock up at the end of a day and leave a problem behind. This explains the higher proportion of claims. When it comes to settlements, the absolute maximum amount of a settlement for a domestic property would be unlikely to exceed the total value of the property. In fact, the vast majority of settlements are for very much less than total value, and the typical cost of a residential claim is in the tens, or possibly hundreds, of thousands of pounds. By contrast, a commercial valuation claim might easily exceed a million pounds; claims for tens of millions are not exceptional and those for hundreds of millions are not unheard of. The total number of claims in this area is lower but the total losses are very much higher.

The matter of professional indemnity insurance within the surveying profession, especially in relation to residential valuation and survey work, has been the subject of much discussion since the 2007 property crash; but resolving the issues which have been raised, and only briefly touched on here, has proved much more difficult than identifying and debating them. It will be interesting to see whether solutions that meet the needs of all stakeholders can be designed and implemented in the coming years to replace the current arrangements which, it is generally agreed, are no longer operating in a satisfactory manner for most of those involved.

Expertise

Possession of a professional qualification and professional indemnity insurance are two basic requirements for the prospective residential surveyor, but underpinning these must be a level of knowledge and expertise commensurate with the type of work which is being offered to clients. The surveyor must be familiar with the nature and complexity of the types of dwellings in the practice area and have an understanding of the area itself. The RICS Guidance Note *Surveys of Residential Property* summarises the likely level of knowledge needed (third edition, p. 9):

- common and uncommon vernacular housing styles, materials and construction techniques; particularly important for older and historic buildings where the surveyor must understand the interaction of different building materials and techniques;
- general environmental issues where information about them is freely available to the public, including flooding, radon, aircraft noise, typical soil conditions, important landfill sites, etc.;
- the location of listed buildings and conservation areas/historic centres, and the implications of these designations;
- local and regional government organisations and structures; and
- an awareness of the socio-historical/industrial development in the area.

If this knowledge is missing for any locations within the area where the surveyor offers services, the gaps should be filled by study and research. If the gaps are directly relevant to a property under consideration then the instructions should be declined unless the necessary research can be completed within the timescales relating to the work.

A surveyor with the stated level of knowledge may be able to provide all three RICS levels of service, but those offering Level Three services – especially where these involve older, complex or historic buildings, those of unusual construction or those in poor condition and needing extensive works – will require a level of knowledge that is not only broader and deeper but also more technical than that needed for reporting on the more common types of urban dwelling. This level of expertise can be gained through involvement in project management, undertaking investigative work, supervising building and repair work, and dealing with applications for Planning and Building Regulation consents, listed building consents and similar activities. Additional study into specific areas, such as building conservation work, can also contribute. This degree of experience will not be acquired overnight, and those contemplating extending their practice offering into Level Three services should consider whether they have the appropriate level of knowledge.

In retrospect, the fate of the attempt by the government of the time to impose mandatory Home Condition Reports was probably sealed by its extraordinary, not to say contrary, decision to ignore the existing expertise of professional residential surveyors and insist on a completely new qualification and training regime for its Home Inspectors – the Diploma in Home Inspection. It offered to train to the necessary standard from scratch those interested in a new career; but with an unrealistically short timescale, it came as no surprise to many that the required numbers achieving the necessary standard did not materialise. Eventually, as successive attempts at compromises failed to meet the necessary objectives, the proposal was abandoned, with the unfortunate consequence for those who had embarked on the training process that the promised work did not materialise. Fortunately, those who qualified with the Diploma in Home Inspection have since been able to offer survey services within the National Energy Services' SAVA Scheme (Surveyors and Valuers Accreditation Scheme), which offers the necessary training and support to its members. Insurance for SAVA reports is provided through the scheme and covers each individual report as part of the report registration cost. In this respect, the basis of the insurance is fundamentally different from the arrangement used by most surveyors, but the protection for clients is effectively the same.

There is no law preventing an individual setting up in business as a 'surveyor', and some do practise as such without membership of a professional organisation or indemnity insurance. They usually make sure that they have no assets so that if they fail to spot defects and are sued, they are found to be 'men of straw' and the aggrieved party is unable to recover any loss. If asked whether they are insured, they will probably say that they are such good 'surveyors' that they do not need to be. That may or may not be true, of course. It might also be the attitude of practitioners who are wealthy in their own right who, while confident of not making mistakes, have sufficient resources to pay compensation if necessary. However, there is nothing to prevent them accepting instructions and carrying out inspections for purchasers provided that the report subsequently prepared is not on a copyright form of one of the professional organisations. Thankfully, lenders are legally prohibited from using unqualified and uninsured 'surveyors' for their valuations for lending purposes.

The Consumers' Association, in its monthly magazine *Which?*, provides occasional articles based on the public's actual experience of dealing with surveyors and on the quality of the reports they produce for those buying dwellings. It has been a consistent source of information whereby surveyors can gauge the public's reaction to their work and should certainly be read by all practitioners, not only for the comments by the buyers but also for the advice the magazine gives to those contemplating engaging a surveyor, what they should require and what they might reasonably expect to get. The existence of unqualified and uninsured 'surveyors' is no doubt one of the reasons why the magazine provides advice to its readers on how to go about choosing a surveyor. *Which?* suggests that the surveyor should be qualified and insured, should be a member of an institution which operates a compulsory arbitration scheme, should practise in the area in which the property is situated and, ideally, should come recommended.

Liability

An individual agreeing to carry out an inspection of a residential property and to provide a report on its condition takes on a responsibility and incurs liabilities in two ways. First, this relates to the person with whom the contract to do the work is made as expressed in the terms of the contract, whether formulated in writing as Conditions of Engagement or by word of mouth. Second, and quite irrespective of the contract, this relates to other persons who may have been expected to rely upon the work should it have been reasonable to anticipate that they would do so. If damage – either physical or financial – is suffered by reason of a failure to exercise reasonable skill and care, a claim may arise by way of breach of contract in the law of contract, or breach of duty in the law of tort in England and Wales or the law of delict in Scotland – in other words, a claim of negligence against the individual carrying out the inspection.

The law covering negligence applies to all individuals who exercise their skills and do work or give advice, and it is irrelevant whether they are qualified or not or whether they are experienced in the particular field. To take an extreme example, if an individual undertakes the work of inspecting residential property despite being totally unqualified and without any experience, the responsibility taken on is the same as for others who have both qualifications and experience. Going further, if the work is taken on for a low charge to undercut others or even for no charge at all, perhaps for a friend, the responsibility again remains the same.

However, for a claim to succeed in contract, tort or delict, the recipient of the report has to have acted on its contents and suffered loss. If the individual carrying out the inspection fails to put into words what had really been intended, that is no defence. The reasonable skill and care requirement extends to how the results of the inspection are expressed in words of advice to the client. The requirement to provide accurate reports often causes as much, if not more, difficulty to individuals who carry out inspections as undertaking the inspection itself. Should a subsequent meeting be necessary to clear up any misconception or misunderstanding in the client's mind, it is most important that the outcome of the meeting is confirmed by letter with an acknowledgment by the client that this has been received.

It will be noted that 'reasonable' skill and care has to be exercised, not that which an expert might employ in similar circumstances. Indeed plaintiffs who succeeded in a lower court in obtaining a judgement for damages on the evidence of an expert, not of a surveyor but that of a structural engineer, found the decision overturned in the Court of Appeal (*Sansom* v *Metcalfe Hambleton & Co.* 1998). The individual carrying out the work has to exercise the same reasonable skill and care as others carrying out the same sort of work at the same time, not some later time. In other words, hindsight does not come into it. This aspect was argued by Lord Justice Brooke in a dissenting judgement in the Court of Appeal case of *Izzard* v *Field Palmer* in 1999. Although he agreed with his colleagues in finding a surveyor carrying out an inspection for mortgage purposes negligent in failing to provide warnings about future liability for charges relating to a flat in a nontraditional block, he disagreed with them on the amount of damages awarded. Other valuers at about the same time had apparently taken a not too dissimilar view, and he considered comparison with those was the proper measure rather than with a figure produced with hindsight.

The responsibilities undertaken do not subsist forever. In England and Wales, if actions are to be brought arising out of reports then in *contract*, to comply with Section 5 of the Limitation Act 1980, they must be brought within 6 years of the *date of the report*; within 3 years if a claim involves personal injury or death; or within 12 years if the contract is under seal – very rare in this type of work. In *tort*, the 1980 Limitation Act, Section 2, prescribes a different timescale in that the 6- or 3-year periods run from the date when the claimant *acted in reliance on the report* by entering into a contract to purchase or sell. This action is considered to be when the damage is suffered, not, as might be thought, when defects are found in the property. To overcome this anomaly for purchasers who did not know they might have a cause of action until defects were found, the Latent Damage Act was passed in 1986. Under this Act, actions – which can only be brought in tort, not contract – have to be brought within a long-stop period of 15 years from the date of the report and within that period, 6 years from the discovery of defects. There is provision for a further 3 years to be added to the 6 from a 'starting date', which is defined as the earliest date on which the claimant had the knowledge required to bring the action and the right to do so.

There was no need for an Act to cover latent damage in Scotland because the anomaly had already been dealt with by way of the two periods for bringing actions laid down in the Prescription and Limitation (Scotland) Act of 1973. This provides for a 5-year limitation period, 3 years if involving personal injury or death, from the time when the claimant could, with reasonable diligence, have become aware of the damage and a long-stop period of 20 years within which all claims must be brought.

This variety of timescales by which claims may be brought against a surveyor makes it difficult for practitioners to be certain when there is no risk of any action in relation to completed work. This is rather less of a concern for ongoing practices than it is for individuals nearing retirement and wishing to hang up their damp meters without the worry of claims arriving years after they have ceased practising. Most contracted work, such as residential valuations, will reach the limitation period after 6 years so the majority of retiring practitioners will take run-off cover to provide insurance for this period. The more cautious, aware that there is a potential for claims under tort for up to 15 years (20 years in Scotland), and especially those whose work was primarily survey work rather than valuations may opt for a longer period of runoff cover, especially bearing in mind that liability may extend to the estates of deceased practitioners. This is a further area where there is potential for improvement in the provision of professional indemnity insurance but, as touched on in the previous section, while the issues are clear, the solutions are not.

When defects costing considerable sums of money to put right are not detected and a claim against the surveyor is successful, the level of damages awarded by the court is often a source of annoyance to the claimant, to put it mildly. Often it is at a sum far below the cost of the repairs because the law only provides damages to cover the diminution in value of the property. The figure is the difference in value between the property as reported on and its value as it should have been reported if the defects had been known about at the time of the inspection. In the case that established this principle – *Philips v Ward* 1956 – the claimant paid £25,000 for a house in 1952, having taken advice. If the advice had been correct, he would have paid £21,000. In consequence, he was awarded £4,000 although the repairs would have cost £7,000 as estimated at the time of purchase. The view of the courts was that the claimant should not be put in a better position than he would have been in had there been no breach of contract or negligence – i.e. compensation for the assessed loss, not punishment of the offending individual – which would have been the case if the cost of repairs had been awarded. If that higher sum had been awarded then the claimant would have been in possession of a house worth more than the £25,000 he paid for it.

There is a certain element of logic in the court's way of awarding damages to successful claimants. On the other hand, it is small consolation to hard-pressed claimants without the necessary financial resources to cope. The Consumers' Association has commented unfavourably on the situation over many years. It cited one exceptional case arising out of a negligent Building Survey where £5,000 damages were awarded against a £12,000 bill for repairs. The problem was compounded when the surveyor became bankrupt and didn't even pay the damages. The response to this criticism is that owners can just sell on to someone else, but this is somewhat facile since the problem giving rise to the sale will be known about and will have a detrimental effect on the price. Although the courts have tended to lean at times towards figures more nearly matching claimants' actual costs, there has been much adverse comment.

Fees

Most of the attributes believed to be necessary for individuals carrying out inspections and reporting on residential property have been considered, but there is one other. It is not too far a stretch of the imagination to say that a further necessary attribute is to know one's worth. In days of severe competition and internet-based fee quotations, responding to enquiries by stating that a fee quotation cannot be provided until the surveyor sees the property is unlikely to produce much work, if any. The enquirer will most likely go elsewhere.

Of course, instructions to inspect and carry out a mortgage valuation will invariably be coupled with the amount of the fee, based on a fee scale set by the lender or panel manager issuing the instructions. Fee quotation will not be involved, and since the instruction may be one of a number received from the same source, the valuer may well be prepared to accept the rough with the smooth. Such an arrangement doubtless bears little relation to the size or condition of the individual properties, but most valuers have traditionally been prepared to accept this situation in return for a steady stream of work. Increased use of panel management firms by lenders and issues around professional indemnity insurance, mentioned earlier, have meant that for many, this previously reliable source of work has effectively dried up, obliging them to focus their efforts elsewhere.

Quoting for business is often a fraught activity in any commercial sphere, and establishing a satisfactory balance between attempts to maximise income while not scaring off too many potential clients may take some time to get right, especially for newly established practices. It seems that the skills that make a good residential surveyor are not those of the natural salesman and many residential surveyors are not particularly good at promoting their practices or themselves. Presumably, if they were better at it, they would be working in different businesses. Nevertheless, to be successful, every business, and the sole practitioner above all, must be able and willing to sell the services they wish to offer to the market. Recommendations may come in time, but satisfied clients will only move house every few years so repeat business, while pleasing, will not keep a residential survey practice afloat. This is not to say that professionalism should be sacrificed but that it is important to grasp the commercial realities of running a practice, recognising that every enquiry from a member of the public is a potential sale and that converting as many of these potential sales as possible into firm instructions means that the practice has a realistic opportunity to flourish. Once the commission has been confirmed, the skill and professionalism of the surveyor becomes paramount to ensure that the service expected by the client is delivered. Until the commission is received, however, the very highest level of professionalism is simply unfulfilled potential.

Later in this chapter, consideration is given to the detailed discussion the surveyor should have when agreeing with the client the level of inspection and form of report the client requires as well as the extent of any special instructions or additional services which may be needed; but when dealing with a preliminary enquiry the surveyor must be able to give an enquirer a good idea of the likely costs for particular types of inspection. The fees charged by each surveyor will differ, even for the same piece of work. This reflects the differing values placed on time and the level of practice overheads – i.e. all that it costs to run an office so that there is an organisation behind the surveyor, irrespective of whether a sole principal or part of a large office, possibly itself part of a larger conglomerate. For small practices in particular, fee levels may be varied deliberately as a means of managing work volumes: higher when work is more plentiful and lower when more instructions are wanted. The Consumers' Association magazine *Which?* once made the facetious comment that charges for a Building Survey depended more on the surveyor than the property, perhaps not realising how true that could be.

Basically there are three methods for assessing charges. There are advantages and disadvantages to each, and the method is chosen according to the circumstances.

1   The first method is by rule of thumb of so much per habitable room, such as bedrooms and living rooms. This method would be suitable for instructions to prepare any level of report and is easy to apply. It has the disadvantage, however, that no two properties with the same number of rooms are exactly the same in character and condition, resulting in quite different inspection times. It might be advantageous to use this method if properties in a particular area are much the same and the client is able to give some idea of the condition. Alternatively, the surveyor could decide that it will take a minimum time to inspect a room whatever the condition. Of course, the rate per room itself will need to be varied according to the type of report being considered.

2   The second method is based on a percentage of the asking price or the disclosed agreed purchase price. A disadvantage here is that a property in a higher-priced area would attract a higher fee than a similar property in a poorer area. It could be argued, however, that the higher fee is justified because if something was missed, the damages awarded would reflect the difference in property values. As a method, it is not particularly well regarded. It has been criticised by the Consumers' Association since, in their view, there is no reason why a higher fee should be charged when the amount of work undertaken could well be much less. Again, if this approach is used, a series of rates will need to be determined for the different types of report.

3   The third method is to charge so much per hour spent on the work. This is acknowledged by most to be the fairest, both to the client and the surveyor. Ideally, the surveyor would agree the rate with the client and, on completion, submit a bill itemising what has been done and the hours spent doing it. Such an arrangement might be acceptable where there is a continuity of instructions and where client and surveyor are well known to each other and there is an element of trust between them. When they are at arm's length, however, this is unlikely to be acceptable because it is too open-ended. Unless some idea of the likely total cost was provided in advance, arrival of the account could well induce at least a disinclination to pay. Most clients need to be quoted a fixed fee on the assumption that the information given to the surveyor about the property is correct. For the surveyor not to be able give a quotation would be considered very unbusinesslike. Every surveyor in practice needs to be able to quote an hourly and daily rate of charge. It can be based on the overall costs of running the organisation, large or small, including either their salary or their expectations if a principal. For a medium-sized office in reasonable premises with appropriate backup staff, it has been suggested that surveyors might be charged at a daily rate of 1 per cent of their gross annual salary

or expectation. To pluck figures from the air and purely as an example, assuming their working day comprises seven hours and their salary or expectation is £70,000 per annum, the daily rate would be £700 and the hourly charge, £100. The percentage figure may need to be adjusted, fractionally, for larger or smaller organisations, but this method has the merit of simplicity and can readily be used in conjunction with the surveyor's estimate of time required to complete the proposed work. An experienced surveyor who knows the housing stock in the area of practice should normally be able to make a reasonable estimate of how long it would take to complete the work in relation to one of the differing levels of inspection and report on properties of various periods, sizes and types. Obviously, the estimated time should reflect every part of the anticipated work – not only the time to carry out the inspection but also the time needed for travelling, preparation of the report and, if appropriate, a reasonable amount of time for discussion with the client either at the property, in the office or on the telephone.

It is important that the work should be remunerative. Low fees have been the bane of the surveying profession over many years, and fees for this type of work have undoubtedly been pitched at unrealistic levels for a variety of reasons. It is often forgotten that the fee level needs to cover not only the time taken on each individual case and the relatively easily calculated office and staff overheads but also time expended on other non-fee-earning activities, such as CPD (Continuing Professional Development), staff training, managing audits, abortive work and (hopefully infrequent) management of complaints. A further consideration – linked once again to the issue of insurance – is an understanding of the risk that the practice is exposed to when deciding to take on each job. This risk will be unexceptional in most cases and covered by the normal level of fee. At times, however, the risk will be significantly greater than normal – whether due to property type or location, high value, variations agreed in the instructions or even something relating to the client – and these higher risks should be priced into the fee quotation. This is all part of the surveyor knowing the true worth of the services being offered. In any event, when preparing to quote a fee, the surveyor should never forget that whatever the fee, or whoever does the work, the responsibility and the liability are the same, and the temptation to cut corners in the hope of saving time and being more competitive can rebound to damaging effect. It is probably best to calculate charges exclusive of VAT, as this can be varied by the Chancellor at any time, but to provide an inclusive figure as well in order to avoid uncertainty.

The next section in this chapter explores in more detail the nature of discussions with the client as he or she decides which type of service is required; but following these discussions, the client may decide to have the standard survey supplemented in some way with additional investigations or in other ways. These might be tests of drainage, of electrical installations or of heating installations, with the provision of long ladders to allow the surveyor access to concealed roof areas, or taking up fitted carpets and floorboards for a closer examination of underfloor construction – always on the assumption that permission will be granted by the seller. It will have to be decided there and then how such instructions are to be met. Does the surveyor have the necessary experience to be responsible for carrying out the tests and reporting on the results, will the work be put out to contractors, in whole or in part, under the surveyor's supervision, or will they be employed by the surveyor on behalf of the client? In the first case, the additional fee for the work can be agreed, exclusive of VAT, and the results incorporated in the report. In the second case, the surveyor may need to ascertain the charges likely to be incurred and obtain the client's agreement to meet them. In the third case, it may again be necessary to ascertain charges on the client's behalf, but it should be made clear that the contractors are employed by the client, report directly to the client and render their account accordingly; the surveyor merely arranges the timing of the inspections, hopefully to coincide with the surveyor's own inspection.

The surveyor should be aware that a range of specialist firms has sprung up in recent years to advise buyers on matters extending beyond those of the normal building services consultants. One organisation will do a check on the neighbours to see if they might cause problems. Enquiries and visits to the neighbourhood will be made at hours other than those when the surveyor is likely to be making his inspection. Another organisation will carry out a check on the locality for matters such as airborne pollutants, contamination, subsidence risk, crime rates and the like.

Surveyors and their clients can also take advantage of the services of woodworm and dry rot eradication and damp prevention firms, many of whom will carry out a free inspection, prepare a report and provide an estimate for work, if ascertainable, or advise on any further investigations they consider necessary. Obviously, they hope to obtain work as a result. However, while their presence provides another useful pair of eyes, the surveyor's responsibility is in no way diminished. The surveyor is the one paid to give the disinterested opinion and advice and this might well extend to commenting on such firms' reports if asked to do so.

Reaching agreement with the client

A crucial aspect of making arrangements for an inspection is to establish exactly what the client requires, bearing in mind the type of inspection that is appropriate for the type of dwelling under consideration. The information given by a prospective buyer not only on what is expected, assuming this is known, but also on the dwelling that may be purchased can be less than generous. It is here that the matter of settling instructions in a precise and structured way comes to the fore, and this is an essential and vital preliminary to any inspection.

Assuming that the enquiry is not from a client already well known to the surveyor, he or she is likely to say simply, 'I am thinking of buying a house. How much would you charge to do a survey?' The surveyor must rapidly establish some key facts regarding the dwelling concerned in order to supply the estimate requested. Vital information at this stage could be limited to location; type of dwelling – detached, semi-detached or terraced house or, should the enquiry relate to a flat, on what floor; approximate age; size in the form of the number of living rooms, bedrooms and bathrooms; condition; tenure and, if leasehold, the unexpired term; whether occupied or unoccupied; plus the asking or accepted offer price. Depending on the replies, further questions might be posed, such as whether there is a garage or outbuildings and if so, whether they are to be included, and whether the dwelling is listed or in a conservation area. As these details are gradually established, it may be that the surveyor already knows the property or at least has a very good idea what it is like. This can be turned to advantage by demonstrating the surveyor's depth of knowledge by making a relevant comment on some aspect or feature so far undisclosed during the conversation.

Once the basic details about the property have been established, it is necessary to ask if the enquirer is aware of the types of report available for buyers. Even if the enquirer has some understanding of the options, it is still worthwhile spending some time explaining the differences and, if appropriate, how the services offered by the surveyor vary from the standard levels defined by RICS. It can be very helpful for the client to see examples of the different levels of report that the surveyor offers. Some practices make samples available on their websites, but handing samples to an enquirer during a meeting or offering to send samples to follow a phone conversation is arguably a more powerful way of selling the services of the surveyor. Of course, the enquirer may have provided details of the type of dwelling on which the surveyor would only entertain carrying out a Level Three Building Survey in which case an alternative would not arise. If it transpires that only a Building Survey is suitable, the enquirer may still require market and fire insurance valuations.

With the above information to hand, and suitably retained as a record of the conversation, it should be possible in most instances for the surveyor to provide an estimate for what the enquirer wants. That, of course, could be the last the surveyor hears from that particular enquirer; if that happens too often, it may be that the surveyor needs to look at the level of the fee quotations being given.

However, should there be an eventual return call with a request to proceed then the opportunity should be taken to have a longer conversation or perhaps arrange a meeting. This is so that the instructions can be fully discussed in order to leave no doubt as to what the surveyor has to do and, just as important, no doubt in the prospective client's mind as to what can be expected for the fee being paid. The enquirer should also be asked to send or bring along any further information about the dwelling – agent's particulars, for example.

It is vital to spend sufficient time with the client to determine which level of service will meet their needs and expectations while also being appropriate for the type and age of the property. If the RICS-branded products are not being offered, the RICS Guidance Note *Surveys of Residential Property* requires all documentation and explanations of alternative survey types to be benchmarked against the three defined survey levels. Variations are quite acceptable but they must be positioned in relation to one or other of the defined Levels so that consumers are able to understand what services are being offered and can compare them with alternatives on offer elsewhere.

When buying a property, private clients are already facing significant expenses over which they have little control, such as mortgage application charges, legal fees and taxes, whereas paying a fee for a survey is discretionary. Faced with a choice between a quite expensive survey (HomeBuyer or Level Two inspection), a very expensive survey (Building Survey or Level Three inspection) or no survey at all, the majority have consistently opted for the nil cost option without fully appreciating the risk they are running of even greater expenses once they move into the property, probably in the erroneous belief that nobody would lend that sort of money if the dwelling was in any way seriously defective. As described in Chapter 1, now that a full range of survey options and prices is available following the introduction of the basic Level One service or Condition Report, experience is showing that when the options are explained briefly but clearly, a significantly greater proportion of purchasers will opt for the mid-range HomeBuyer survey than when it was the lower-priced option. It may not be logical but the psychology of the marketplace seems to be playing a helpful part.

During this meeting or telephone call, the surveyor should attempt to see matters from the enquirer's point of view. In this way, it is possible to empathise with some of the apprehensions which assail most people about to embark on probably the most expensive purchase of their lives, particularly if they are first-time buyers, and thus help the enquirer to reach the necessary decisions. Trading up to a much more expensive dwelling or downsizing from a family home to a flat, much smaller house or cottage can produce similar anxieties. Successfully assisting during this process will help to convert an enquiry to a paying client.

Most prospective buyers by this stage have spent a lot of time looking at dwellings within their price range and have settled on one that more or less meets their requirements. Although an element of compromise usually enters into the decision, the dwelling could be at the top end of their financial resources and they will no doubt be desperately anxious for a favourable report. However, the surveyor tactfully needs to enquire whether the dwelling has been looked at sufficiently, preferably more than once, for the clients to be satisfied on its visible condition and that they have the resources, with the aid of a loan if necessary, to deal with any obvious defects. It is quite surprising how viewers, cheerfully bundled around by owners and their agents, can have a very uncertain idea of what they are proposing to buy, and further questions by the surveyor can be vitally necessary for both parties to put matters into perspective. However, given some reassurance that the clients have decided that this is the dwelling for them, the surveyor can then go on to enquire if there any aspects to which the surveyor should give particular attention, such as whether the clients wish to carry out any minor alterations on which the surveyor can comment as to feasibility – for example, the removal of a partition – without the clients necessarily incurring an increased fee.

How does the surveyor respond to clients who wish to attend at the end of the inspection to discuss the surveyor's findings? Is this something the surveyor welcomes or resists? Depending on the level of service being discussed, there is no right answer, but if the surveyor is willing to do this then the additional time involved will need to be reflected in the fee charged.

Is the client's financial situation so critical that consideration should be given to curtailing the inspection if a major problem requiring urgent work is unearthed, thus saving a proportion of the fee? Do the clients wish to be summoned to the dwelling for an explanation should this transpire? During a discussion of this nature, the surveyor will be able to gauge quite a bit about the character of the clients. Have they been through all this before and know precisely what they want from the surveyor or do their hands need to be held at all times? The surveyor will no doubt react accordingly but should seek to avoid a patronising 'I know it all' attitude with the client.

Having established what the client wants, it is important for the surveyor to make quite clear in discussion what can and cannot be done. Some clients have difficulty in recognising that surveyors do not have X-ray eyes and that they are not able on a visual inspection, for any level of inspection, to pull things apart to see what is behind or to take up fitted carpets, laminate flooring, floorboards or decking to see what is underneath. All that the surveyor can do if there are grounds for thinking there is something nasty behind the woodwork is advise opening up and a fuller investigation. Even though all the things a surveyor will do and what cannot be done in someone else's property will be set out in the terms and conditions of engagement sent or handed to the client – and which the client is expected to sign as having seen and accepted – it is important to go over them verbally since the significance of the written word is not always fully appreciated and certainly not always read as closely as it might be.

During the discussion, the surveyor will need to find out as much additional information as possible about access to the dwelling for the inspection. The precise address, the name, address (postal and email) and telephone number of the owner, and the name and the same particulars of any agents involved are required. How is access to be arranged – direct with the owner or through the agents? If the dwelling is unoccupied, where can keys be obtained? Are all the services connected and turned on? For unoccupied premises, it is important for the surveyor to ensure that security systems are deactivated before arrival or at least to be provided with accurate information for dealing with them. Few things are more wasteful and time-consuming than setting off an intruder alarm when entering an empty property; plus there is the attendant noise and possibly even the arrival of police to contend with, though the latter happens surprisingly rarely.

Further discussion will no doubt settle on a provisional day for the inspection; the likely day for the delivery of the report; details of to whom and to where the report is to be sent; and whether the report is to be sent only by email or whether hard copies need to be posted. Details for making payment will also need to be agreed. Depending on the established routines of the practice, this will be when the signed terms and conditions of engagement are returned (essential before the report can be issued), on receipt of an invoice which will be submitted with the report, or even at some later time after any subsequent additional services have been completed.

Most prospective clients are happy to accept standard terms and conditions, at least once the inevitable limitations of an inspection have been explained, and accordingly will agree on appropriate fees for 'proper' inspections. Sometimes, however, a surveyor will be asked, perhaps by a well-known client who provides an amount of other profitable work or, worse, by a relation or friend, 'Would you mind having a quick look at a house I am thinking of buying and let me know if it's OK?' It is highly unwise to take on such a request even if done for a friend or relation for no fee in the surveyor's spare time, irrespective of whether the outcome is to be reported either orally or in writing. Negligence cases in tort as far back as the 1950s and 1960s (*Sincock v Bangs* 1952 and *Sinclair* v *Bowden Son & Partners* 1962) decided that even on a brief inspection, the surveyor was expected to indicate the presence of woodworm, dry rot and settlement, however slight the manifestations. Even the tempting offer, 'If it seems OK, I will ask you to do a detailed survey' shouldn't be taken. Experience shows that the second instruction never comes and the buyer proceeds headlong to purchase, leaving the surveyor at considerable risk. Far better to point out to friends and relations that their interest would be better served elsewhere by having a 'proper' inspection carried out by a surveyor taking full professional responsibility for opinions in a written report. In the case of valued clients, who really ought to know better, pressing for full instructions to do the work properly is probably better than a flat refusal. If the client is sufficiently valued, it may even be worth making a commercial decision to do the work properly while charging a significantly discounted fee or even for a nil fee. This at least enables a proper inspection to be completed. In reality, the potential liability is no different from the lottery of a quick look around and a verbal report, but the risks of missing something significant will be much reduced.

Another request by a potential buyer which needs to be treated with considerable care could be phrased thus: 'The present owner has handed me this report on his house with a view to speeding up its sale. Can you check it over for me?' Going over another's work in these circumstances amounts to transferring the responsibility for the accuracy and the opinions expressed in the report from the person who prepared it to the person checking it. Accepting such an instruction should not be countenanced in any circumstances. Since the Contracts (Rights of Third Parties) Act 1999, the surveyor endorsing such a report is liable in negligence to any buyer acting on the contents, and this liability is extended for the benefit of any lender who takes the dwelling as a security for a loan. There is no way of excluding this liability and this is as it should be.

Making arrangements with the occupier

When making arrangements for the inspection appointment with a vendor or occupier, it is a good idea to explain what the process will involve and how long it will take while also offering reassurance about how little disruption it will actually involve for the occupiers. Often the vendor will say, 'Goodness, I had no idea it would take so long' – the first intimation that life might be made difficult. However, while remaining polite, the surveyor should be firm and not accept, say, an appointment for 4 p.m. if a three-hour inspection is anticipated, even more so if it is wintertime. Weekday morning appointments are ideal but, unfortunately, it has to be accepted that the odd awkward appointment might have to be entertained – though only as long as the surveyor's capability to do the work is not impaired in consequence.

When the appointment is being made, it is necessary to stress that access to all parts of the dwelling is required and not to assume that this is understood by the occupier. Keys for locked rooms and cupboards need to be produced and if, for example, the basement, the top floor or an individual room is sublet, the occupant must be notified by the owner so that the accommodation can be inspected.

Unless instructions are received to inspect an unoccupied dwelling, it has to be admitted that the surveyor is not always a very welcome visitor. The inspection could well bring to light defects not known about previously that are likely to lead to a reduced offer. The surveyor, therefore, may have a fairly hard task initially to overcome what may range from a slight coldness to almost downright hostility. Seldom will the welcome be warm and if it is, such is life that the surveyor is likely to become wary of possible manipulation into missing signs of defects. A reasonable formality is probably the best reception to have, particularly if accompanied by a remark such as: 'I am off out for the day. Tea and biscuits are in the kitchen and don't forget to lock up and put the spare keys through the letterbox when you go.' Unfortunately this doesn't happen very often.

Once at the dwelling, the surveyor must be careful not to cause any damage; but if an accident occurs, the owner must be told and the surveyor must be prepared to pay for the reinstatement. The idea of fudging the issue should not be entertained. A particularly relevant court case, *Skinner v Herbert W. Dunphy & Son* 1969, concerns a surveyor who, having lifted the cover to a drainage inspection chamber, could not replace the cover as he found it. The frame and surround had broken on lifting the cover, and he was later sued by the owner's wife who, it was held, had not been warned of the hazard and tripped over it some days later. Damages and costs of £1,500 were awarded for the injuries sustained, and it can well be imagined what the damages would be if the case had arisen in this more litigious age. Even if the damages were paid by the surveyor's professional indemnity insurer, it would, nevertheless, amount to a claim which could easily have been avoided with the exercise of more consideration, which would also prevent any impact on the insurance renewal premium. Even leaving the front door on the latch while trotting in and out can be dangerous. One contractor was held liable for the loss of valuables removed by an unnoticed intruder in just such circumstances, and this could just as easily happen to a surveyor.

Needless to say, when the inspection has finished, the home should be left as it was at the start and either the occupier thanked for their patience or an unoccupied property left securely locked. When asked, as will frequently occur, 'Does it pass?' by a nervous vendor feeling like an examination candidate, some bland observation of a very general nature is usually the best response along with some helpful information about the likely timescales before the report can be produced and be with the client. If any broad reassurance is offered, it is surprising how even the most simple statement will be misunderstood and quoted back at a later date, bearing little relationship to the message which was intended. The only exception to this is when something posing a serious danger is observed, such as a partially blocked gas flue or a dangerously leaning retaining wall. These, and the risks they pose, should be pointed out to the occupier and also to the estate agent so that remedial action can be taken without delay to prevent an injury or death.

3

Equipment and personal safety

Equipment

It is for good reasons that a survey has been likened to a forensic examination. The same structured and analytical approach is required to ensure that the underlying causes of any problems have been properly identified. The technologies available to assist the surveyor in this task are now increasing, and for specific investigations of particular building defects, some quite sophisticated equipment is now available. The objectives and limitations of pre-purchase surveys, however, restrict the opportunities for detailed investigation so the equipment routinely required by the surveyor is relatively basic (5).

Undertaking a survey of a dwelling is primarily an exercise in using the eyes and comparing the visual observations with what the surveyor's training and experience lead her or him to expect to be seeing in a property of that age and construction. Other senses naturally come into play: surfaces are touched, tapped and knocked; sounds are assessed for solidity or hollowness; and certain distinctive smells – dampness or rot – indicate potentially serious problems. The mind will be constantly interpreting these as well as visual observations. The various pieces of the surveyor's equipment assist in this process but, to borrow from another context, the most important item available to the surveyor is the apparatus between the ears.

In the first edition of this book, the authors expended much ink trying to reconcile a series of sometimes conflicting lists from different sources setting out the equipment which the surveyor should have available for the inspection. The demise of the Home Information Pack simplified matters by removing what had been the main source of potential confusion. The *RICS Building Survey* Practice Note issued in 2012 followed established custom by attempting to provide a definitive equipment list. The 2013 RICS Guidance Note *Surveys of Residential Property*, on the other hand, takes a rather different and more flexible approach, helpfully grouping the equipment into four principal categories with a fifth category of additional equipment normally associated with Level Three surveys.

5 The absolute basic minimum equipment: ladder, torch, moisture meter, tape measure and clipboard. It is difficult to envisage completing a survey without using any of these, but a full range of equipment must be at hand for use whenever needed.

One of the categories in the 2013 RICS Guidance Note relates to health and safety needs. There has rightly been increased recognition that health and safety is just as important a consideration for surveyors as for any other working person. Many activities of the residential surveyor are self-managed, frequently as a lone worker, so there is an added imperative to take this subject seriously. Rather than touch on it briefly here as part of the surveyor's equipment, this important matter warrants wider consideration and is covered in the section 'Personal safety' at the end of this chapter.

The other three main groups defined in the RICS guidance comprise equipment used simply to see or gain access to areas which would otherwise remain hidden, equipment for taking measurements and equipment used to record the observations made during the course of the inspection. The individual items in each of these groupings are considered in turn.

Equipment to assess remote or concealed areas

Binoculars or a telescope

These are essential for viewing chimneys, roofs and any other part of the dwelling that needs to be viewed from a distance, including gutters and the upper parts of walls. Zoom systems can be useful but 8x magnification is the minimum and more than 10x will be difficult to hold steady. Selection is a matter of personal choice but investing in a good-quality instrument with good light-gathering optics will pay dividends on gloomy winter days.

Surveyor's ladder

A surveyor is expected to be able to gain access to roof spaces and building elements up to 3m above floor or ground level. A traditional four-section surveyor's ladder or a telescopic ladder will suffice for most purposes. Putting old socks – clean ones, it is suggested – on the ladder ends will help to avoid marking decorations. The surveyor should be mindful of the safety issues around using ladders at all times but especially when working alone and whenever the ladder needs to be used at its fullest extent.

Torch

A surveyor without a torch for seeing into poorly lit areas is unthinkable. The main requirement is for good, even illumination at the furthest reaches of the roof space so that defects do not go unobserved simply due to inadequate light. The carrying of spare batteries and bulbs should be taken as read, and having a spare torch is also worth considering. The failure of the torch without spares being available, besides making the surveyor look foolish, can be more than just irritating. Personal preference again comes into play, some advocating the hands-free benefits of head torches with others preferring the floodlighting of a million-candlepower rechargeable spotlight. A torch that stands up on its own as a lantern has advantages as does one that will survive being dropped onto concrete from a height. Arguably, better lighting can be obtained in a roof space by means of a protected lamp and a mains extension lead to a lamp or socket on the landing below. The owner or occupier should, of course, be asked before doing this as some can be touchy about electricity consumption. Even when this is the preferred method, a torch will still be needed at properties without an electricity supply or in the furthest reaches of large roof spaces.

Lifting equipment

Something to provide a better element of leverage than the fingers, and to save the nails in the process, is required for lifting drainage inspection covers and loose floorboards – if required by the level of service being provided and always provided lifting and replacing can be done without causing damage. The idea of inspection chamber covers is that the occupier of the dwelling should be able to lift them if required. Domestic covers, even of cast iron, are designed to be lifted by one person, preferably by hooks round the handles, although these are usually broken – in itself a defect. Normally the surveyor's weapon of choice will be selected from a bolster, a claw hammer, a crowbar, a large flat head screwdriver or often a combination of these. A pair of large screwdrivers, working side by side, will usually prise up the edge of most chamber covers. A crowbar provides the best leverage for more serious lifting and is also useful for clearing the mud which tends to accumulate around the lifting handles and the edges of covers, even when they are in good condition. A bolster and claw hammer, supplemented by a crowbar if necessary, are usually effective with loose floorboards.

Screwdriver, bradawl or handheld probe

A penknife is probably the most useful and effective form of handheld probe. A judiciously placed blade makes an almost indiscernible mark on timber but instantly yields reliable information about the soundness or degree of decay present in the area being tested (**6**). The round profile of a bradawl or, even worse, the rectangular head of a screwdriver make much more obvious and damaging marks. A penknife blade can also be used to scrape surfaces, poke mortar and judge the depth of cracks. The widespread installation of uPVC replacement window frames has very much reduced the need for the surveyor to studiously check every window sill and frame for evidence of deterioration; but areas of vulnerable timber, whether outside or inside, are still present and should be checked. Should there be any concerns, surveyors can be reassured that anti-knife legislation allows the carrying and use of knives for work purposes, and in any event, folding knives with blades no longer than 75mm – which will include most penknives – are not covered by the legislation. A small selection of screwdrivers will occasionally come in handy but it is only on Level Three surveys that the surveyor will be required to open panels or roof hatch covers that have been secured. Unscrewing light switch covers to inspect electrical wiring used to be a common practice, but this no longer forms part of any pre-purchase survey and should be left strictly to qualified electricians.

Mirror

A handheld mirror is invaluable for seeing around corners, under floors, behind panels or into crevices. It needs to be small enough to manoeuvre into awkward spaces but large enough to allow a torch to be held close to the eyes and the light reflected into recesses being examined. A 100mm-square mirror is a good compromise and some form of protection will help to stop it being damaged. A mirror on a telescopic arm can be particularly useful in some circumstances.

Meter cupboard key

A standard meter cupboard key will enable the gas and electricity meters of more modern dwellings to be seen. These can sometimes indicate the date of construction, or possibly the date the meter was replaced, and may give some clue regarding the standard of the installation.

6 The surveyor was not guilty of removing this large section of a badly rotted doorstep, but the deeply inserted sharp probe, in this case a penknife, indicates the extent of the problem. This timber is beyond repair. Discrete examination with such an instrument can be extremely revealing about the condition of timberwork.

Equipment for taking measurements

Measuring dimensions

No single device will conveniently meet all requirements for measuring dimensions during a survey. A tape measure 30m long with a hook on the end is required when building exterior or site measurements are needed. A 3m retractable steel tape can be carried in a pocket and is useful for measuring wall thickness or the size of roof timbers. A 5m or a 7.5m retractable steel tape is less convenient but is long enough to measure rooms easily. A handheld laser measure is convenient for internal room measurements as long as there is an opposing wall to 'bounce' the laser off. The 2m boxwood rod which folds handily into six sections is less popular nowadays but has its devotees. It is good for measuring ceiling heights and has the added advantage of enabling the surveyor to tap ceiling surfaces for loose areas of plaster.

Crack gauge or ruler

Strictly speaking, any item that measures in millimetres will suffice for measuring crack widths. Nevertheless, a crack gauge is vastly superior for the purpose as it provides a vivid illustration of how cracks of various widths appear on plaster or rendered surfaces and even on rougher surfaces such as brickwork. Its use will avoid reliance on subjective descriptions such as 'slight', 'moderate' or 'severe' (**7**).

7 A simple but very useful crack gauge, comprising a plastic ruler with crack widths marked on it, provides a ready guide when assessing the significance of cracks.

Levels

It is surprising how sensitive an experienced surveyor can become to floors being out of level while simply walking across them, but a spirit level gives an objective indication. A digital spirit level is even better as it will show the angle of any slope without having to resort to trigonometry. A long level of 1m or 1.5m is needed for floors and to check verticality of short lengths of walls. A small level is useful for checking window sills and door frames. Marbles, ball bearings or golf balls are often mentioned as helpful for checking floor levels. These may assist with identifying the direction of slope on floors with hard and level surfaces, but unlike a spirit level, they cannot be used to work out the angle of any fall and they can be useless on carpets or uneven surfaces.

Plumb bob

As with floor levels, the eye of the experienced surveyor becomes accustomed to noticing when the vertical corners and planes of walls are out of true, but an objective check should be made. A long spirit level may be used on short sections of wall but a plumb line is useful for checking longer sections and also for identifying unevenness and bulges over the height of a wall. Taking precise measurements with a plumb line is not easy when working alone, especially outside if there is any wind, but hung from an upper-floor window, it can give a helpful indication and can often provide a better check on the verticality of a wall than the eye alone.

Inclinometer

Increasingly, especially with extensions to existing buildings, roof slopes need to be checked to see whether the slope is adequate for the roof covering. A digital spirit level or a basic inclinometer will avoid the need for calculations. If these are not available then a simple spirit level and an adjustable set square, or alternatively a folding carpenter's rule with protractor-like divisions on the brass hinge, can be used instead.

Electronic moisture meter

There are different ways to test walls for dampness, and several different technologies are represented by the variety of moisture meters available. Each has benefits and drawbacks. The easiest to use – and also the most prone to misinterpretation by the unwary – is the electrical resistance meter. This is characterised by having two sharp steel prongs, the ends of which are pressed into or against the material being tested. These meters are small, light and give an instant reading so they are the most suitable for the type of survey envisaged by this book, but it is essential that users understand how the meters work, what the readings indicate and the circumstances when false readings can arise. Used properly, they are very effective and an essential tool for the surveyor.

Equipment for recording observations

Developments in information technology mean that the recording equipment used by a surveyor no longer simply comprises clipboard, paper, pen and pencil, although these continue to have much merit in their own right. Personal voice recorders are favoured by some and tablet-based software packages are increasingly being used (**8**).

Whatever the personal preference of the surveyor, the requirement is to complete an accurate and comprehensive record of the property and the surveyor's observations at the time of the inspection. The need to back up any electronic notes should not need to be mentioned, and there is much merit in producing hard copies of anything recorded electronically, including photographs, and storing them separately. Written notes can be stored in traditional files but even these are vulnerable to damage by fire, flood and other more conventional hazards, so the provision of suitable long-term storage for these valuable original documents should not be neglected. Ultimately, if a case should end up in court, it is important that all notes taken on site should be available and that they should be contemporaneous – that is, dating from the time of the actual inspection and not a later reconstruction. If original notes are incomplete or missing, it will make mounting a defence to a claim very difficult, if not impossible.

One of the first things recorded at the start of the inspection should be the orientation of the building. This will assist in assessing its exposure to the prevailing weather as well as the likely effectiveness of any solar panels – or the potential for having them installed. Using simplified compass directions is more helpful when recording aspects of the property than such descriptions as 'front left' or 'right side elevation' which might be open to misinterpretation. They can also be used in the report as long as the orientation is suitably simplified for the benefit of the reader and the simplification is expressly stated at the outset. It used to be necessary to carry a magnetic compass to determine the actual orientation but many digital devices can now provide the necessary information with built-in GPS.

Digital photography is now so ubiquitous that it warrants a special mention. Digital cameras are so easy to use that they are virtually indispensable. Mobile phone cameras are convenient and some produce good-quality images, but even a

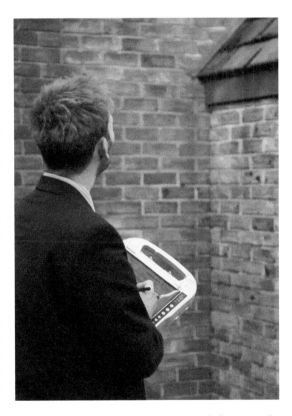

8 Tablet-based technology is increasingly being used. Originally it was solely utilised for mortgage valuations and asset management applications but survey-related software is increasingly being adopted. Many doubt that it will ever replace the trusty clipboard, but time will tell.

small dedicated camera will usually provide better results. The images can be an aid to reflective thought, and if the camera has an optical zoom (not a digital zoom) and the quality is good enough, enlarged images can even reveal more details than direct viewing with binoculars. Flash photographs taken in dark areas such as cellars or roof spaces can sometimes give a better overall impression than is revealed by torchlight. Conversely, capturing useful images of some subjects, such as minor cracking to plasterwork or the interiors of cavity walls, can prove irritatingly elusive. Strangely, areas of dampness on walls or ceilings can be more apparent in a photograph than in life. This can be something of a mixed blessing and is sometimes rather unnerving when comparing images with notes of on-site observations.

Photographs should only be used to supplement the notes taken on site and not as the sole record on which the report will be based. Consent to take images should always be sought from the owner and any refusal, as occasionally occurs, must be respected. The judicious inclusion of photographs can greatly enhance the final report as long as they are of good quality and properly illustrate the text. Simply scattering a random selection of images or using poor-quality photographs will only detract from the completed report. Digital images can also be invaluable in circumstances after the inspection and report have been completed, either to answer queries or perhaps, if a challenge should arise at a later date, to establish what may or may not have been accessible or concealed at the time of an inspection.

Additional equipment for Level Three services

The fifth group of equipment listed in the RICS Guidance Note *Surveys of Residential Property* is described as those 'normally associated' with Level Three services, but some – for example, large screwdrivers, hammers and mirrors on extendable poles – could be just as useful for Level Two surveys. At least this list, unlike the 2012 *RICS Building Survey* Practice Note, doesn't include spare batteries and bulbs in the optional list. These are essential and have no place on a list of optional extras.

Other items on the 'additional' list are more understandable. Borescopes, moisture meter accessories such as insulated deep probes, metal detectors and equipment to allow more accurate measurements of building distortion, such as quickset levels and measuring staffs, will not routinely be used during a pre-purchase inspection but may need to be available if the surveyor is to offer services at Level Three. The 2012 *RICS Building Survey* Practice Note even mentions a thermal imaging camera as worthy of consideration. This is certainly not yet a standard piece of equipment for pre-purchase surveys, but as the price of such sophisticated electronics becomes increasingly accessible, making these instruments less specialist, the more experienced practitioner will start considering whether or not they can make a worthwhile contribution to the professional services being offered.

Equipment – further thoughts

Care should be taken to ensure that moisture meters and other electronic and digital equipment are maintained in good working condition. Regular calibration checks should be made in accordance with manufacturers' requirements and records of these checks should be kept. When the entire equipment list is considered, it is remarkable that in most respects – the main exceptions being the thermal imaging camera, the handheld metal detector and the digital camera – the items remain little changed from what most surveyors have routinely carried since the 1960s. Removing the electronic moisture meter would take the list back some decades more. It is not without reason that surveyors are considered a conservative bunch.

One further item that doesn't fit into any of the earlier categories but which should be considered an essential part of every surveyor's toolkit is a handy guide to assist with the identification of trees. There are numerous suitable publications but it is more important to have a useful pocket version readily available for use on site than a comprehensive volume sitting on the office bookshelf.

Over the years, every surveyor will also gradually acquire a collection of additional tools that have proved their worth in the course of daily investigations. Most will be favoured variations of equipment in common use by all surveyors but some will be particular to individuals who would not be without them. At least one surveyor always carries a piece of white blackboard chalk to help conceal inadvertent marks on surfaces. In their first edition, the authors recommended a length of stout wire and a box of long household matches to check for blockages and adequate airflow through airbricks, essential for ventilation of suspended timber floors.

The second edition author has found two additional items particularly useful. First, a small magnet is helpful to distinguish between plumbing run in steel, to which it will be attracted, and that run in copper, which has no magnetic effect. Second, a supply of disposable shoe coverings will placate vendors who would otherwise insist on shoes being removed to protect their carpets. Whilst understanding the sensibilities of house-proud vendors, this does pose obvious practical difficulties in conducting an inspection which involves repeatedly walking in and out of a building and climbing a ladder to the roof space. Furthermore, acceding to this request arguably raises a health and safety issue about inspecting an unknown environment whilst wearing only socks. Providing sensible protective shoe coverings should enable the surveyor to remain shod.

Personal safety

It should be clear from the discussion about necessary attributes for the role as a residential surveyor in the previous chapter that physical fitness is a prerequisite. For many years, the capability to clamber where necessary to gain access to the more inaccessible parts of a building was all that was required to complete the work. Little regard was paid to the risks posed by, for example, large drops on the far side of low parapet walls while edging around roofs along narrow parapet gutters; exploring the furthest reaches of excessively hot roof spaces on a summer day; or crawling through shallow voids below floors to inspect joist ends. Obviously the individual would 'take care' but little objective assessment of the risks of any particular situation was ever undertaken. Occasionally the potential hazard would impose itself on the surveyor's imagination. The prospect of climbing a ladder into a loft hatch located directly above a stairwell might have given pause for thought, especially if the landing floor was slippery and the loft hatch cover was heavily glazed to throw borrowed light onto the staircase. It was relatively rare, however, that an area would not be inspected simply because access was considered unsafe.

It is an important part of the surveyor's duties to report to the client on any significant personal risks which the building or its surroundings might pose to the occupants or their visitors. These are examined in Chapter 29 but the surveyor is now also expected to consider the extent of any risks to which he or she might be exposed whilst conducting the inspection. Until relatively recently, this aspect of the inspection process justified only the briefest of mentions in RICS Guidance Notes and Practice Notes or within text books, but increasing awareness of health and safety issues generally, combined with legislation and regulations imposing responsibilities and duties on both employers and employees, have raised the profile of these matters.

The overriding legislation is the *Health and Safety at Work etc. Act* 1974 but regulations under that legislation have been regularly introduced, such as the *Personal Protective Equipment at Work Regulations* 1992 and the *Management of Health and Safety at Work Regulations* 1999. Meanwhile, a series of guidance notes have been issued by the Health and Safety Executive, some of direct relevance to many surveyors; for example, *Health and Safety Guidance on the Risks of Lone Working – INDG73* and *Driving at Work – INDG382*. The bar was raised for many employers when the *Corporate Manslaughter and Corporate Homicide Act* 2007 came into effect and replaced common law with criminal law in that an organisation can be found guilty of an offence where an employee's death is due to a gross breach of a relevant duty of care owed by the organisation to the deceased. After relying for some time on a skimpy few pages on the subject, RICS issued the very informative and helpful Practice Standard *Surveying Safely* in 2011. The whole document provides sobering reading for both employers and employees, still probably relatively few of whom realise just how far their respective responsibilities extend. For residential surveyors, Section 6 'Visiting Premises and Sites' is especially relevant and should be regarded as essential reading.

There is no need to repeat the RICS Practice Note but a couple of key points bear emphasis. First, surveyors should establish a procedure for undertaking a formal risk assessment at the start of an inspection and, when necessary, repeat the process when embarking on any more hazardous parts of the inspection. This need not be very time-consuming but it should be structured and the results recorded in the site notes. The RICS Practice Note has suggestions for the ways in which this might be arranged. Just as important, however, is that the results of the risk assessment should not simply be recorded and no action taken; where appropriate, they should lead to actions that mitigate higher-level risks. For example, not only should adequate protective equipment be available to the surveyor but this should actually be worn when appropriate. A high-visibility jacket, safety helmet and protective footwear will usually be required when entering a construction site. They may not be needed for the majority of domestic inspections, but gloves and a face mask are probably the minimum requirement for most roof spaces. It is easy to succumb to the attraction of 'just having a look' without going back to the car to don overalls or a face mask, but there is a cumulative risk to health of repeated exposure to dust and vermin droppings, not to mention uncertainty about whether asbestos may be present in a roof space.

Ladders, in particular, should be treated with respect. When ladders are being used, they should be properly secured to prevent slipping or at least positioned to ensure that if any slip should occur, it will be limited and not lead to the ladder falling, possibly taking the surveyor with it or perhaps leaving the hapless individual stranded in the roof space (**9**). While on a ladder, the temptation to stretch just that little bit farther to see or touch something is often very great but should be resisted. Ladders should also be checked periodically, especially those with some form of automated locking system, to make sure they remain in a physically sound condition. Because ladders are fundamentally of robust construction, they tend to be taken for granted, but minor flaws or damage can deteriorate if ignored and the consequences of any failure are likely to be serious.

Second, there should be robust arrangements in place to ensure that the working location of the surveyor is known, that the expected time due back at the office or at home is known and that procedures are in place for dealing with a circumstance where a surveyor's well-being may be in question. On one inspection, the surveyor fell down an inspection pit and was severely injured and left unattended for a whole night because he wasn't expected back that afternoon. The surveyor may carry a personal alarm and should certainly have a mobile phone, though phone signals are not always reliable. The question for the surveyor to answer is: 'If I didn't return from the property, who would know I was missing and how would they be able to find me?' Accident or sudden illness can incapacitate anyone, and in either event, a long delay before a search takes place is far from desirable.

Occasionally the risk to the surveyor will not be from some inanimate feature of the building, but from a living occupant. The most common source of difficulty is untended dogs in dwellings where owners are at work, but other animals may pose difficulties at times and, on occasion, a human occupant may cause concern. Most vendors accept the presence of a surveyor as a

9 When the two 'safety' bolts on a loft access ladder have been replaced respectively with a nail file and a bunch of keys, as here, the surveyor is wise to decline use of the facility and revert to using the surveyor's ladder in the interests of self-preservation!

necessary part of selling their property and put up with the inconvenience and intrusion with a greater or lesser degree of grace. Dealing with difficult occupiers is just a part of the job for most surveyors, but on rare occasions, the situation can become unacceptable. Whatever the source of the difficulty, the surveyor may decide that the most sensible course of action is to leave. The inconvenience of a last-minute postponement is not to be measured against personal safety in a threatening situation or the risk to reputation in a potentially compromising one. An explanation must be given to the client and, if possible, arrangements made to complete the inspection in more amenable circumstances. The overriding needs of the parties involved will usually mean that a way will be found to overcome the particular difficulty.

PART 2

The exterior

4

Pitched roof coverings

The inspection

Given reasonable construction and freedom from damp, the ravages of insects and untoward disturbance from settlement, the structure of a roof will last for hundreds of years, but this is not the case with roof coverings. Depending on the type of covering – and there are many – the quality, the fixings and the degree of pollution in the atmosphere, renewal may be required in anything from 30 to 100 years. Even if the covering material itself will last for 100 years or longer, the fixings will probably corrode in perhaps half that time, requiring the roof to be stripped and the components sorted and then relaid, possibly with a modification of details and improvements to the underlay.

The approaching prospect of major expenditure on a roof can sometimes be the motivation behind an owner deciding to move. Accordingly, it is essential for the surveyor to form an assessment of the current condition and to consider, if necessary, whether re-covering using existing material or entire renewal is essential in either the short or long term. This is an important aspect for whatever form of report the inspection is being carried out.

Since few roof coverings are likely to have a life of more than 100 years, the surveyor's knowledge of the age of the dwelling can be helpful in determining its life expectancy if the covering is contemporary with the structure. For example, in the majority of cases it would be unusual to find a great deal wrong with the actual material covering the roof of a dwelling built since 1980. Having said that, there will admittedly be occasional exceptions and there may be incorrect use of coverings in relation to the pitch or other faults such as poor detailing or fixings. However, the generality is sound: in properties of this age, the covering material itself will not usually be defective. On dwellings built between 1960 and 1980, however, it is a different matter. Defects are less likely to be found if the dwelling was built in the private sector, which was more inclined, even in those days, to favour more traditional materials and methods. For public sector dwellings, however, it was a time of much innovation, from which roof coverings were not exempt. Although it now seems incredible, much housing was roofed in what was thought of, even at that time, as a temporary material – three-ply bituminous felt – on both flat and pitched roofs. Most of these dwellings will, of course, have been reroofed by now, but it is still necessary to consider very carefully how this was done. Was it a repeat operation or was a longer-lasting material employed? With pitched roofs, was the new covering appropriate to the angle of the pitch and was the roof strengthened?

Dwellings built between 1940 and 1960 will probably have roofs that are traditional in shape with a steeper pitch and, in consequence, more sharply angled hips. But that was a time of shortages and substitutes. Concrete products in particular were much used, some more successfully than others. Those covered with tiles, however, may now – along with those built between 1920 and 1940 – be showing signs of the need for re-covering, depending on the quality of the original clay or concrete tiles used. Some will already have been re-covered, presenting the surveyor with a variety of tile shapes that show the full range of what has been available in comparatively recent times, some guaranteed by their makers for periods of up to 60 years. Asbestos cement slates, which came into use during the interwar period, have proved of variable quality. Many have already been replaced but those still in use can only have a limited remaining life (**10**). Dwellings built between 1900 and 1920, if not reroofed already, are likely to require re-covering very soon while there are relatively few built before 1900 that have not already been dealt with. Such simplification is not, of course, valid for dwellings built, say, before 1870. It is to be expected that such dwellings will have had their roofs radically attended to at least once and possibly twice since the original date of construction.

Ideally, to obtain a complete appraisal of an existing roof covering, a close examination of both external and internal surfaces of the covering should be made, but this is rarely possible. Taller dwellings of the eighteenth and nineteenth centuries, with their flat fronts topped by a parapet with a gutter behind and rooms in the roof space, can provide relatively easy access through the dormer windows to view at least part of the roof surface close up. Perhaps best of all for access to see all the pitched roof surfaces are those dwellings where butterfly roofs were provided in the latter part of the nineteenth century. Given access to the roof for clearance of snow from the central valley gutter and for necessary repairs and maintenance, they might have a doorway approach via a short flight of steps rather similar to the better type of access to flat roofs. On the other hand, they might just have two trapdoors – one in the ceiling of the top floor and the other in one of the sloping sections of the roof – giving access to the outer surface via a roof space of restricted height.

With more recent buildings, when it is within the limitations of the surveyor's ladder, it should be possible to look closely at the upper surface of the lower courses of the covering material on bungalows or single-storey extensions to taller dwellings; and occasionally a balcony or dormer window will enable a closer inspection to be made at a higher level. If single-storey extensions are covered with the same material as the main roof slopes at higher

10 The original asbestos cement slates on the left of this pair of houses have reached the stage where, like those on the right, they ought to be replaced. As there are no signs of sagging to the right, the roof structure was presumably strengthened to take the heavier replacement interlocking concrete tiles. The roof light suggests some form of loft conversion, possibly undertaken at the same time.

level then all to the good, but often they are not. 'Catslide' roofs provide another opportunity for close inspection, as do cases where, with older dwellings, there are internal facing roof slopes or safe access to a flat roof from the interior. The danger with flat roofs lies where there are no parapets or balustrades, and even when these are present, they can be a hazard if too low. As in all matters of access, the surveyor needs to use common sense as to whether it is safe and sensible to use the available means, perhaps in conjunction with the surveyor's own ladder. There are times when it will be and times when it won't, and there may not be the chance to complete the inspection if unnecessary risks are taken! Even where opportunities for external access are available, the surveyor must still use binoculars to examine all parts of the roof exterior that cannot be inspected close at hand but which are visible from a distance.

As to the underside of the roof covering material and the fixing components, close inspection may be possible on properties built prior to about 1950, except where roofs have been boarded – a relatively few cases in England and Wales although the use of boarding as an underlay has always been common practice in Scotland. Since then, however, the provision of an underlay of roofing or sarking felt to keep out wind-driven rain or snow has been provided on practically all new roofs as well as during roof re-covering work, so the opportunity to see the underside is now fairly rare. Advantage should be taken, however, of any situation where there is a tear or gap in the felt to take a peek at the underside. This can sometimes be possible where a ventilating pipe is taken through the roof slope, for example.

On the assumption that neither surface of the pitched roof covering is available for close inspection, what can the surveyor tell from a sight of the external surface of the roof covering through binoculars? It should be possible to go some way at least towards identifying the material and type of covering used. To take, at one extreme, some of the rarer examples first: it should not be very difficult to identify sheet materials where these have been used to roof a dwelling – copper by its green or brown patina, lead by its dark grey colour, zinc by its lighter grey shade, each with its own characteristic cappings or rolls down the slope. Dark grey material with flat-welted seams ought to suggest roofing felt, and there should be no problem in identifying where, even more rarely, corrugated asbestos or corrugated iron or thatch have been used. What can be said about condition requires further consideration of each material, but these materials account for but a small proportion – probably only about 5 per cent – of roofs covering dwellings.

It is when it comes to what covers the vast majority of dwellings and identifying what tiles and slates are made of that doubts can begin to form. Tiles cover 75 per cent of the total housing stock and can easily be distinguished from slates, covering about 20 per cent, by their shape, thickness, texture and colour; and also one from another by whether they are plain or profiled. Plain tiles, however, can be of hand- or machine-made clay or machine-made concrete. From a distance, they can appear indistinguishable. In rare cases, tiles might be mistaken for shingles – a long-traditional version made of wood. Profiled tiles, again of long tradition, are also hand- or machine-made of clay and they too can be machine-made of concrete. Accordingly from a distance, while distinguishable from slates, it is not possible to be certain from which material tiles are made. Nor may it be possible in all cases to give a name to the shape, unless the surveyor is familiar with tile makers' catalogues. These can sometimes reveal different names for similarly shaped tiles, and the surveyor should also be aware that it is possible to come across pressed steel sheets which look remarkably like profiled tiles from a distance (**11**).

Similar doubts can be generated when thin flat slates are seen as the covering. Natural slates – split from metamorphic rock – can be copied by moulding resins with powdered slate waste so that they are virtually indistinguishable from a distance, presenting the same riven texture, irregular edges and colour. Worryingly realistic looking 'slates' made with reconstituted rubber from car tyres have even been produced (**12**). Such good copies, from an appearance point of view, are only some of the more recent artificial slates to appear on the market. Earlier versions, some made of Portland cement and asbestos fibres, although flat and of a dark grey colour, could usually be distinguished from natural slates by their mechanical, uniform shape, their straight, sharp, thin edges and their unnatural very smooth and somewhat shiny appearance. Even earlier, before attempts were made to actually copy natural

11 At a quick glance, these roofing tiles could be mistaken for profiled concrete when in fact they are in the form of coloured pressed steel sheet; see also **64**.

12 Artificial 'slates' made of reconstituted rubber from car tyres at a demonstration unit at the Building Research Establishment. Surprisingly convincing in appearance, they can be bent with a finger. It will be interesting to see how durable they prove to be on extended exposure to the elements, but car tyres don't weather very quickly.

slates for replacement work, asbestos cement slates were available in colours such as russet brown, blue and the natural grey shade of asbestos. Since the raising of health concerns in the 1970s, they have not been used, but they often provided the roof covering over the chalet-type bungalows fashionable in the 1930s because of their very light weight. When seen, they are highly unlikely to be mistaken for slates in any other material and they will be showing signs of their age, indicating a need for replacement. Since most artificial slates can now be supplied with matching accessories for ridge, hips, valleys and roof ventilators, the inclusion of these may be the only features distinguishing artificial from natural slates.

Once again, the surveyor should be aware that concrete tile manufacturers produce a flat, lightweight, slate-coloured, interlocking tile. From a distance and at a quick glance, these too give the impression of a slate roof. A second look, however, will show that they are thicker and too uniform in shape, texture and colour to be mistaken for either a natural or artificial slate-covered roof.

So-called slates – usually called 'stone slates' but more like stone slabs and sometimes called 'tile stones' – were the traditional roof covering in areas where the local non-metamorphic stone formed the main building material. They continued to be so until the product of the massive mines and quarries of North Wales, with the help of the canals and railways, tended to oust them on grounds of cost in the 1800s. Whereas natural and artificial slates normally range from 5mm to 8mm in thickness, stone slates at their least are twice that in thickness and can range up to over 30mm. Through binoculars, that difference coupled with the contrast in texture and colour is sufficient to set them apart. However, where there has been sufficient demand, local stone slates have been copied using Portland cement and crushed stone moulded to give a texture that at a distance makes it virtually indistinguishable from the real thing.

While there will, therefore, be times when a surveyor may be confident in identifying the material used to cover the roof of a dwelling, there will be other times, in the absence of facilities for both close external and internal inspection, when it would be unwise to be dogmatic. Phrases such as 'appears to be ... when viewed through binoculars from ground level' are quite legitimate for inclusion in a report if that is the only vantage point available, irrespective of whether the roof looks sound or in need of repair. If the latter happens to be the case, the next person inspecting could well be another surveyor or roofing contractor with a long ladder, able to be certain; and this could be embarrassing, at the very least, to the initial reporting surveyor should the report have been too dogmatic.

With a lack of any facilities for close inspection but with reasonable confidence in identifying the material used, the sight of an apparently sound roof with no sign of damp stains internally on surfaces not decorated for some time, the surveyor can report favourably without further ado. Where surfaces immediately below the roof have been redecorated recently, it is necessary to be much more circumspect since temporary roof repairs and redecoration, adequate in the short term, could have been carried out specifically for the purposes of sale. Checks should then be made to see whether the pitch and other features are appropriate for the covering and are satisfactory, looking very carefully and using a damp meter on timbers in the roof space at places where there could be a possibility of damp penetration.

In cases where the roof of a dwelling is covered in either tiles or slates, the sight of many missing, slipped or broken items, or in the case of slates, many refixed with 'tingles' (the narrow strips of lead or zinc used to re-fix slipped slates) is indicative, depending on the number, of the need for either minor or major repairs. A few flaws on a roof covering, not particularly old looking, might be due to isolated incidents of wind gusting or the clumsy feet of aerial installers and be straightforward to rectify. On an older roof, it may be the beginning of a more general deterioration, likely to accelerate as time goes on and worthy of further investigation. Circumstances will dictate how the surveyor phrases the report in such a situation, but it would be wise to recommend a closer examination.

On the other hand, in the absence of a close inspection and with little possibility of ascertaining the likely cause of failure, the sight of many defective components requires the surveyor to recommend – again, quite legitimately – that further investigation should be undertaken to determine the cause of the defects and the extent of any necessary remedial work while also warning that that major expenditure is likely to be required. With binoculars, it will probably not be possible for the surveyor to tell, for example, whether tiles or slates are delaminating and require complete replacement or whether, when the underside is obscured by felt, it is the fixings which have corroded and

13 A recently re-covered roof in plain machine-made tiles where, in the absence of any new damp stains below, the surveyor can be reasonably confident in reporting. The adjoining roof will shortly also require re-covering.

14 In the absence of facilities for close inspection, the surveyor cannot do otherwise than suggest a need for close investigation to check that the remaining old tiles on the roof of the 1930s semi-detached dwelling on the left are not going to need replacement shortly. They have all been renewed on the other house of the pair as well as elsewhere on the estate, suggesting that this might have been a better course to follow than the patching carried out here.

15 At times, the surveyor may need to be creative to see parts of a property. The surveyor's ladder is not only useful for gaining access to roof spaces but can also be used to check any suspect external areas within safe reach, such as single-storey flat roofs, gutters or conservatory detailing. Here, a ladder is being used to inspect roof slopes on either side of a concealed central valley that can only be seen from a landing roof light. The view may be limited but that is better than not seeing anything of it at all, although the surveyor here also needs to consider the risk of undertaking this activity at the top of a staircase. If the risk is regarded as too great, it is important to keep notes explaining there was a conscious risk-related decision not to inspect. Such a decision can be justified whereas simply not realising there might be an opportunity to inspect cannot.

the repair will involve relaying rather than wholesale renewal. Either will be a major expense, but relaying less so than replacement.

Given access for close inspection to both external and internal surfaces, but primarily the latter, it ought to be possible for the surveyor to draw conclusions about the soundness of the covering and the reasons, if applicable, for disrepair. Sometimes a document of guarantee or possibly an estimate for repairs may be offered. While certainly of interest, these should be regarded as irrelevant until the surveyor has decided from personal experience what advice should be given. Having formed an opinion, as far as practicable, about the qualities of the material which has been used, the surveyor needs to look closely for any emergent signs of degradation, however slight, both in the actual covering used and in the fixings and battens. It is always possible, more likely perhaps on renewal work than on new build, that inferior product or inferior fixings have been used, less than ideal practices followed or corners cut. With the benefit of full access, the surveyor ought to be able to identify shortcomings of this nature and, if necessary, include appropriate advice in the report.

Tiles

By far the most common form of roof covering which the surveyor is likely to encounter is tiling since it is estimated that throughout the UK, something like three-quarters of all dwellings have tiled roofs. Two-thirds of these tiles are of concrete, this being the cheapest form of satisfactory semi-permanent roof covering for the last 50–60 years, the remaining third being of clay. Clay, moulded and fired, has been used to make tiles in the UK for many hundreds of years. Clay tiles lost out in favour of slates for around 140 years from about 1780 to 1920, then resurfaced in popularity for a further period until overtaken in the 1950s by the first really successful concrete tiles. Both clay and concrete tiles have always been, and continue to be, made in two forms: plain and profiled.

Unlike the makers of concrete tiles, some of whom currently guarantee their products for 50–60 years, the makers of clay tiles are more reticent, perhaps because the processes for producing both handmade and machine-made tiles from natural materials are less capable of fine control. While the makers of clay tiles naturally extol their virtues, pointing out that many seen can be well over 100 years old, guarantees in excess of 30 years are unusual. Whereas British Standards for the manufacture of concrete tiles have existed since the 1950s, there is no British Standard for the manufacture of clay tiles – only for the size and performance by testing of plain, not profiled, tiles.

Plain tiles

Plain tiles are rectangular in shape, 265mm × 165mm or thereabouts, with a slight camber lengthwise and sometimes also a cross camber, holed to be secured with wooden pegs or nails at one end and in most cases with nibs for hanging on battens. Being comparatively small, plain tiles are adaptable – unlike the profiled variety – to roofs of relatively small and complicated shapes, and they can also be hung vertically on walls when tiles shaped at the lower end and bearing such self-evident names as arrowhead, spade, club or fishtail are often used. They are laid with a double lap so that there are always three thicknesses of tile over the point of support (**20**). Every nibless tile needs twice nailing and those with nibs require nailing at least every fifth course below 60° and every course at a steeper pitch, along with every tile at the crucial points of verge, eaves, ridge, hips and valleys.

In order to prevent the entry of windblown rain and snow – a more likely occurrence when tiles with a cross camber are used because of their better, less mechanical appearance – the surveyor should look for the underlay of boarding or roofing felt which should be employed, nailed to the rafters below the battens, on all tiled roofs. Ideally, counter battens should also be provided when boarding is used.

If a close look at the tiles is possible, the surveyor should be able to decide by scraping the surface whether they are of concrete or clay. Early concrete tiles had an application of coloured granules to the surface that can usually be scratched off even if they have not already been partly removed by weathering, giving a patchy appearance to the roof. Although unsightly, this weathering does not usually affect the roof's performance. Later concrete tiles are usually through-coloured and either smooth- or granular-faced, but they can still be distinguished close up from clay.

Handmade clay tiles are usually distinguishable from their machine-made cousins by irregularities of shape, both as to thickness and camber as well as to size and the rougher texture produced by sand moulding. Though machine-made tiles can be provided with a sand-faced finish, this too can usually be scraped off, unlike the texture produced by sand moulding. Much more typical of machine-moulded tiles is the regular appearance, both as to texture and colour, and often just the one-way camber. Usually they lack that variety of texture, size and colour that make the handmade product so much more suitable for use on older dwellings.

Bearing in mind that the pitch of the tile when laid is always less than the pitch of the roof structure, the pitch of the latter for plain tiles – whether of clay or concrete – should never be less than 40° and preferably should be around 48° to 50° or even more, as on many early tiled roofs. Where the basic roof pitch is shallow, there is always a danger that when, for example, sprockets are provided, rainwater may not clear rapidly enough to prevent moss forming or, in winter, to prevent it soaking into the more porous tiles and accelerating frost damage at the eaves, a frequent cause of tiles delaminating.

Efflorescence of salts from within the tile itself has been known to cause the underside of the tile and the nibs, if any, of both clay and concrete tiles to erode and crumble. Unduly porous tiles lacking a cross camber are more prone to this defect as the salts carried in solution to the bottom of the tile migrate by capillary action towards the head of the tile where the solution evaporates, leaving little more than a white powder. It was the failure to spot this white powder, which had fallen on to the top of the ceiling joists in an unfelted roof space, that led to a surveyor being found negligent in one of the earliest reported cases of surveyor's negligence, *Last* v *Post* 1952. If the dwelling is one in an estate, a glance at other properties might indicate whether a total renewal of tiling has been found to be generally necessary, whatever the possible causes of delamination.

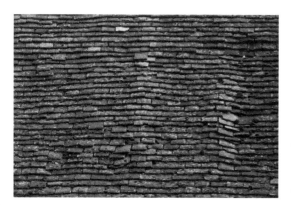

16 A roof of plain tiles where localised repair and replacement has been carried out with varying degrees of success but where complete re-covering cannot be postponed for much longer. The accumulation of moss is due to the proximity of trees and will interfere with free flow down the slope, encouraging water retention and capillary attraction between the tiles and increasing the risk of frost damage and probably blocked gutters.

17 Too much disturbance to the lie of plain tiles can introduce gaps that will compromise the effectiveness of the covering if allowed to become worse, as with this extreme example. Careful examination might indicate the possibility of retaining the main structural elements but strengthening may be necessary and packing out the rafters, often possible with less severe unevenness, is unlikely to adequately realign these slopes before the covering can be relaid.

18 Plain tiling in serious need of attention. The obvious hole requires patching urgently but a number of other tiles are deteriorating badly and should be replaced at the same time to prevent dampness penetrating in other areas in the near future.

19 Shaped tiles are not only used for hanging on walls, whether timber framed, brick or block. They make an attractive roof, as shown here. When used in this way, the run-off to the gutter is facilitated by a few courses of plain tiles at the eaves.

Where slipped tiles are present, close examination from a balcony or dormer window may confirm that they are delaminating from one or other of these causes and that those still present are crumbling away and will meet the same fate fairly shortly, thus indicating that total renewal is required. On the other hand, if slipped tiles show no signs of delamination then it is likely that the fixings have corroded or the battens have rotted. It may not be possible, in the absence of full exposure, to be sure which of these possible defects is

20 The overlapping arrangement of the nibbed plain clay tiles laid on boarding was revealed as this 1930s roof was stripped prior to re-covering, necessitated by loft conversion works. The original covering would otherwise have been capable of many more years of service, subject only to minor repairs.

21 The regularity of machine-made tiles, showing disturbance here where they are beginning to delaminate.

22 The underside of a handmade plain tile roof covering secured with pegs. Signs of efflorescence can be seen on the tiles, and elsewhere some of the pegs have worked loose, leaving some hanging on one peg only.

the cause. If the roofing felt on the underside of the tiles has not been laid loosely enough but has been stretched tightly across the top of the rafters, it will prevent drainage of windblown rain from the underside of the battens. If tightly stretched felt is found where tiles have slipped, it would suggest rotted battens as the possible cause. Rot could also be the cause of failure in the fixing of old tiles. Very often, early handmade tiles were nibless and relied on wooden pegs for their security. An old custom to improve weathertightness, particularly if the lap was less than ideal, was to torch with lime plaster and hair the underside of the tiles, covering the pegs in the process. While torching helped to hold the peg and tile in place, it frequently became wet and led to rot in the peg. In either of the above cases, of fixings corroding or battens rotting, it will be a case of stripping the roof, sorting sound tiles and providing and fixing new felt and battens, and relaying the tiles.

Profiled tiles

Roofs covered in profiled tiles have just as long a history of use in the UK as those covered in plain tiles. S-shaped clay pantiles, probably introduced and imported from the continent, were developed as a one-piece tile from the flat or curved undertiles and curved overtiles which made up Roman and Spanish tiling of long tradition. Being roughly twice as wide and one and a half times as long as a plain tile and laid at a lower pitch with only a single lap, they have always been much more economical in use. Against this advantage, however, is that they do not lend themselves to roofs of complicated shapes because they are more difficult to cut, providing problems in forming satis-factory joints at abutments to walls and at projections through roofs, and they are less tolerant of unevenness in the structure below.

As well as the traditional pantile shape, clay tiles also came to be made as single units with either one bold roll or a pair of narrower rolls so that when laid, an impression of Roman tiling was produced. These are known respect-ively as single or double roman tiles (**32**). Manufacture in concrete as well as clay began in the early 1900s, leading to the development of the wide range of interlocking tiles available today.

Finishes can be smooth or sand faced in various shades and also, when glazed, highly coloured, as on clay tiles. In shades of green, purple and blue, their use was popular on some of the rendered and white painted dwellings of the 1930s. Such tiles are still available from the continent with an extended colour range, but it may be difficult to match both shape and shade if replace-ments for tiles from the interwar period are required, although scouring building reclamation yards can prove fruitful. A few of the shapes of profiled tiles available are illustrated here (**23–35**) along with some of the colours that are still seen on dwellings from the 1930s.

Profiled tiles, because they are made from the same clay and concrete materials as plain tiles, suffer from the same flaws in manufacture and the same effects from weathering, particularly frost damage. Even so, the main problem encountered with profiled tiles is found where they are used in circumstances not appropriate to their shape. When they are used over splayed bay windows, over small entrance porches, at the sides of valley gutters and at the junction with hips, the necessary cutting to make them fit is often crudely done; this can remove the interlocking anti-capillary grooves, carefully provided by the makers, which normally help to keep them in place and keep water penetration at bay. Sometimes even the nail hole is removed, leaving a section of tile to rely on its weight to hold it in position and vulnerable to high winds or physical disturbance when gutters are cleared.

Furthermore, the very shape of profiled tiles means that when they abut a hip tile or ridge piece, much mortar has to be used to fill in the spaces

23 Traditional clay pantiles finished with half-round hip tiles. This shows that whenever other than the simplest shapes are adopted, there is always awkward cutting and filling in with mortar, which has a tendency to shrink and fall out – in this case, to the ridge tiles.

24 Replacement traditional-style pantiles on a restored early timber-framed dwelling.

25 Interlocking concrete tiles on the left-hand side of a pair of semi-detached local authority houses of the 1920s, with the original interlocking clay tiles still present on the right-hand side.

26 Shallow-pitched roof of the 1960s with a characteristic profiled concrete tile, widely used at the time and ever since.

between the two. If the mortar is too strong, it usually shrinks excessively and leaves gaps. If galleted – when slips or pieces of tile are inserted – they, in consequence, will become loose and slip down. If too weak, the mortar might not hold the galleting and will let the water through. In recent years, to help overcome these problems and also to provide ventilation to the roof structure,

27 Striking blue glazed pantiles, characteristic of the 1930s. If replacements or complete re-covering should be necessary, owners should not be fobbed off with a tale of 'you can't get them now' as enquiries can often prove fruitful. It may, of course, be a requirement to match should the dwelling be listed or in a conservation area.

28 Glazed clay Roman tiles, coloured both green and blue, from the 1930s, used as copings to the parapets of these flat roofs to art deco-style semi-detached houses. Unfortunately, the curved glass, steel-framed casement windows – the principal characteristic feature, along with the white walls and flat roofs, of this style – have been removed from the bay window of the house on the right and the curved sections of walling have been crudely filled in with brickwork, totally altering the appearance.

29 To even out the flow and lessen the risk of overshooting the gutter in heavy rainstorms, pantiles are sometimes provided, as here, with a few courses of plain tiles at the eaves.

30 Natural variations in handmade tiles mean that perfect alignment is difficult to achieve, especially with pantiles. As long as the differences are not too pronounced, in which case there is risk of damp penetration, these minor discrepancies add character to the resulting appearance.

manufacturers have produced a range of fittings, some in plastic, for use as ridge pieces. The plastic versions come in a limited range of colours and while, initially, they achieve their objective, the matching to the tile shade needs care and their longevity is far less predictable.

32 Old, broken and patch-repaired profiled clay double Roman tiles which point to a roof covering of limited life.

31 Profiled concrete double Roman tiles. As can be seen, each tile has two flat sections separated by a complete roll with the interlocking joint two-thirds of the way round the next roll. These are heavily marked by lichen and algae and there is extensive moss growth spoiling the appearance, which although harmless in itself, requires removal and a surface treatment to inhibit growth because it encourages moisture retention.

33 A very old clay pantile roof, extensively patched with cement in a very rough and ready manner and now in need of complete renewal.

34 Profiled concrete tiles to the simply shaped roof of a house in the form of economical double pantiles. The verge to the gable has no overhang, leaving the detail with little protection and a rather mean look, but at least the tiles at the edge are provided with clips to help resist wind uplift.

35 Cement fillets are often provided to profiled tiles at abutments but lead has been used here. However, the cutting of a groove in brickwork for the top edge in lieu of stepping into raked-out joints is not satisfactory, being much more liable to work loose; and neither is the dressing down on to the profiled tiles adequate.

Slates

Natural slates

Even though a roof may be identified as one of the 20 per cent or so of all dwellings covered with natural slate – as distinct from the artificial stone or asbestos cement varieties – the surveyor should be aware that there are good and bad slates. The good can last for hundreds of years but the bad for very few.

Natural slate, locally mined or quarried and capable of being riven – i.e. split as distinct from sawn – has long been a traditional roof covering in many of the areas where metamorphosed rock occurs. Its use expanded mightily in the late eighteenth century and became almost ubiquitous through the nineteenth and early twentieth centuries when slates from the huge quarries in North Wales could be transported relatively cheaply, first by sea and canal and then by rail, to most parts of the country. Thin Welsh slates roofed most of the dwellings built during this period. Welsh slates are now so expensive that when it is found necessary to renew the roof covering, a cheaper alternative is usually selected. There are still many dwellings remaining, however, with their original slate roofs, provided and installed according to the custom of the time, still unfelted and, accordingly, well ventilated.

The custom of the time looms large when it comes to roofing in natural slate since there is really no scope for improvement in the basic component. Slates have always been sold in many different sizes – up to 30 or more – and usually in three thicknesses between 4mm and 8mm; but apart from being provided with either one or two nail holes at the head or centre, they remain thin and flat with none of the cambering, anti-capillary grooves or interlocking features of manufactured tiles.

All slates eventually succumb to disintegration – the poorer quality, sooner; the better, later. They can be prone to splitting due to strong wind uplift, particularly at verges, hence the preference for slate and a half rather than half slates at this point. They also suffer the depredations of those clambering over roofs to install TV aerials, solar panels and the like. Being a natural material, sometimes the cause of disintegration can be inherent due to the presence of impurities. Iron pyrites, for example, present in the slate when mined, oxidise in the atmosphere into sulphur dioxide. This causes a brown stain and when combined with rainwater, turns into sulphuric acid which attacks the calcium carbonate component of slate, turning it into mushy calcium sulphate.

Even if iron pyrites are not present in the slate, pollution in the atmosphere combines with rainwater to descend as acid rain. Because slates lie flat against each other, the acid is drawn along the underside of the slate by capillary action to continue its attack, particularly in the area of the nail holes, long after it has stopped raining. The signs of such degradation are a lightening of colour on the edges of the slates and powdering, flaking or blistering of the surface, initially much more pronounced on the underside. It is essential, therefore, for the surveyor to keep a sharp eye open for these symptoms.

The effect on the edges of the slates is to separate the layers making up the thickness, allowing water to penetrate, which will cause yet further disruption on subsequent freezing.

Where good-quality slates have been used on a dwelling, it is much more likely that the fixings will fail before the slates themselves. This may be after a life of 40 to 60 years though it may be longer depending on the degree of pollution and exposure and the quality of nails. Nails eventually rust and a strong gust of wind lifting up the slate will cause the nail to break. If single nailed, the practice in some parts of the country, the slate will slip down the slope. If double nailed, it may hang on for some further length of time, perhaps a bit askew, until another gust breaks the second nail. The sight of many missing, slipped or broken slates is an indication of trouble of this nature. So too is the sight of many slates refixed with tingles – strips of lead or zinc nailed to a batten and turned up over the tail of the slate to hold it in position when slid back into its former place. Repairs of this type can often be continued for many years – an option chosen by many owners – but entire relaying will eventually be necessary.

Purely temporary palliatives will occasionally be found applied to slate roofs in an effort to postpone the otherwise imminent renewal of the covering. These include laying hessian, canvas or other mesh material and bonding it with bitumen to the top surface of the slates, a process often described as 'Turnerising' after a previously widely used proprietary version of this treatment. Unfortunately this not only spoils the external appearance of the slates but it also renders all slates on the treated slopes wholly unsuitable for reclamation and reuse. Applying a rendering or slurry to the top surface is also a traditional remedy used in some localities, but all these surface treatments restrict ventilation and are likely to induce rot in the battens. These treatments are very much a last resort and their only arguable advantage is that, as long as the supporting timbers resist deterioration, the surface can be periodically patched or renewed to further postpone the eventual renewal of the covering. The surveyor, finding such a treatment at a property, can only recommend that the client has the roof stripped right back to the structural timbers and completely re-covered.

As external Turnerising treatments have gradually fallen from favour over the last few decades, other methods of holding the slates and nails of unfelted roofs in place have been promoted. One method is to apply adhesive blocks to fix slates to battens and, if necessary, battens to rafters. Another involves spraying the whole of the underside of an unfelted slope, between the rafters but including the battens, with a foam-like bonding agent that quickly sets and secures the slates in place. Like an external surface treatment, this risks causing rot in the now-concealed battens, prevents any subsequent localised repairs and renders the existing slates incapable of reuse in the future. These treatments often come with guarantees of effectiveness for many years and promise to avoid the need for fairly frequent renewal required by the other methods. But to the surveyor, they are a clear indication that a roof covering has reached the end of its effective life and needs to be completely renewed. Depending on the replacement covering selected, the roof structure may also require additional strengthening prior to re-covering.

Since the 1980s, there has been a rapid growth in the international trade in roofing slate with Spain becoming the predominant source of roofing slate in Europe. Imports of slate from further afield, including the Americas and the Far East, are also now established. Unfortunately, by the turn of the century, it was clear that a significant proportion of imported slates were proving less durable than anticipated. Research by the Building Research Establishment (BRE) found that only about a third of imported slates met the requirements set out in the British Standard, which had applied to British slates since the 1940s, and some were showing signs of failure after only a few years of exposure to the elements. In 2004, the British Standard was superseded by a European Standard that imposed a different classification system more applicable to the greater range of sources now available. The onus is now on the user to select a quality of slate suited to the locality and building where it will be used but, of course, none of the relevant information applicable to a batch of slates is likely to be available to the inspecting surveyor unless inspecting a newly slated roof.

More often, however, the surveyor will be presented with a natural slated roof that is neither particularly new nor old looking and on which no information is forthcoming from the current owner of the dwelling. If there is no access for close inspection to either upper or lower surfaces, the scope of advice which the surveyor can provide in various circumstances is covered in the section at the start of the chapter, headed 'The inspection'. With the opportunity for close inspection of either surface, the surveyor will be able to examine and check a great deal more based on knowledge of the material and its characteristics when used as a roofing material in the conditions of exposure and pollution relevant to the area in which the dwelling is situated.

There is also the factor of second-hand slates to be considered. For some years, roofing contractors have been replacing the natural slate roof coverings of dwellings built in the nineteenth and early twentieth century with other materials. Broken slates from the stripped roof end up on the skip as rubbish, but sound slates are usually carefully removed and sorted for second-hand sale for repairs or to those who would rather see their roof covered or re-covered with slates than an alternative material. Many second-hand slates will be sound and suitable for reuse, but others may well be approaching the end of their useful lives.

With post-1980 slate-covered roofs, much depends on the source of the slates as well as the standard with which they have been laid. It may well be that many years of service will be obtained from imported or second-hand slates, but after very close inspection and in the absence of any signs of premature failure, the surveyor can only describe the condition of the roof

36 Sometimes the presence of roofing felt will hide the underside of a natural slated roof, leaving some of the surveyor's questions unanswered. In its absence, much information can usually be gleaned. By the damp stains on the purlin in this case, it looks as though the renewal of the roof covering could not have come too soon, but the surveyor will need to enquire about the quality of the renewed slates and workmanship. The slight sagging correctly given to the felt between rafters prevents accumulations of dust and debris and allows drainage of windblown rain or snow to clear from the upper edge of the battens. Tightly stretched across, wet rot can be expected in the battens in time.

37 It is still possible to view the underside of many natural slated roofs unencumbered by roofing felt and as originally laid. These slates and their fixings are still providing service after some 100 years, though it cannot be long before relaying will be necessary as the nails, visible where driven through the battens, are very rusty. The white material, which has fallen away in places, is torching to reduce the risk of penetration by dampness, more usually applied beneath clay tiles than slates. Despite this hindrance to inspection, there are indications of a lightening of colour at the edges of a number of slates, suggesting that some, if not all, will need replacing.

38 Slipped and missing natural slates due to nail sickness and slippage of an inadequately tacked lead flashing at the change of pitch to this mansard roof. A novel way of refixing slipped slates has been used – the slate is turned head to tail and refixed through the single nail hole by a screw and washer to the batten between the two slates in the row below. Not recommended!

39 The presence of replacement coverings on the roofs of surrounding properties of a similar age is a usually a good indication that a roof covering is at the end of its life, but the condition of this slate roof leaves no room for doubt. If a heavier replacement covering is used, the structure will probably require strengthening. The slate detailing to the chimney and parapet walls is unusual but when well executed, can be effective. Lead flashings, such as with the property on the right, would be preferable to cement fillets when the roof is re-covered.

40 The sight of many lead or zinc tingles on an otherwise sound natural slate roof indicates that the fixings are corroding and that the time for entire relaying approaches, provided the slates are still sound, or complete renewal of the covering if they are not.

at the time of inspection. Hopefully, the abuses occurring prior to the European Standard have ceased so the residue of problems will reduce in time. But meanwhile, surveyors should highlight the risks associated with slates imported during this period. Beyond that, there is little more that the surveyor can do other than explain that there is no guarantee of durability with natural materials.

Artificial slates

A glimpse on the roof of the earliest type of lightweight artificial slates from the first half of the twentieth century, namely those made of asbestos cement, uncoloured or in shades of red or green, need not detain the surveyor for long. As they have not been manufactured for many years and were never considered to have a very long life anyway, despite the evidence that some have lasted for well over 60 years, they are likely to be extremely brittle and very much due for replacement. While there is not considered to be any danger to health from the presence of asbestos cement slates on a roof as long as they remain undisturbed and their surface condition is firm, a special warning needs to be included in any report where the slates are very old and particularly where they are showing signs of flaking, powdering or crumbling. The asbestos fibres released could pose a threat to health, giving even more reason to recommend their total replacement as a matter of early priority. It is not necessary to use a licensed asbestos contractor to remove asbestos cement roofing slates, but it is a requirement to use a 'competent contractor' – a straightforward requirement likely to be met by any specialist roofing contractor who is a member of a reputable trade association.

The same care needs to be taken in regard to those asbestos cement slates manufactured later in the twentieth century specifically as a replacement for natural slates, coloured accordingly and at roughly the same weight as thin Welsh slates, if not slightly lighter. Those found to be flaking, becoming powdery or crumbling need to be removed and replaced with a material unlikely to become a health hazard. As when replacing natural slates, new battens and roofing felt will be required and there is the possibility, depending on the choice of replacement covering, that the roof structure may need to be strengthened.

Artificial slates developed later, made from Portland cement and other materials, such as crushed stone, pulverised fuel ash, glass fibre, polypropylene or polyester resins using crushed slate as a filler, must be dealt with as found. If comparatively new and showing no signs of premature failure, the surveyor should press the present owner for any evidence there may be as to when the slates were installed, their make and type, and whether they came with a guarantee – preferably one backed by insurance in case the manufacturer goes out of business. Although there is no British Standard covering all types of fibre-based artificial slates, some have obtained Agrément Certificates and makers may guarantee their products for particular periods. Such may provide some reassurance to a prospective buyer, but the surveyor should point out that some of the versions of artificial slates did not prove to be satisfactory and that although no signs of degradation may be visible, long-term satisfaction in use has yet to be demonstrated.

Weathering on less than entirely satisfactory artificial slates will begin to show at the edges as either a lightening or darkening of the original colour, an early giveaway if damaged slates have been used. In addition, those slates provided with an applied surface coating will begin to lose this sooner or later depending on the quality of the manufacturing process.

If a roof covering is sufficiently recent and the surveyor is able to find out the name of the manufacturer, it may be that a copy of the fixing manual and specification for the particular type of artificial slate can be obtained or perhaps downloaded from the manufacturer's website. Any guarantee is invariably subject to the installation being carried out in accordance with the manufacturer's recommendations. The surveyor will then have the opportunity, within the limitations of the visual inspection, of verifying that with regard to the dwelling's exposure category and pitch adopted, the correct size slate and lap has been used, the correct fixings have been employed and any special fittings such as ventilators, ridge, hip, valley, verge or eaves pieces are of the same make and have been installed properly. The extent of such further research will naturally be determined, and probably limited, by the type of report being provided. If there are serious doubts regarding the installation in the surveyor's mind, however, it may be necessary to recommend more detailed investigation, which might include seeking advice from the manufacturer.

41 Asbestos cement slates, available before the health concerns about asbestos, are not likely to be mistaken for slates of any other material. By now, they are bound to be due for replacement, yet many are still found – though not usually as heavily covered with moss as this. The roof structure will require strengthening if a heavier replacement covering is used, such as with the adjacent property, but fibre cement slates of a similar weight are available.

42 When newish, some types of artificial slate have a distinctly shiny appearance, as here. There is a noticeable absence of clips to the verge, which should have been provided to prevent uplift; but a roof space ventilator is visible and there is an adequate lead flashing to the parapet wall, complete with lead clips for holding down. On the other hand, a narrow width of slate has been left out, which means that at one point there is the thickness of only two slates instead of three, with a corresponding greater risk of damp penetration.

43 Someone was swindled here as the artificial slates are beginning to slip down, revealing roofing felt stretched tightly across the rafters but an absence of the battens to which slates are normally fixed. It is hard to see what was meant to hold them in position and clearly the whole roof needs urgent re-covering.

44 Fibre artificial slates beginning to bow on the south-facing slope of a bungalow. Will it be progressive, the single slipped slate being an indication of things to come? Slates on the north-facing slope are unaffected, suggesting that further hot summers could cause more damage.

Stone slates

Of all the materials available and used for covering roofs, none is more bound up with the location of the dwelling than what are generally called stone slates, sometimes known as tilestones. For example, it is almost inconceivable to find a dwelling in the Cotswolds with a roof covered with stone slabs from Horsham in Sussex. They would look quite out of place, being a different, much darker colour than the local stone and much larger and thicker than the stone slates common to that area. Not only that but the practice of forming a roof of stone slates is more often than not a matter of long tradition in the area where a particular stone occurs. To pursue the example further, stone slating in the Horsham area presents a formal regular patterned appearance of slabs about the same size and thickness whereas in the Cotswolds, slates of many different sizes and thicknesses are used and the usual practice is to lay them in diminishing courses from eaves to ridge. Furthermore, the weight of the stone slabs would have made it likely that the cost of transporting such material for any great distance beyond the immediate locality would have been prohibitive for any but a very important and exceptional dwelling.

All this suggests, as do so many other aspects of inspecting buildings, that the surveyor should take the trouble to become familiar with the materials typically used in the area where the surveyor practises and even the craft practices adopted to turn those materials into the components which eventually form parts of the buildings which characterise the area. Certainly, the practice of stone slating is one of those cases.

Unlike true natural slates which are always riven – i.e. split from a block of metamorphic sedimentary rock compressed and fused by heat – stone slates are derived from limestone and sandstone sedimentary rocks that have not been metamorphosed. However, they can be turned into comparatively thin slabs either by being split with a hammer or sawn. Also unlike natural slates which are impervious, stone slates are to an extent porous and hence have to be much thicker to take in a certain amount of rainwater and then allow it to evaporate.

Limestone slates are normally split along their bedding plane by a hammer and chisel. They tend to be creamy greyish brown in colour – the brownish shades encountered in Oxfordshire, Gloucestershire, Wiltshire, Somerset and Northamptonshire and parts of Worcestershire and Lincolnshire; the greyish shades, in Cornwall, Wales, Cumbria, and Leicestershire and parts of Lancashire, Dorset and Scotland. They tend to be somewhat thinner than sandstone slates, usually from 12mm to 20mm thick, and used in small random sizes, irregular in shape and thickness and rough in texture. These are all considered good qualities to produce an attractive roof by skilled labour at – because they are thin – a steepish pitch of about 50° to 55° and often of comparatively complicated shape.

Some sandstone can be split to form stone slates, but more often it is sawn to produce regular slabs, which are quite large, sometimes up to 75mm thick, in beige to very dark brown shades. Sandstone roofing is typically encountered in Yorkshire, Lancashire, Derbyshire, Northumberland, Herefordshire and Sussex and parts of Cheshire, Durham, South Wales and Caithness in Scotland. As the slabs are thick and large, the pitch provided is usually much lower, at around 35°, with the slates laid in even regular courses on plain simply shaped roofs.

Inevitably, roofs covered in stone are, in the main, likely to be old and on dwellings that are either listed or in a conservation area. The unevenness, mottled colour and perhaps somewhat unkempt appearance are what attracts and makes such dwellings desirable. Therefore, a new owner will be required to maintain and conserve rather than alter that appearance. A surveyor's ready condemnation of such roofs, and unthinking advice to replace with something prim and proper, according to modern practice, is inappropriate. It should be understood that such roofs, constructed and maintained over the years by craftsmen according to the needs of local conditions of weather and exposure, have often lasted for hundreds of years without the benefits of modern practice. Nevertheless, items wear out and things do go wrong so that new owners may need to dig deep into their pockets to put things right in order for the roof to provide trouble-free protection from the elements for some years to come.

The current owner may, of course, claim that any necessary work has already been done. If so, the surveyor should make enquiries about the nature of such works and, if possible, see any Building Regulation or listed building consent, estimates, details of the work carried out, bills and receipts together with any guarantee which may have been provided. If there is no access for close inspection to either the external or internal slopes, there is some reassurance to be gained from finding no apparent defects visible and no signs on oldish decorations of damp penetration to top-floor ceilings. Where top-floor ceilings have been newly decorated, there is all the more reason to ask questions as this could be an indication that the dwelling has been redecorated preparatory to sale, perhaps with a view to disguising evidence of former or even existing problems.

Signs of defects on external slopes of stone slates, when viewed either from a distance or close up, can include cracked, broken, loose or missing slates, delamination or an over-accumulation of moss. There may also be undue settlement in the slopes or the ridge or in the supporting structure, something which is more likely to be a problem with the larger and heavier sandstone slabs than the more accommodating, smaller limestone slates. It should, however, be evident from an inspection of the roof space whether the roof structure needs strengthening or, if movement is progressive, restraining. One of the problems is that the battens become overstressed if there is undue settlement, sometimes causing them to split but in any event distorting and loosening the securing peg, causing slates to slip down. It may well be that if the condition of the roof is so bad that the slates need relaying, some judicious packing out may be necessary to improve the future lie of the slates.

As the thinner slates, in all forms of stone slating, are usually laid nearest to the ridge, these are more likely to split due to wind uplift, particularly on the side furthest from the direction of the prevailing wind. These should be examined as closely as possible and ridge pieces should be scrutinised for breakages. Lead flashings are less likely to be encountered because they are difficult to fix into stone walls, other than ashlar, so cement fillets are likely to be present at abutments. These should be checked for shrinkage cracking, as should ridge pointing and verge pointing on exposed gables.

When porous stone is laid flat as a coping or at an angle as in roofing, there is always the danger of salts migrating as rainwater evaporates towards the exposed edges. The concentration of salts then affects the binding material of the stone, which becomes friable leading to the edges crumbling away. This is a more pronounced characteristic of some stones than others.

Roofs covered with stone slates, particularly those near trees, can acquire considerable quantities of moss and lichen. While these do not cause damage in themselves, they can interfere with the rapid run-off of rainwater, allowing it to accumulate and correspondingly increase the effects of pollution which will cause more pronounced damage to the stone below. Should the growth of vegetation build up in the side joints, the edges of the slates will be more affected by being in longer contact with polluted rainwater. If the growth is pronounced, advice needs to be given for the vegetation to be scraped off rather than removed by the use of fungicide, which will probably cause unsightly staining if used.

As with all roofing materials, an inspection of the underside, if at all possible, could reveal whether lamination – either inherent or caused by salt action – is taking place and whether this might have affected or be about to affect the area of slate around the peg holes. Should this area become friable and the holes enlarge in consequence, the pegs could become distorted and eventually snap, allowing the slate to slip down. If nail fixing is present, it probably indicates a later roof or relaying at some time in the past, and the nails will need careful inspection for signs of corrosion. Torching may be found as this was a fairly common provision for stone slating. If torching is loose and friable, but there are only a few defects in the roof, its renewal may conceivably extend the life of the roof since, apart from its original purpose of keeping out windblown rain and snow, its presence helps to hold stone slates in place.

As mentioned previously, the surveyor should be aware that the concrete industry has been producing artificial stone slates in reconstituted stone for many years in sizes, shapes and textures to match the traditional varieties for the various limestone and sandstone areas. This awareness is necessary so that, during an inspection, the surveyor can correctly identify what has been used as a roof covering, but also in recognition that artificial slates may possibly be suitable as an alternative material when complete replacement of a defective natural stone slate covering is required as the cost of renewal in natural stone may be beyond the reach of either a present or prospective owner. Some artificial slates are better in appearance and weathering qualities than others, but in their own locality, surveyors will hopefully be able to find out – perhaps from the local Conservation Officer – what are considered to be most suitable for the area concerned. With experience, the surveyor may also be able to come to a view regarding the likely level of cost for the work that may be required although such an opinion will usually only influence the content of the written report rather than be specifically quoted in it.

45 A roof covered in the limestone slates commonly encountered in the Cotswolds, Northamptonshire and the Purbeck area of Dorset – smaller at the top and progressively larger down the slope. Such stone slates lend themselves to steeply pitched, more complicatedly shaped roofs with gables, valley gutters and the like. The irregularities are a characteristic of the areas where such limestone occurs, and replacement with a roof of more formal appearance would be inappropriate and considered undesirable.

46 The plain, shallow-pitched, simply shaped roof of this timber-framed 1400s house is here covered with sawn slabs of the Horsham stone of Sussex. The regular appearance with courses lining up and the slates of a comparatively even thickness is typical of roofs in the sandstone areas of Yorkshire, Cheshire, Northumberland, Caithness, Hereford and South Wales. The term 'tilestones' does seem more appropriate than 'stone slates' for these components. The gap left between slabs is intentional to assist run-off.

47 The gradual increase in the size of the stone slates towards the bottom of this slope is clear, as is the need for some repair and replacement work. Streaks of lighter discolouration to the wall indicate several places where water from overflowing and leaking gutter joints has been washing the wall clean.

48 A roof covered with stone slates under repair, showing the stout timbers required to support the weight.

Thatch

It is estimated that there are between 50,000 and 60,000 dwellings in the UK with roofs covered in thatch, cared for by some 800 to 1,000 thatchers. These dwellings occur mainly in country areas of East Anglia, the South and the West Country but can crop up in many other areas, even city suburbs. Thatch is one of the oldest forms of roof covering, and as it is also one of the lightest of the traditional forms, it is particularly suitable for dwellings where the walls are of unbaked earth – such as cob or clay lump – or where they comprise the less substantial examples of timber framing, which are not strong enough to carry the traditional heavier forms. No thatch seen today can be more than 80 to 100 years old, so its presence on a roof does not invariably indicate a dwelling of any great age and other evidence needs to be sought to establish an approximate date of construction. Indeed, the revived interest in vernacular forms of construction has encouraged the use of thatch on a surprising number of new dwellings. In appropriate settings, thatch not only helps individual newly built properties to blend into older surroundings but can also assist efforts to give recent larger-scale developments, with their less regimented and more random layouts, the appearance of having grown organically.

Nevertheless, many thatched dwellings are of some age and subject to listing or included in a conservation area where any proposed change of covering may require consent and might be met with opposition. This would be so even though the disadvantages of thatch – the ravages of birds for nesting material, infestation by vermin, the danger of fire and the correspondingly higher costs of insurance – may prove burdensome to an owner. The availability of grants in certain circumstances, the good insulation and soundproofing qualities, and the undoubted attractive appearance could well be compensatory factors influencing a decision to purchase or persevere with ownership. It does no harm to advise prospective purchasers of these pros and cons along with other necessary routine precautions, especially those relating to fire risk.

Statistically, homes with thatched roofs are no more likely to catch fire than those with conventional roofs; however, if they do, the results are often rapid and spectacular. It seems worse, mainly because thatch fires in unprotected properties usually cause severe damage, often complete losses. Statistics show that 70 per cent of fires in thatched homes are caused by appliances that burn solid fuel, and the installation of a wood burner or a multi-fuel appliance requires great caution due to the extreme temperatures generated over prolonged periods. A specification for a fire barrier in thatched roofs – known as the Dorset Model – has become established, operating on the basis that the thatch can be made sacrificial in the event of a thatch fire as long as there is an unbroken barrier which prevents fire spreading back into the rest of the property. This specification can only be met with new-build properties or when a roof is re-covered, but with existing properties, other measures are possible to reduce fire risk. In all cases, external chimneys, including the pot, should terminate at least 1.8m above the ridge and there should be a loft access hatch of at least 600mm × 900mm to allow access for firefighting. It should be recommended to occupiers that they install interconnected fire alarms in the roof space and in each floor level as well as extra protection for wiring in the roof space, and that they not use recessed lighting in ceilings below the thatch or external floodlighting below the eaves of the thatch. Other sensible precautions include having an external mains tap with a hose fitted that will reach all parts of the roof and, of course, not being away from the property on Bonfire Night. Further helpful advice on this important topic is readily available from numerous sources and should be researched in detail if thatched properties are likely to be inspected.

In the past, many different types of grown material were used for covering dwellings ranging from rushes, reeds, heather, bracken or broom to the straw obtained on harvesting rye, flax, oats, barley or wheat. While examples of all these may still be found on a local or a regional basis, thatching is now mainly carried out in either water reed, combed wheat reed or long straw. The reeds, 2m to 3m long, obtained from the Norfolk Broads and the marshes and the margins of rivers elsewhere in East Anglia and further south and west have long been considered the strongest and best material for thatching. Cutting and collection has always been arduous and they are difficult to work with, but demand still outstrips supply from these sources. This makes water reed more expensive than any other material for thatching. The quality of reeds has also been affected by the run-off of pollutants from farmland into the water. Consequently much reed is imported from Eastern Europe and the Middle East.

In areas where arable crops are grown, where conditions for reed growing on a large scale did not exist, long straw was the most common form of thatching material in use. There is a considerable difference in longevity, however, between water reed and long straw. A roof thatched well in water reed could last a lifetime in the drier parts of the country while a roof of long straw, even of good quality and closely laid, would only last half that time. In the warmer and wetter parts of the country, if short in stem, ravaged by birds and the weather, it might only last 15 years. However, modern growing methods and modern farming machinery have rendered straw, unless specially grown, somewhat less useful for thatching purposes in normal circumstances; it grows too quickly to the detriment of its strength, it is too short and the straws are crushed by combine harvesters.

Consequently, for thatching dwellings, it is now necessary for wheat to be grown over an extended period to produce longer and stronger stems. When cut, these can then be used in the traditional way as long straw, where the stems are about 1m long and the ears, stems and butt ends can all appear in the finished surface of the thatch. However, when it is cut and specially combed, it is then known as 'combed wheat reed', a distinct misnomer as wheat is in no way botanically related to water reed. The special combing leaves the straw clean and unbroken, ready for use with all the butt ends laid in one direction. Using good-quality material and with a skilled thatcher, a very close similarity in appearance to water reed thatching can be achieved with a life somewhere between the longer life of water reed and the shorter life of long straw.

All thatched roofs keep the rain out because there are a great number of small components to receive and transfer it down a long series of nearer horizontal than vertical stems to the eaves. Whatever the pitch of the roof structure, the stems are always at a shallower pitch. To follow the axiom of the smaller the roofing component, the steeper the pitch – small tiles for steep pitch and large slates for shallow pitch – the steeper the pitch for thatch, the better; 50° to 55° is preferable and this certainly should not be less than 40°. Shallow pitches should be viewed as more prone to earlier biodegradation as the water is in contact with the thatching material for longer. This is one of the reasons why long straw thatched roofs, which traditionally tend to be provided with less steep roof structures, need more frequent replacement. This is so even if long straw is replaced by wheat reed without, as is often the case, the pitch being altered; though admittedly the replacement material will provide a somewhat longer life. The steep pitch coupled with the innate property of permeability helps the thatch to resist wind pressure and avoids the need to weigh it down with ropes and stones as is sometimes necessary in very exposed positions; although securing with steel chicken wire or plastic mesh is often desirable, particularly towards the end of its life.

Thatched roofs tend to have an overall thickness of about 300mm to 600mm, rain penetrating initially to a depth of around 50mm to 100mm before running off at the extremity of the eaves in a band of about the same width, from where it can get blown back against the wall below. Consequently a deep overhang of 400mm to 500mm is required. Gutters and rainwater pipes are neither traditional nor necessary and, indeed, gutters would require extensive brackets and need to be of inordinate size to be effective. However, some provision by way of channelling should ideally be present for dealing with the rainwater when it hits the ground. A good check on whether the thatch drains satisfactorily is to see whether after rainfall, the underside of the eaves is reasonably dry near the wall head on the side sheltered from the wind. Rainwater tends to drain from a roof in a consistent pattern, and over time, this results in the formation of a series of darkened triangles along the base of the eaves. A helpful rule of thumb is that the larger the area of discoloration under the eaves, the older the thatched covering and the deeper into the thatch the rainwater is penetrating. It follows that the further the darkening is from the elevation the better.

Thatch performs best on simple, uncomplicated roofs and any changes in pitch to sweep over top floor or dormer windows should be gentle. For this reason, such windows should be spaced reasonably far apart. Chimney stacks are best sited on the ridge as back gutters are not easy to accommodate.

As with all roof coverings, the surveyor should endeavour to provide the client with a reasonable idea of how long the present covering will last before entire replacement is necessary, together with what maintenance work ought to be carried out in the meantime. In view of the different lifespans of the materials used for thatching, identification of the type of thatch is important. In the case of long straw with the shortest lifespan of around 20 years in the wetter and warmer parts of the country and perhaps 40 years in the very driest, this should not be too difficult. The method of laying is different, leaving lengths of stem visible rather like a traditional style of haircut where the hair is brushed down smooth to lie flat – strokeable, in fact, like a cat. This is not the case with either combed wheat reed or water reed unless it is in very worn and battered condition when it will need renewal anyway. The identification of water reed, with the longest life of 80 years in the driest parts of the country and 40 to 50 in the wettest, as distinct from combed wheat reed, with a lifespan of 30 to 40 years according to location, is more problematic as they are very similar in appearance when laid. With water and wheat reed, only the thick ends or butts of the stems show so that roofs appear similar to a cropped hair style – close, neat and tidy, gently pattable or, as one writer put it, much as a hedgehog is enveloped in its spines.

Location could, at one time, be an initial guide in that a thatched dwelling in East Anglia was more likely to be covered in water reed whereas in the West Country, the covering was more likely to be of combed wheat reed – locally called Devon or Dorset reed in the appropriate areas. With the importation of water reed, this is no longer a reliable guide; and neither is the presence of the traditional patterns along the ridge and down the slopes, typically associated with water reed, necessarily a distinguishing feature since these can all be closely matched in wheat reed. What is needed to determine which of the two types is present is a close examination. If it is at all possible to pull out some stems which remain unbroken, signs of ears and a length of about 1m would indicate combed wheat reed whereas the presence of flowers, long stem lengths of around 3m and a coarser texture would be evidence that water reed had been used as the covering.

Confirmation of the surveyor's assessment of the type of material might be gained by questioning the current owner, though with the usual reservations as to veracity. However if, for example, a receipt can be produced for examination, this could reinforce the assessment and, furthermore, provide a date for when work was last carried out. Contradictory evidence and any elements of doubt in the surveyor's mind should be explained to the client and advice given that an experienced local thatcher should be commissioned to inspect and report.

If the surveyor is satisfied about the identity of the material that has been used, some consideration should then be given as to the remaining life of the two principal components of a thatched roof, namely the thatch itself and the ridge feature. The ridge of a thatched roof should be looked at first as this is the most exposed part, receiving the most battering from the weather and deteriorating faster. It is for this reason that an essential item of maintenance for thatch is the renewal of the ridge capping; this can be expected to be necessary at least once and perhaps several times in the roof's life, depending on the type of material used on the main slopes. Where water reed is the main covering, the reeds are too stiff and brittle to be bent over the apex where the steep slopes meet and some other material, usually sedge – a coarse, pliable, grass-like, water-loving plant – is employed. Sedge is the most durable but, alternatively, tough grass or straw may be used, all normally laced with willow or hazel spars, providing a contrast in appearance with the reed-covered slopes. Such materials are not, in themselves, as durable as water reed and need to be renewed every 10 to 12 years. The same form of capping is added to roofs of combed wheat reed. Straw is sufficiently pliable to be bent over the apex so if long straw is the main covering, the same material can be used for the capping to give greater protection. In this case, the appearance of ridge and main slopes will match in colour and texture.

If the ridge is showing signs of patchiness or any decorative shaping is disintegrating and generally of an uneven and unkempt appearance, particularly at the top of south-facing slopes, it is undoubtedly due for renewal. Neglect to do this at the appropriate time is a distinctly false economy. Deterioration in the ridge releases pressure on the thatch beneath, which can then be washed down the slope, perhaps to be blown away leaving the upper part much more exposed than before.

If the remainder of the thatch is seen to be in good condition then re-ridging, together with an overall examination to tighten the fixings of any loose sections and the cleaning off of any moss, may be all that is required for another 10 to 12 years when the same process may need repeating. At the penultimate time of re-ridging, or perhaps at the one before, it may be thought advisable to secure chicken wire or fine plastic mesh over the whole surface of the thatch, which helps to hold it in place when its strength is weakening towards the end of its life.

However, it may be that renewal of the ridge capping coincides with the need to renew all of the thatch. Signs that this is due are distinct discolouration at the eaves, the pattern of the fixings showing through, considerable loss of shape to any decorative features or patterns, and areas of weathering where an unduly high proportion of stems of water or combed wheat reed can be seen on the slopes. As with all types of thatch, general untidiness, unkemptness, uneven and distinctly more pronounced patterns of run-off showing in certain areas are all signs of degradation. Intruders in the form of rats, mice and squirrels do damage from the inside of the roof space and this can show up as many short, chewed lengths of thatch lying around. From the outside, damage by birds in spring can be extensive, in particular on long straw roofs where they pull out whole stems to build nests. Covering with netting does not always act as a deterrent and if wildlife becomes trapped, the damage can become aggravated. Patching with long straw is sometimes carried out to put off the day when entire re-covering becomes vital, but this shows up very clearly and is only an indication that the time for entire renewal is not too far away.

As with damage to the ridge capping, defects are likely to be far more pronounced on south-facing slopes where the thatch dries out too much but also, conversely, where areas are too sheltered from drying winds. For example, where a thatched slope is sheltered by a bank of partially overhanging trees,

49 Thatch is an unmistakeable covering for a pitched roof, although to believe that it is only found on old dwellings in country areas is disproved by this example on a suburban development, built in the 1990s but already showing signs of wear.

50 A roof in good condition thatched in long straw. The thickness of the thatch at the eaves is due to the practice of only partially stripping back the previous covering before thatching over the top, resulting in the total thickness of the covering gradually increasing each time it is re-covered.

51 This thatched roof has the classic smooth curved 'tea cosy' appearance which distinguishes it from the 'sharper' lines of hips on roofs thatched with water reed. The long stems that characterise long straw thatching are visible on the slopes.

52 The velvety appearance of the finish to this slope and the clearly visible cut ends of the straw mean this roof has been thatched with 'combed wheat reed'; this is, in fact, the same material as long straw but the straw has been aligned, or combed, so that only the butt ends face outwards. Long straw has a mixture of heads and butts so the finished appearance is not quite as smooth.

an undue accumulation of moss can build up, preventing drying out and accelerating the natural biodegrading process through increased fungal activity. Thatch prefers dryness to moisture; hence it will usually last much longer in the less humid eastern regions than the damper areas of the west.

53 The thickness of some thatch can be more fully appreciated on a gable such as this. Here, it can be seen to be approximately 800mm thick on this timber-framed with brick infill cottage. The thickness is due to the practice of thatching on top of old thatch – a not uncommon method.

54 This pair of semi-detached cottages could not illustrate more clearly the contrast between a thatched roof in good condition and one showing signs of wear. The condition of the one on the right is far from dire but it won't be long before the attention of a thatcher will be required.

55 This roof now requires re-covering. The surface has become so worn that the hazel spars used to secure the thatch to the roof have become exposed and gullies are starting to develop where rainwater runs down, increasing the rate of deterioration. The protective wire covering now stands proud of the ridge and all that remains of the ridge decoration is the wire outline of two birds, each containing a sad lump of straw.

56 It would be false economy to ignore the wear and loss of shape to the ridge at the top of this thatched roof. Renewal of the ridge and cleaning off the moss will give it another lease of life, and as it is already covered in a fine wire mesh, it should give years of further service.

57 Familiarity with local forms of vernacular construction, in all their varieties, is a key requirement for residential surveyors. Although there has been a revival of interest in thatched roofs in recent years, it is unlikely that many surveyors will be called upon to inspect a property with a heather-covered roof, even in more remote locations. Those practising in areas where this or any other unusual material might be encountered, however, should ensure they have a proper understanding of the material and principles of use before being tempted to advise.

Shingles

It is conceivable that the surveyor will at times encounter a roof covered in wood, either as boards laid horizontally or vertically or, much more likely, in smaller rectangular sections known as shingles – in effect, wooden tiles. This is likely to happen more rarely than would have been the case in the past.

Wood shingles are another form of roof covering of long tradition. The cedars of the Middle East provided suitable wood in early days and the oak used in medieval times for timber framing and boat building in many parts of the UK also proved suitable, as did other hardwoods such as elm and teak. Indeed, the roof of Salisbury Cathedral was at one time covered with shingles. After the many fires in towns during the twelfth and thirteenth centuries, shingles were perceived, along with thatch, as a fire hazard and prohibited in many urban areas. These bans coupled with shortages of suitable timber and the ready availability of other roofing materials saw a substantial decline in their use. Elsewhere, in North America and Scandinavia, their use continued in popularity because softwood in those locations had always been found to be equally suitable; for example, larch, pine and redwood but, in particular, western red cedar. The importation of the latter to this country in the early twentieth century saw an increase in the number of roofs covered in shingles, particularly in rural areas.

Western red cedar is particularly suitable for use in shingles because of its close straight grain. It starts off as reddish brown when first laid but turns to a darkish silver grey tone on weathering. Shingles are very light in weight, something to the order of a tenth of the weight of tiles, and accordingly may be found on some of the less substantially built bungalows of the 1920s and 1930s, themselves often having light timber frames.

Subject to treatment with a fire retardant, shingles are an acceptable material for covering roofs under the Building Regulations, with certain limitations on distance from the boundaries and other nearby buildings, in the same way as with thatch. They need, however, to be treated in addition with a preservative to prevent insect attack as the natural repellent inherent in western red cedar can get leached out under the heavier rain conditions likely to occur in the western parts of the UK.

As is the case with other materials, the surveyor presented with the sight of new or near new shingles on a roof needs to seek assurances as to the species of wood used and the treatments applied to improve fire resistance and for preservation purposes together with any guarantee or at least an indication of the anticipated length of life. Use of a less suitable species, such as larch, may reduce life expectancy down to 20 years or so, particularly if used in the wetter parts of the country. Western red cedar shingles, on the other hand, might be expected to last something in the order of 50 years if appropriately treated. Good-quality shingles – often known as 'edge grained' – are obtained from quartered logs, alternately rift sawn since shingles taper down from tail to head. They are sawn outwards from the centre of the log at right angles to the annual rings so that the lines of grain are roughly equidistant apart and ensuring that the fibres run continuously for the whole length of the shingle. Cheaper shingles can be produced by sawing tangentially to the annual rings, leaving the rings far from equidistant. When laid, this can lead to early shrinkage in width and warping and splitting when exposed to the elements as well as being more liable to decay.

Where a roof of shingles is neither new or near new and can only be viewed from a distance, signs of warping or splitting coupled with an undue gap between the shingles could indicate that the cheaper flat-sawn variety have been used. If the opportunity for closer inspection exists, such defects may be more easily ascertainable and signs of decay or undue erosion on the exposed surfaces may also be seen. Warping can also occur if, on laying, the shingles are nailed too closely together with a gap of less than the 6mm normally employed to allow for swelling. If this is so, the defect will probably be apparent much sooner than, for example, decay or erosion.

Shingles are normally centre nailed but, even so, wind uplift may occur if they are not very securely fixed. Two nails are needed for each shingle and it is preferable for them to be ring shanked for greater security. Because of their lightness in weight, there is a greater need than with other roof coverings for adequate tying down of wall plates and the roof structure – a point to be checked if there are any signs of disturbance.

58 Terrace houses with rooms half in and half out of the roof space, showing the characteristic colouring and appearance of roofs covered in shingles. The porch roof and the house on the right have been re-covered with natural slates.

Because the run-off of rainwater from a shingled roof is acidic, signs of defects may be apparent in rainwater gutters and pipes if any type of metal fitting has been used, as can also be the case with flashings against or around any projection through the roof surface. Such projections are ideally avoided, but if they are essential, a protective bituminous coating will help though plastic fittings are preferable.

Traditionally, shingles were nailed to battens with no underfelt in order to allow a free circulation of air, and it was not common for felt to be provided. If it is provided nowadays, then counter-battening is essential to allow moisture to run down below and between the shingles. Therefore, the sight of underfelt from the roof space should lead the surveyor to enquire whether counter battens were employed – that is, if it is not possible to check at a tear in the felt whether this was done.

Surveyors should be aware that artificial shingles can be made from wood particles bound together with Portland cement. Although thicker and heavier than wooden shingles, these are similar in appearance to Western red cedar shingles but are claimed to have better resistance to the spread of fire and a longer life expectancy. They have been available abroad for some time and it may well be that opportunities for their use here in similar circumstances to wood shingles will appeal to developers and their designers. If encountered, a surveyor will need to include advice and warnings as he or she would in the case of other new materials brought into use without evidence of long-term satisfactory performance.

Other forms of pitched roof coverings

Occasionally, the surveyor will encounter forms of material covering pitched roofs other than tiles, slates, thatch or shingles. The metals copper, lead and zinc can be used on the slope, the two former often used to crown features such as canopies, porches, balcony roofs and turrets; but their main purpose in domestic building is as coverings for flat roofs on the more expensive dwellings. Their characteristics are the same and the detailing is not wildly

59 The characteristic green patina of sheet copper is unlikely to be confused with any other material when used on the slope, as here.

60 The differences in these roofs mirror the development of this school site. The buildings at the rear and on the left are the oldest and these are roofed with copper. When new buildings were constructed (centre and rear right), they were roofed with green-coated metal sheets to blend with the original buildings. Closer examination reveals that not only is the match not exact but the sheets on the newer roofs run from ridge to eaves in a single length without the necessary joints in the copper covering. It seems that when the most recent addition was built (front right), there was less of an effort to provide a match.

61 Much of the roofing of this development of light gauge steel-framed houses is not readily visible from ground level, but sufficient can be seen for the surveyor to realise that the covering is some form of sheet metal. The most likely material is coated steel but, unless information is available from an occupier, enquiries will need to be made to confirm this. The distinctive characteristics of lead, zinc and copper are absent and indeed the use of such traditional materials would be unusual on a development of this type.

different when used for either purpose. These will be dealt with in Chapter 5 under the respective headings for each metal.

Profiled steel sheets are available in a variety of designs intended to replicate conventional roofing tiles, and these can be sufficiently convincing to trip up the unwary. These are especially popular on mobile homes, intended to give an air of permanence, but they are also used on conventional housing (**64**), including by at least one New Town development authority. Their lightness means they are also useful when covering over flat roofs with pitched structures. As their manufacturers typically only offer a life of around 25 years, however, they should not be regarded as a long-term roof covering. Other

62 A lead roof on a slope, showing the characteristic near round batten roll side joints. Whereas for lead flat roofs, the joints across the fall need to be in the form of a drip with a change in level not less than 50mm and preferably more, this is not necessary on a slope. Lengths cannot be too long or the stresses of thermal movement would lead to splitting.

63 Colour-coated steel panel roofing to a modern block of flats. The surveyor will hope that more details about the roof will be available from the management company or their agent but, at times, will be disappointed.

64 At first glance, this bungalow roof seems to be covered with interlocking concrete tiles, but closer examination reveals terracotta-coloured steel tile sheets. The best clue is at the gable end where, instead of conventional detailing, a metal profile covers the edge of the sheet covering. Subtle clues are present once suspicions are aroused – for example, joints between sheets and even dents may be visible – but this is certainly a potential hazard for an inspector in a hurry. This roof covering may help to distinguish former BISF houses (see **286** in Chapter 14).

65 Corrugated asbestos on the roof of a two-storey house probably dates from the 1950s. Clearly, the house is in need of renovation and a more permanent form of roofing that is less hazardous to health is required; but how will the structure below need to be adapted? Costs will be increased by the need to comply with the regulations for the removal of asbestos.

66 The corrugated 'iron' – in fact, steel – to this roof is showing signs of rust and needs a new protective coating. The walls are also clad with the same material but these have been maintained to a better standard.

67 One wonders what was said about the roofs to prospective purchasers of long lease flats in this block when it was new, perhaps 15 to 20 years previously. The reflective surface felt is not lasting well and is already patched above the right dormer. The flashing to the separating wall is looking suspect and the one opposite has already had to be replaced with other material. All lessees are going to have to dip deeply into their pockets for a new roof covering very soon; for a long-term solution, perhaps tiles on the sloping surfaces with asphalt or a metal on the flat sections above would have been infinitely preferable.

materials that may be encountered, such as corrugated steel and corrugated asbestos, can be discounted as being wholly inappropriate covering materials for dwellings and should be advised for renewal in a more permanent form.

Much the same can be said for bituminous or other forms of felt, including felt made in the form of tiles, sometimes incorrectly and confusingly known as 'shingles' in the building trade. Failure with the early types of built-up felt roofing and the workmanship involved has long been evident although a new generation of more durable reinforced bitumen membranes (RBMs) are now available. These are becoming widely used on flat roofs and are dealt with in the relevant section in Chapter 5, but they are not yet generally accepted as a covering for pitched roofs, particularly in domestic situations where the simplicity of structure more often found in commercial premises is usually not achievable. The use of such coverings on outbuildings such as garages is a different matter since defects do not take on such a critical importance. Clearly, however, their condition needs noting along with whether their renewal is required.

Solar generation installations

While considering pitched roof coverings, it is also necessary to briefly consider solar generation installations, which are increasingly encountered. The majority are retrospective installations, requiring disturbance of the existing covering, but in some cases, the panels are an integral part of the roof surface (**68**). Integral panels obviously overcome the inevitable problems associated with interfering with an existing covering. It seems possible that integral panels may become an expected part of the specification for new-build dwellings, but they are not economical for existing properties unless the roof covering already needs to be replaced so they are unlikely to predominate for the foreseeable future.

The surveyor is not expected to report in detail on a solar installation, but there should be a willingness and capability to make some basic observations and to report matters which come to the surveyor's attention when examining the building. To avoid the embarrassment of making a simple factual error, the surveyor must be able to distinguish which type or types of solar installations are present on a roof. The most common are solar photovoltaic (PV) panels which generate electricity. These have a distinctive grid pattern with a dark blue reflective surface. Similar in appearance but with a plain, more matt black finish are solar thermal panels that generate hot water. To the casual observer, which the surveyor cannot afford to be, it is easy to confuse these two types; but they perform different functions and the associated internal fittings are completely different, one being connected to the electrical installation and the other to the plumbing. An alternative method of generating hot water is from evacuated tube collectors, which are distinctively arranged side by side in sets of a dozen or 20 or so, occupying a similar area of the roof slope as a thermal panel. It is not particularly unusual to see a single roof with both electricity and hot water generation equipment (**69**).

Although PV panels are relatively recent introductions to the urban scene, the technology they represent is well established and very durable. It is more likely that internal fittings will need to be replaced before the panels themselves so there need be no concerns about their longevity. By contrast, solar thermal generating systems contain liquids and may be vulnerable to leakage, evidence of which might be seen on a roof slope, and the vacuum in evacuated tubes can fail, leaving a distinctive white appearance (**70**). Solar installations naturally operate most effectively where the sun shines most strongly, so they are better on south-facing slopes and more efficient in the south of the country than the north. As the roof slopes they are fixed to start to face further away from the south and towards the east or west, they operate progressively less efficiently until slopes which start facing more northerly are unsuitable for installation. Similarly, any shading will adversely affect their efficiency. PV panels, in particular, are very sensitive to even partial shading as the electrical connections mean that shaded areas can affect cells in unshaded areas, significantly reducing the electricity being generated. The surveyor should report any significant matters that come to attention such as obvious defects or, as has been known, panels in wholly inappropriate locations.

All types of solar panels form part of the dwelling's services so a comment will need to be made in that section of the report, and there may be planning and legal considerations which should be referred to the client's legal representative. There are also a number of matters relating specifically to the building that the surveyor should consider when solar installations are encountered. Panels are typically fitted with a 25-year timescale in mind so it is important to consider the general state of the roof covering. Ideally it should be in a good condition but that is not always the case; if it needs repair or will clearly need to be re-covered within that 25-year period, the client should be warned that the presence of the panels will create additional complications which will need to be taken into account.

Panels are usually secured to frameworks fixed to the supporting structure beneath the roof covering. With single lap tiles – typically interlocking concrete tiles – this is a relatively straightforward process which involves easing individual tiles out of place to expose the timbers beneath, screwing the fixings to the underlying rafters and then replacing the tile after it has been slightly cut, notched or filed down to neatly incorporate the projecting fixing (**71 and 72**). It is inevitable that as workmen scramble over the roof to carry out this work, tiles will be damaged or broken and these will need to be replaced. Once the framework is secure, the panels are fixed and the roof covering beneath is effectively invisible and cannot be inspected, so it is important that only reputable installation providers are used to ensure that all necessary repairs are carried out and work has been properly completed.

Much greater problems are likely to occur with double lap coverings such as plain clay tiles or slates. Exposing the roof timbers beneath these coverings is much more difficult and time-consuming than with single lap coverings, and making good around the fixings is awkward and can be complicated, ideally involving individual lead flashings (**73**). The amount of collateral damage requiring repair and replacement can also be significant. Indeed the execution of high-quality installations can be so demanding that some providers simply refuse to install panels on roofs with such coverings and instead concentrate solely on those which are so much quicker and easier to complete to a satisfactory standard. Unfortunately, as with so much in life, there is an inferior quick fix available which fills this gap in provision. Instead of carefully removing and replacing the slates or tiles, it is possible to drill blind straight through the covering into (hopefully) the supporting roof timbers and make the fixing directly, waterproofing the resulting hole in the covering with a silicone or mastic sealant. Although not recommended, if properly carried out, this can provide an effective support for the panels although the long-term durability of the waterproofing may be questionable. However, it will come as no surprise to learn that unfortunately, yet again, installers adopting this process do not always operate at the highest level, and there are examples of drill holes for the fixings missing the rafters completely so the framework is only secured to battens, the fixing seals not being adequate and so on.

The difficulty for the surveyor is that once the solar panels are in place, there is very little that can be seen to assess the standard of the supporting installation. Sometimes it is possible with binoculars to see fixings beneath the edges of panels and this may help an assessment. Where possible, checks should also be made from within the roof space to see whether there are any indications of substandard installation, damp penetration or other damage. As double lap roof coverings are more likely to pose difficulties, they should always be checked particularly carefully (**74**).

68 An integral solar thermal panel for hot water generation. This was fitted as part of the roof covering on a new-build dwelling, avoiding the disadvantages of retrofitted installations. Thermal panels typically have a dull, matt finish, unlike the distinctive grid-like pattern visible on PV panels for electricity generation.

69 An array of PV panels for electricity generation along the bottom of this slope with evacuated tubes above – an alternative to thermal panels, as in **68**, for hot water generation.

70 An array of solar evacuated tubes, one of which has the characteristic white colouration indicating that the vacuum has failed. Fortunately, individual replacements are available.

71 Fitting the supporting framework for a retrofitted installation is relatively straightforward with single lap coverings. Brackets are fixed to the roof timbers and only minor making good is needed to allow the tiles to sit flush afterwards.

72 After the metal framework has been fixed to the brackets (top) and the wiring prepared and fed beneath the roof covering back into the roof space (centre), the panels are fitted in place and connected to the wiring.

73 Once PV panels are in place, only the very outer supports can be seen, and then often only with great difficulty even with binoculars. This fitting onto a slate roof appears to have been carried out in an exemplary manner with lead flashings around the projecting bracket base. It seems reasonable to assume the remainder of the installation has also been finished to the same high standard.

74 There are several missing tiles apparent near both upper and lower fixings on this double lap, plain tile roof. If the PV panels were not present, every surveyor would consider this unsatisfactory and report it. Simply because the panels are there does not make the situation any better. The situation should still be reported although remedying the faults will be more difficult because the panels will need to be removed to allow access. This is part of an eight-panel array, so how many more areas are likely to require similar repair?

5

Flat roofs

The inspection

Flat roofs are normally regarded as those with a pitch of below 10° and, in the main, their coverings differ substantially from those used to cover pitched roofs. While those used to cover flat roofs can be used satisfactorily to cover pitched roofs, the reverse is by no means the case. Because water is in contact with the flat roof surface for much longer periods, conditions are much more onerous and coverings have to be impervious whereas those for pitched roofs can afford to absorb a proportion of water for it to evaporate when the rain ceases. The components which make up most pitched roof coverings are small and capable of accommodating and adjusting to a limited amount of movement in the structure below, far more so than those used normally to cover flat roofs. Ideally flat roof coverings are seamless and have no joints but the only material in use now which fulfils that ideal is asphalt. All others are delivered to the site in sheets and have to be jointed appropriately as they are laid according to the type of material.

Traditionally, flat roofs have been covered with lead, copper, zinc or mineral felt. Mineral felt, widely used until the 1980s, can perform well if laid to a good standard and subsequently protected, but it has acquired an understandable reputation for premature failure. It remains an economical covering for garden sheds and the like, but with a relatively short lifespan – 10 to 15 years is often quoted – it has now been superseded as a suitable material for long-term use on dwellings by modern reinforced bitumen membranes (RBMs) with polyester reinforcements and possibly polymer-modified bitumen. Mineral felt will still frequently be encountered for assessment during inspections, but when a replacement covering is required, modern alternatives, which are far more robust and less susceptible to degradation and have a service life of up to 50 years, should now be used instead.

Being unable to see the construction is, of course, the bugbear on providing suitable advice on the condition of flat roofs. As far as possible, however, the surveyor should attempt to determine the nature of the construction and the decking on which the roof has been laid because the differing constructions give rise to different problems; an awareness of the actual construction will greatly assist in diagnosing defects.

A cold roof or cold-deck roof has insulation placed directly on or just immediately above the ceiling and below the roof deck. This is the usual arrangement for pitched roofs and lofts but it creates a risk that moist air will condense on the underside of the roof covering, hence the requirement for ventilation of roof spaces. Vapour control barriers can also be fitted below the insulation but the practicalities of preventing condensation in cold-deck flat roofs are such that they are now often converted to warm-deck roofs.

With a warm-deck roof, or sandwich roof, the insulation is located above the roof deck and below the covering with a vapour control layer placed between the deck and the insulation. The waterproof covering is placed on top of the insulation. On an inverted or upside down roof, the insulation is placed above the waterproof membrane and paving slabs or decking are used to protect the insulation and provide a surface to walk on, but also prevent wind lifting the insulation. It has become popular for new apartments to have their own balconies or terraces but it can often be impossible to see or inspect much of the insulation or waterproof covering without lifting the protective slabs or decking on the surface, although it may be possible to identify whether the roof is of inverted or warm-deck design.

Considering the approximate age of the property might be thought to be a help, but it does not provide any degree of certainty. Cold-deck flat roofs are likely to be found on dwellings built before the 1960s. If the covering has been renewed and there is no ventilation, there is no telling for certain whether the roof has been upgraded to either a warm deck or inverted warm deck or left as it was, possibly with dangerous consequences. On the other hand, the presence of ventilators, whether old or new, does provide the assurance that ventilation has been provided – essential if a cold-deck roof has been retained. It is best for the surveyor not to speculate on what might be and to confine statements to whether construction is hollow or solid and whether or not ventilation is provided, with appropriate warnings of the possible consequences if not.

Apparent failure of a flat roof covering can usually be attributed to one or more of four causes: constructional moisture, physical failure, condensation or damp penetration actually originating from elsewhere. Those undertaking pre-purchase inspections are much less likely to encounter problems associated with constructional moisture, but an awareness of this possibility should be borne in mind if problems are found during the first years after construction, especially when a roof has been formed above solid construction such as poured concrete slabs or wet screeds laid to falls. Physical failure can occur when thermal or differential movement stresses the covering material and causes splitting. Condensation can cause more problems in buildings than seemingly more likely failures in coverings. The problems of providing effective ventilation to cold-deck roofs with the increased levels of thermal insulation now required mean that warm-deck roofs are now recommended in all cases to overcome condensation issues. Leaks blamed on a defective roof covering can often be due to a failure in a different building component. The surveyor should not overlook the need to check the usual suspects, such as parapets, copings, damp-proof courses, windows, roof lights, gutters, and even service pipes within the roof void, when trying to identify the source of dampness below a flat roof. A helpful checklist of areas to consider when inspecting flat roofs is given in **Box 1**.

Although the surveyor's ladder should allow access to the upper surface of most single-storey structures with flat roofs, it is better for safety reasons to commence the inspection internally to endeavour to establish whether the roof is of hollow or solid construction and whether subject to leakage. A sharp tap on the ceiling with the surveyor's measuring rod, or a handy broom handle, will indicate by a resonant sound that it is hollow or a dull thud that it is solid. If the roof is poorly insulated, there might be pattern staining on a ceiling not recently redecorated, giving an indication of the spacing of joists. In the majority of cases for domestic premises, the construction will be found to be hollow, and if the structure lower down matches the rest of the dwelling then it will probably bear a distinct relation to the floors elsewhere. This is likely to be true even if the structure is found to be solid, as it might well be in the case of a flat roof to a block of flats.

A consideration of the general condition both internally of the ceiling and externally of the roof should be made next to help towards deciding whether it would be wise to walk on the roof. Undue neglect, damp stains on the ceiling and an aged covering with visible defects when viewed from the ladder in the case of a single-storey structure, would indicate that it would be unwise to do so. Hollow timber-constructed flat roofs have been known to collapse when warmth, damp and a lack of ventilation provide ideal conditions for outbreaks of dry rot. The failure to warn a prospective purchaser of an upper maisonette in a converted house of the lack of ventilation to a hollow flat roof, and the possible consequences of damp penetration through badly constructed upstands without angle fillets or cover flashings to the felt roof, led to a surveyor being found negligent in the case of *Hooberman* v *Salter Rex* 1985. It transpired that chipboard had been used as a decking with the vapour control layer placed on top instead of underneath, resulting in its disintegration through condensation and extensive dry rot in the timber structure. It is noted that the surveyor was found negligent for failing to warn the purchaser of possible consequences derived from what he could see, not for failing to report on unsound construction that he could not see.

Nevertheless, a close visual inspection of the ceiling can provide crucial information. Damp stains have already been mentioned but cracks in the plaster can be caused by deflection in the structure. This could be due to inadequacy in its strength due to, for example, the use of joists of insufficient depth, the formation of a trapdoor without proper trimming or the placing of heavy items on the roof that were not envisaged at the time of construction – perhaps a

---

Box 1 Flat roof inspection checklist

The RICS Information Paper *Flat Roofs*, second edition (2011), section 1.4, lists the following areas where it is advisable to pay attention when inspecting the covering of a flat roof:

(a) roof lights and roof light upstands;

(b) parapet walls and copings;

(c) verges, eaves and edge trims;

(d) abutments with adjoining roof features (such as pitched roof slopes and walls);

(e) movement joints;

(f) expansion joints;

(g) ventilators;

(h) vent pipes to above ground drainage systems;

(i) roof vent pipes;

(j) steps and changes in level;

(k) flashings;

(l) handrails, gantries and fire escapes;

(m) lightning conductors and aerials;

(n) parapet gutters;

(o) ventilator and air conditioning equipment;

(p) patent glazing;

(q) pedestrian areas/promenade tiles, etc.;

(r) method of surface water discharge and the route it takes; and

(s) the number, location and size of rainwater outlets.

Source: copyright RICS – reproduced with permission

---

cold water storage cistern or a tank – or the formation of a terrace. Many flat roofs of timber construction are supported on wood plates set halfway into one-brick walls, which are very much exposed to driving rain. The rain can penetrate the brickwork and causes either wet or dry rot in the plates with consequent deflection in the structure. Compounding the defect, deflection is a frequent cause of failure in the covering itself.

Deflection might be a permanent condition and the surveyor's ladder will facilitate a scan across at ceiling level to see whether or not the ceiling beneath is bowed. If the deflection is intermittent from foot traffic or the occasional heavy fall of snow, the ceiling is more likely to be merely cracked and without permanent bowing because the structure will recover once the loading has gone. Of course, the presence of bowing does not necessarily indicate a defect in the supporting structure. If the key to a lath and plaster ceiling has failed, the ceiling plaster may, depending on the circumstances, be bowed to a greater or lesser extent and even possibly in danger of collapse. A gentle push with the palm of the hand should indicate whether the key has failed or not. Whether to do this must be at the discretion of the surveyor who ought, first, to obtain the vendor's permission, having explained the possible consequences.

Deflection in the supporting structure will often result in ponding because the fall, which should have been provided on construction to counteract those permissible amounts of deflection allowable in all forms of design and construction, will be compromised. Even on seamless asphalt, ponding can cause problems by encouraging a build-up of moss or other organisms – particularly near trees – which can obstruct outlets. In extreme cases, ponding can allow a build-up of water to such an extent that flashings to pipes or stacks are covered, resulting in damp penetration. With sheet materials, the build-up of water can breach the joints themselves, particularly if blown by the wind. For sheeted roofs that have stuck or welded joints, the persistent presence of water will soon reveal any faults in the workmanship.

There was a time when it was thought better to construct seamless asphalt roofs without a fall and to form a reservoir with a weir to control the level of water so as to obviate the effect of the sun on the material. Certainly for domestic purposes, excessive deflection, obstruction of the weir and drying up in periods of drought have tended to provide greater problems. The provision of a proper fall and a light-coloured reflective surface to the material are now considered preferable.

The effect of relatively intermittent ponding will still be evident on the dry surface of a roof covering by the whitish colour of the deposits left behind by the slow evaporation of a quantity of water. More persistent ponding will reveal itself by the vegetation which has become established. It will be encouraged and more firmly embedded if the covering is rough rather than smooth; for example, mineral surfaced felt or the more corroded metal surfaces.

Finally, the increasing demands of the Building Regulations will have implications for any works where a covering is being renewed. Indeed, at the time of writing, if more than 25 per cent of a given roof area is being re-covered, there is a requirement to improve the thermal insulation to a stated level where reasonably practicable. The additional cost will be offset, to some extent at least, by energy savings, but the client should be made aware of the potential implications when recommendations are being made to renew a roof covering.

When reporting on specific flat roof covering materials, the surveyor should be particularly wary of venturing too far from observable facts into unsupported speculation. Clients should, by all means, be warned about the uncertainties of concealed areas and the potentially serious risks associated with any degree of damp penetration; but the surveyor who does not have direct experience of specifying or supervising works relating to flat roof coverings or who has not researched the material in question may need to exercise a degree of caution in reporting. Any sweeping judgements – whether of condemnation or approval – are liable to be exposed as superficial and of little value, if not seriously ill-informed, if an expert should become involved at a later date for whatever reason. Familiarity gained through regular encounters with a particular material, such as mineral felt, will enable the surveyor to differentiate between examples that have been laid well or laid badly, and those which are in better or worse condition. However, inspections of a roof covered with copper, for example, are likely to be much more infrequent; and the challenges and implications of reporting on an extensive area of lead-covered flat roof, perhaps on a large period property, are very much more serious than when considering a lead covering to a modest 1930s bay roof. This is an area where the surveyor may need to be particularly conscious of the limits of his or her professional competence and be prepared to seek specialist advice when venturing into unfamiliar territory. The RICS Information Paper *Flat Roofs* (second edition, 2011) is a first-class resource for information on coverings, their defects and the possible remedies.

Mastic asphalt

Naturally occurring bitumen mixed with sand has long been used for waterproofing purposes in the Middle East, but mastic asphalt did not come into use as a covering for flat roofs in the UK until the middle of the nineteenth century. Initially, the material would have been entirely the imported natural mineral rock found in parts of Europe and containing a percentage of bitumen. Another bitumastic product laced with mineral material was found in a lake in Trinidad and refined and imported. Later, the distillation of petroleum was also found to produce bitumen and this derived product, when mixed with a limestone aggregate, produced a synthetic mastic asphalt. Cast into blocks, it could be delivered to the site in the same way as the natural product. The synthetic product is modified with the addition of polymers to avoid cracking in colder climates and strengthened against indentation with further mineral additives to produce a surface suitable for foot traffic.

A surveyor inspecting a flat roof covered in asphalt will not be able to tell whether the material is natural or synthetic or how it may have been modified since there is no observable difference when laid. On the inspection of a new or near new dwelling, it might be possible to examine the specification; but even then, there is little possibility of checking that the work has been carried out in every detail. It is thought that the natural might be more long-lasting than the synthetic, but any difference there may be is materially insignificant since it is usually the detailing rather than the material itself that lets the covering down. Well laid to a sound specification, a life of 50 to 60 years may be expected from asphalt – sometimes even longer.

On delivery to the site, the blocks of asphalt are heated in a cauldron to become a thick paste which is spread, usually in two coats about 19mm thick, on an isolating sheathing membrane to separate it from the structure below – either a timber decking or a cement screed on concrete. Both decking and structure should be as firm and stable as possible although no movement joints are required as the asphalt will take up any slight imperfections and accommodate any slight movement.

The fact that the blocks of asphalt have to be melted down to produce a spreadable material provides a clue to its main weakness. Laid on surfaces no more than 10° from the horizontal, it will perform well, only softening under the effects of hot sunshine and not expanding as do the metals used for flat roof coverings. However, laid at an angle – particularly the 90° required at abutments to walls, chimney stacks, roof lights and pipes – in the full glare of the sun, it will sag and creep downwards by gravity and its own weight – a fault commonly found. Angle fillets of 45° with lathing to provide a key at the junction of horizontal and vertical surfaces are therefore essential along with, where there is brickwork, joints well raked out to a depth of 20mm. Furthermore, a reflective white or silver-coloured paint should be applied to the asphalt before a flashing at least 75mm deep, preferably of lead, is well tucked into the brickwork and dressed down over the top of the asphalt. To reduce the effects of solar gain, an inverted warm-deck roof, where insulation is placed above the asphalt and also included at the abutments, is a preferable type of roof construction for asphalt, along with a hefty type of ballast to hold the insulation down against wind suction. Inverted roofs have only come into use since the 1980s, and unless details of the specification at the time of covering should be available, the surveyor will only be able to report on the surface material and any visible areas of asphalt, which are likely to be fairly limited.

Blistering of the asphalt is another defect encountered fairly frequently, though much more so if the structure below is of concrete rather than timber. It is due to entrapped moisture which expands on warming by the sun, pushing up the material. The fitting of small dome-hooded ventilators at the time of construction is the way now adopted to anticipate this problem, but if not incorporated at that time, it is possible to have them inserted at a later date should blistering appear. The normal method of poulticing with hot asphalt to carry out repairs can be used to remove a small section of roof covering for their insertion. Small blisters can be left undisturbed, but blisters larger than 150mm^2 indicate a more serious problem that should be investigated and remedied. If nothing is done, there is a possibility that the blister will expand further until it bursts, with almost certain consequential damp penetration.

Deflection in flat roof construction has already been mentioned, and this might be the cause of any crazing seen on the top surface of the asphalt where excessive trowelling on laying has brought a more fluid and less dense coating of bitumen to the top. This is usually of no consequence and can be distinguished by its patterned effect from cracking through the total thickness brought about by movement in the decking as distinct from the structure. If expansion occurs in the decking, this could produce an upward pressure and induce longer, straight cracks in the asphalt.

Cracks in the asphalt, other than crazing, together with indentations caused by foot traffic perhaps in excess of that originally envisaged can be repaired. Also repairable are skirtings which have lost their adhesion to any projections through the surface of the asphalt. There will come a time, however, when the extent of the visible defects coupled with a worn, degraded appearance of the general surface will suggest to the surveyor that the client should be advised that the covering be renewed. Warnings of what might be found if there are already signs of damp penetration need to be given. It would then be up to the client to decide whether to continue with further uneconomic repairs or to follow the surveyor's advice to renew.

75 Surface crazing to a mastic asphalt roof is usually brought about by over-trowelling bringing the mastic content to the surface. It is generally of little consequence. That is clearly not the problem with this roof, which has deep cracks and multiple blisters. Asphalt has the advantage of being suitable for the repair of localised defects when the rest of the surface is sound, but that is patently not the case with this covering.

76 Cracks in the asphalt surface are usually the cause of damp penetration, as here. Any movement in the supporting structure can often cause straight cracks and needs investigating. The badly and repeatedly patched projecting parapet indicates persistent problems. Total renewal is required, coupled with warnings about what might be found underneath.

77 Areas of patching to this asphalt roof indicate past problems, but grass growing along the line of the more serious crack at the change in levels shows that the patching has been ineffective. More thorough remedial works are required, including resolving the problem of differential movement between the separate supporting structures above which the roof has been laid.

78 Asphalt skirting to a flat roof beginning to sag downwards through the action of the sun and pushing the flashing outwards due to a lack of clips, which will eventually cause damp penetration. This is a case where the upstands could be renewed because the near horizontal asphalt surface is sound and intact.

79 An asphalt roof in a very sad state – the cracking, plant growth and peeling edges all indicative of neglect and the need for replacement.

Lead

Along with zinc, lead is the most likely metal to be encountered on flat roofs over dwellings. It is heavier, softer and more malleable than zinc and, therefore, more likely to be found as the material chosen for covering porches and canopies of complicated shape as well as balconies and dormer windows. The difference in cost between lead and zinc usually finds the former provided on the better quality of dwellings, but less so nowadays than in the past. The difference in cost is justified by the difference in the length of life – something like 100 years for rolled lead (formerly known as milled lead), which is perhaps double that of zinc. In earlier times, lead was cast on sand in sheets which came out much thicker and, when reversed, produced a rougher texture for exposure to the weather. This could last, astoundingly, to the order of 400 years depending on the amount of pollution in the atmosphere. The use of cast lead is nowadays generally confined to historic buildings. Lead is also produced using a continuous machine cast method but this is not manufactured to British Standards.

Although a shiny silver when new, rolled lead soon develops a patina of lead carbonate when exposed to the atmosphere, which turns it to a medium-grey shade but provides the metal with a strong degree of protection against pollution. Lead has a relatively high coefficient of expansion, which has to be allowed for in the design of joints. Side joints to sheets are usually accommodated in near round wood batten rolls in the direction of the fall. Standing seams, although possible, are too easily damaged on flat roofs and, if encountered, will usually be found to be so. Lengthwise joints, so as not to interrupt the fall, are welted drips and involve a slight change in level – not less than 50mm and preferably more. Any resulting build-up of windblown leaves, moss or grit can cause flow problems towards outlets and perhaps compromise the joint.

The need for a change of level at lengthwise welted joints can cause design problems in long flat roofs and even more so in parapet gutters where there is limited height. This sometimes results in detailing which is less than ideal and intermittent problems of damp penetration, particularly after heavy falls of snow. A critical examination of the detailing is, therefore, essential in such circumstances so as to warn the client of possible problems. The detailing can be checked against best practice available from the Lead Sheet Association, which can be obtained from their website (www.leadsheet.co.uk). Checking details is facilitated by the malleability of the metal, which permits it to be lifted at rolls, drips and flashings to see whether the detailing is sound. While checking the detailing is relatively simple, checking that lead of the specified code or thickness has been used is less straightforward unless delivery notes and invoices are available when a new or near new dwelling is being inspected.

Over the years, the continual intermittent movements under the effects of hot sunshine can induce metal fatigue and the development of ripples, buckling and, in due course, splits and cracks. These can be repaired by burning in new lead, but the frequency of their occurrence will eventually render repeated repair uneconomic and renewal will be required. This may also involve renewal of the decking.

Despite the protective coating of lead carbonate, which gives good protection from normal atmospheric pollution, lead is still vulnerable to chemical attack from a variety of sources that can cause pitting and corrosion. Acid rain from a badly polluted atmosphere can cause problems, but aggressive run-off can originate from a surprising range of materials and these can be concentrated on specific points of a roof as water flows downwards. Acids are present in the run-off from copper, perhaps from an adjoining roof or a lightning conductor; cedar shingles or oak details, possibly in sills; from other timbers treated with preservatives; and from moss, lichen or algae which can be present on otherwise innocuous tiled or slated roofs. The remedy in any of these situations is either to remove the offending materials or to replace the lead roof with a material that will not be affected by the acidic run-off. A further alternative is to provide a sacrificial flashing which will need to be renewed periodically but which will protect the main covering.

Even the alkali in mortar attacks lead unless it is has been protected on both sides with a coating of bitumen. This protection was often not provided in the past, leaving the tucked-in section at the top of flashings vulnerable to

80 Ridges have developed as repeated thermal movement has caused the lead on this bay roof to buckle. As movement has continued, splits have then opened up on the top and along the full length of each ridge. There is surprisingly little evidence of damp penetration in the room beneath but the supporting timbers are clearly at risk.

attack, often resulting in a split at the right angle bend. This can be compounded by the weight of the metal preventing its recovery after sagging
downwards in hot weather.

Damage from foot traffic – much more likely from flat roof use in summer
when the metal is hot from the sun but also from maintenance operations at
other times – can cause indentations and eventually holes to form unless the
roof is provided with duckboards or some form of raised platform or terrace;
these are seldom found.

Condensation on the underside of a lead roof can rapidly result in serious
pitting and corrosion of the underside surface and ultimately to complete failure,
necessitating complete renewal. Obviously, this can be difficult to detect if
there is no access to the roof void, but if a lead roof covering is being renewed
then it is essential to ensure that detailing of the insulation, ventilation and
vapour barriers are properly specified to ensure that this does not happen.

Zinc

Zinc came into use as a covering for flat roofs in the latter part of the nineteenth
century as a cheaper substitute for lead. It will frequently be found on the flat
roofs of the back additions, or offshoots, of dwellings constructed in that period
or during the early part of the twentieth century. Many zinc-covered flat roofs
on fairly old dwellings have had to be renewed perhaps several times, and the
surveyor may occasionally encounter a new or near new zinc covering.

Like lead, zinc is a shiny silver when new but is far less malleable. When
laid, it soon develops a carbonate, grey in colour, which gives it some protection
against pollution but which is by no means as effective as the carbonate of
lead since it is not as dense and adherent. The life of a zinc sheet is generally
proportionate to its thickness. A sheet of 0.8mm might be expected to last 40
years in average urban conditions and longer in a rural or a marine environment.
On a sloping roof, it will also tend to last longer than the same thickness on
a flat roof.

Very much like lead, only to an even greater extent, zinc is vulnerable
to acid run-off from surrounding materials, including copper, moss, lichens
and algae. The same remedial actions are required – namely removal of the
source or replacement of the zinc with a more suitable covering. Acid rain
can also be a particular problem if the roof is in close proximity to a source
of pollution producing significant quantities of sulphur dioxide.

The jointing of sheets follows somewhat similar principles as those for
lead since the coefficient of expansion is about the same. However, the battens
for the side joints are more rectangular than the near round shape used for
lead as the material is that much less malleable, and they are provided with a
capping over the turned up sides of the sheets. These tend to be displaced
rather than scored or damaged by foot traffic. When this happens, as it often
does, there is a serious danger of water penetration. Not only that but even

with the caps in position, there can be damp penetration where a heavy fall of snow reaches the top of the capping, starts to melt and penetrates the roll joint since it is formed in a less effective way than it is with lead.

Although zinc is not as soft as lead and therefore more difficult to work, it has the advantage that it does not creep or sag in the same way. Nevertheless, where it is turned up, the sharper angle formed tends to hold polluted rainwater for longer and it is often at these points that the material eventually splits. In reverse so to speak, at the top of flashings, the sharp turn for tucking in to the brick or stonework is also prone to splitting due to alkali attack from the mortar used for pointing unless the metal is coated with unsightly bitumen to protect it – a precaution not often taken because of the effect on appearance. Often if there are design difficulties, both in flat roofs and in gutters, long lengths of zinc will be soldered together. This forms a weak spot that will frequently be the first place where a split will appear.

81 A fairly new zinc flat roof showing the characteristic near rectangular batten side joints and cappings and, in the background, a generously sized flashing.

82 A zinc-covered flat roof, probably halfway or so through its life although the uneven areas are a concern. The defects in the decking, allowing ponding of rainwater, are the places where pollution will affect the metal to the greatest extent and where holes and splits will appear first of all. It may be that the essential clearing of the build-up of snow – always a bugbear with flat roofs of zinc and, to a lesser extent, lead – has not been carried out as diligently as it might have been. This can allow damp penetration under the cappings at the side joints and lead to rot developing in the decking, so it would be worthwhile checking the affected areas. In the background, a flashing is visible against a chimney stack, but the covering over the top of the adjacent parapet wall (top left of photo) leaves much to be desired.

83 Zinc on a flat roof, fast approaching time for complete renewal as the pattern of the decking is showing through, not helped by the presence of rusting metal objects. The poor-quality capping to the parapet is a crude attempt to overcome the problem of cracked cement rendering. A joint effort by both owners to do the job properly would have been better.

Over time, the patina becomes whiter, brittle and more crusty, less able to accommodate movement – either in the structure or in the metal – from temperature changes, and what is left of the metal begins to corrode and split. While soldered patch repairs can be carried out in the relatively early stages, this soon becomes uneconomic and total renewal will be required. The sight of attempts at such repairs or patching with strips of felt and bitumen are sure signs that the time for renewal has arrived.

Copper

The least likely covering to be encountered for flat roofs over dwellings nowadays is copper. This is because its initial cost is high and, accordingly, it does not feature at the top of the list of choice for developers, even for houses on upmarket estates. Its use is more likely to be found on individually designed houses for the wealthy or as part of or as a renewal on a listed dwelling that has historical associations, anticipated to be a feature of the landscape for at least another 100 years or so – the probable life of the metal itself. It is less suitable for a marine atmosphere.

Although more expensive than lead, copper does have certain physical advantages. For one, it is very light in weight – about one-fifth the weight of lead when used for the comparable area of roof. For another, it expands under hot sunshine at only about half the rate of lead or zinc. Both of these factors tend to favour its use on steeply sloping surfaces, such as domes and cupolas and the roofs of balconies, since the structure can be lighter and the metal does not sag or creep to the same extent. These advantages are also useful when it is used on a flat roof. When new, copper is a reddish brown colour (**84**); but what clinches it for designers of visible features are the attractive shades of pale green (**86**; see also **59** in Chapter 4) and also brown or even greyish patina, depending on the type and degree of pollution in the atmosphere, which develop over the first 10 to 15 years of exposure.

The patina comprises a layer of copper carbonate, which protects the metal from attack by the acids in the atmosphere. The development of the patina is a slow process but can be accelerated by the presence of salt in the air of coastal regions. While alkalis do not affect the metal, it can, however, be subject to severe deterioration through electrolytic action if, for example, nails or screws of any material other than brass are used for fixing or other metals are used in conjunction in any way. Copper is also vulnerable to corrosive run-off from asphalt surfaces and the characteristic patina can be disfigured and stained by run-off from metal finishes above the copper so, if appropriate, rainwater from higher levels should be drained away from copper surfaces.

Although not so subject to expansion and contraction as lead or zinc, allowance nevertheless has to be made for what is likely to take place under the hottest sun and, accordingly, side joints on flat roofs are in the form of batten rolls. As with zinc, they cannot be near round in shape since neither metal is as malleable as lead. The rolls are often made triangular in shape although round top, undercut or even more decorative shapes are also used to improve appearance where a flat roof can be seen from upper-floor windows. While standing seams, or stand-up welts, can be used for inclined roofs not subject to any form of traffic, staggered double lock cross welts are usually adopted for joining sheets end to end and a form of drip is only required in parapet gutters.

The surveyor should be aware that the run-off from copper roofing can damage zinc, steel and aluminium so valleys and gutters of these materials collecting run-off below copper roofs or claddings will be affected. Undesirable replacements of original cast iron gutters are a possible source of trouble in this respect and may need replacing, perhaps in plastic if the original is thought too expensive. Run-off from copper can also severely disfigure stone and brickwork, the staining being almost impossible to remove. The aggressive qualities of the metal do, however, have one upside in that it does not sustain organic growth, so moss, lichens or algae will not be found adhering to the

84 New copper on an indoor demonstration panel, showing its characteristic reddish brown colour, a standing seam joint used on sloping roofs and features, and a square batten roll joint used on flat roofs.

85 The surveyor will need to establish what material has been used here for roofing and cladding and as a capping to the brick on the end parapet wall of this recently completed block. The makers probably thought it might pass off as copper at a quick glance, but it is far more likely to be of coated steel or aluminium which have a much more limited lifespan.

86 A flat roof covered in copper, about 50 years old, showing batten roll side joints and the characteristic green patina. It is adjacent to a flat roof covered in mastic asphalt but arranged so that there is no run-off to give rise to unsightly stains. The asphalt roof has the typical dome-hooded ventilators to reduce the risk of blistering.

surface. It is still very necessary, however, to keep copper flat roofs clear of leaves and any other form of debris or deposits that accumulate. Small areas of copper roofing can be repaired as a temporary solution but, as with all metal roofs, signs of undue wear, buckling, splits or patch repairs are indications that the time for total renewal is fast approaching.

Aluminium

There was a vogue in the 1960s and the 1970s for the use of aluminium as a covering for both shallow-pitched and flat roofs, despite its high cost, but it is unlikely that the surveyor will now encounter many dwellings with roofs covered with this material. Most have already been replaced.

Bare aluminium has a comparatively slow rate of corrosion but it rapidly becomes whitish and rough on exposure, resulting in a poor appearance made worse by a propensity to pick up dirt and collect pollutants which, of course, accelerate its corrosion. Anodising the surface in colours suited the styles of the time and, in theory, lengthened the life by producing a layer of oxide on the surface of the metal. However, the anodised coating is brittle and easily damaged on installation and fixing so the length of life was not significantly extended beyond the 20 to 25 year span of the unprotected metal, which is attacked by both acids and alkalis and is in even greater danger of corrosion if in contact with other metals. Frequent washing or periodic painting have been put forward as ways to extend its life but not many owners wish to incur this liability either physically, by way of payment to contractors or through a service charge.

Other problems also materialised: condensation below the metal could affect the decking and the need to tie down the covering very securely because of its lightness was not always appreciated with the result that quite a few roofs were stripped off by wind suction. It is extremely difficult to successfully solder aluminium on site and components are usually joined under workshop conditions; but large-scale profiled aluminium sheets, which are still used for commercial properties, do not usually suit the relatively small scale of most domestic properties. It is, accordingly, unlikely that aluminium will be encountered as a roofing material on new or near new dwellings. If it is, the surveyor needs to be very prudent in the advice given since experience indicates that the longevity found in most other metals used for roofing does not extend to aluminium, despite the material being covered by British Standards.

Built-up felt and other membranes

Traditional built-up roofing membrane felts have an expected life of 10 to 15 years but much more durable roofing membrane roofing systems are now available and do not deserve to be condemned out of hand. RBMs are the direct descendant of felts but their reinforcements and bitumen coatings are far more robust. The term RBM refers to the higher-performance membranes, typically polyester-reinforced (with a life of 15 to 30 years) and ideally also having elastomeric bitumen (which may achieve a 50-year life when professionally installed). RBMs are often described as 'high performance' but this is rather misleading and it is better to describe them by their reinforcement and coating type. The term 'roofing felt' is not appropriate for such membranes and should be reserved for the traditional, poorer-quality material.

Identification of the particular membrane in use is crucial in deciding upon the likely durability and will help in assessing the feasibility of any necessary repairs. Occasionally, the surveyor will be fortunate and find original specifications or invoices to be available. Examination of an aged cut-edge may allow the surveyor to differentiate between glass tissue, organic fibre or polyester reinforcement. If a piece can be flexed, the coating can be inspected for cracking to differentiate oxidised bitumen and elastomeric bitumen coatings.

Built-up felt

It is, of course, true to say that as far as any roof covering is concerned, it is only a matter of time before renewal is required. Someone purchasing a dwelling, given this warning, will at least be somewhat consoled if it is followed by an assurance that the need for re-covering is not likely to arise within, say, the next 10 to 15 years. Unfortunately, where traditional built-up felt roofs are concerned, that reassurance is never really possible.

Most traditional built-up felt roofs, usually found on extensions, back additions or dormer flat roofs, were provided in the first place on the grounds of cheapness. When found covering whole dwellings, they are often on buildings dating from the 1960s and 1970s, a time when it was fashionable to embrace alternative designs and forms of construction (**88**). Many of these failed very quickly and have since been removed and replaced with pitched roofs, but quite a few still remain with renewed coverings. Others on new or near new dwellings could be a case of a developer chancing something to keep down costs. Where encountered, more often than not, the poor materials and even poorer workmanship will inspire no confidence. The defects which the surveyor is likely to encounter are generally some, frequently all, but only rarely none of the following:

1   Blistering of the layers, often due to moisture entrapment.
2   Embrittlement and cracking of the covering due to the exposure of the oxidised bitumen content of the felt to solar radiation.
3   Splitting and cracking due to laying in cold weather, which can split the sheets as well as causing a lack of adhesion.
4   Rucking, buckling and splitting due to differential movement with the decking or insulation material or poor bonding on laying, perhaps due to a lack of or inadequate cleaning of surfaces before laying.

5   Ponding, either due to inadequate falls, sagging of decking between joists or any of the above faults interrupting the flow. This can rapidly seek out weaknesses of adhesion at joints, particularly if manufacturers' recommendations for overlaps have not been followed.

6   Inadequate solar protection to the surface either by omission or failure of reflective coatings or by the washing away of materials such as stone chippings applied for the purpose.

7   Mechanical damage from maintenance traffic, particularly where surfaces are uneven.

8   Missing or inadequate flashings, for example, a lack of angle fillets to upstands and abutments, particularly around flues.

9   Poor detailing.

While some of these faults can be found in relation to other types of roof covering, most are particular problems for built-up felt roofing. The main difficulty with this covering is that while the material itself may be reasonably satisfactory, with a life of maybe 15 to 20 years or so, the fact that the material is in sheets and laid to overlap means that it relies on the workmanship of bonding for its durability. It is difficult to ensure that this work is done correctly or satisfactorily and flaws will not necessarily become apparent immediately. When they do, maybe in under 10 years, it may be too late to prevent widespread damage, and the flaws may prove to be so extensive as to make repair uneconomic.

Repairs can be successful if a defect is localised or the source of any problem can be identified. They can even be undertaken using a more durable

87 There is no mistaking the appearance of a bituminous felt-covered roof even, as here, in good condition. Sure indications are the colour, the surface texture, the stuck down side joints and where the lengths cut off from the rolls overlap. Apart from the quality of the material itself, it is on the integrity of the joints that weathertightness depends.

88 It is unusual to find a single dwelling with a shallow-pitched roof covered with mineral felt, and those encountered probably date from around 1970, like this one. The present covering looks relatively recent and in good condition, but the surveyor must be careful neither to assume all is well nor to condemn the covering outright until further investigation has revealed more details of the work carried out, including exactly what material has been used – hopefully at least an RBM.

RBM. But attempting to locate the origin of damp penetration through a built-up felt covering can be time-consuming and frustrating to the point where it may be easier to simply accept defeat and strip the old covering and replace it with the longer-lasting RBM. Even when presented with a flat roof covered with new or near new traditional built-up felt, the surveyor should be wary of giving any view as to future performance because so much depends upon the standard of installation.

To repeat the warning given at the very start of this chapter, however, it is no longer sufficient for the surveyor to assume that any non-metallic flat roof covering is traditional poor-quality built-up felt and simply sound gloomy warnings about life expectancy. Care needs to be taken to identify which material is present before making a definitive pronouncement. RBMs and polymeric single-ply materials are very much more durable than traditional felt and, correctly installed, the top-range materials should have a service life of up to 50 years, so it would not be appropriate, so to speak, to tar them with the same brush.

Reinforced bitumen membranes (RBMs)

RBM roofing is usually constructed by bonding two or more layers of RBM, although single-layer RBM systems are available. Coverings are usually applied by either 'pour and roll', where hot bitumen is poured as an adhesive in front of the membrane as it is unrolled, or by 'torch-on', where the bitumen on the underside of the roll is heated to create a small amount of molten bitumen as the roll is laid out. 'Torch-on' has a number of advantages for small areas or where access is restricted, but the materials require careful handling and extra fire precautions are necessary.

89 A question mark over many bituminous felt roofs is the quality of the supporting decking. If the decking material is vulnerable to dampness – and early decks always were – the inevitable sagging which follows any damp penetration serves only to collect more water above the point of weakness, creating a vicious cycle with inescapable consequences. The catastrophic failure of this porch roof illustrates the problem perfectly.

Polyester-based and elastomeric roof coverings are now in common use and since 2007 have been covered by the European product standard BS EN 13707: 2007 and by the British Standard BS 8747: 2007. The British Standard BS 8747: 2007, *Reinforced Bitumen Membranes for Roofing – Guide to Selection and Specification*, provides guidance on which membranes are appropriate for varying types of flat roof.

Where isolated faults are found in a built-up RBM roof and where the membrane itself is still in a satisfactory condition, patch repairs can usually be carried out. If there are numerous faults, however, or where the membrane is becoming more seriously degraded, it will be necessary to renew the covering, unless it is practicable to overlay the old covering.

Polymeric single-ply roofing

These comprise a single sheet in which synthetic polymers are the main constituent, but there are a number of different variations on this theme and product development actively continues so further types can be anticipated. The specific material determines whether joints are formed with hot air, solvents or adhesives, and the nature of the deck determines whether the covering is loose laid and ballasted, mechanically secured or fully adhered to the deck. There is no British Standard but Agrément Certificates are available and there is a harmonised European Standard for single-ply waterproofing, BS EN 13956: 2012. These materials have a life expectancy of more than 25 years when properly installed.

Surface treatment for maintenance is not usually required, but as these materials are typically less than 2mm thick, they are more at risk from misuse or impact damage such as being walked on. It is rare to find visible defects and expert advice should be sought if problems are encountered or suspected. The Single Ply Roofing Association will assist if the manufacturer cannot be identified from the product itself or from paperwork relating to the installation. If a leak should occur, one clear advantage of single-ply over built-up membranes is that electronic leak detection works extremely well in most cases and locating the source of the problem is much more straightforward.

Paving, decking and lantern lights

Flat roofs, particularly in closely packed urban areas, encourage use for outdoor gatherings and some, in their normal finished state, are more suitable for this purpose than others. Metal-covered roofs (lead, zinc and copper) should not be used because of the risk of damage to their projecting side and lengthwise joints. Asphalt is more encouraging because of its clear surface, but at the time of year when such outdoor gatherings are likely to take place, it is at its softest and can be damaged by pointed heels and the legs of chairs and tables. Membranes, even the stronger modern ones, are not robust enough to withstand anything more than very light traffic, and in warm weather, bituminous membranes share the same disadvantage as asphalt. The provision of suitable lightweight paving tiles is sensible to reduce the incidence of such damage and, if in a light colour, does much to combat the effect of hot sunshine. By taking the paving as near as possible to any abutment and dressing a lead flashing as far down as possible, good protection can be provided.

Roof decking, as distinct from paving, on the other hand presents the surveyor with the difficulty that he or she cannot see either the roof covering or the way the decking is supported. The decking can be raised above the projections from the surface of metal roofs, or it can bring to level the falls across the surface of asphalt or membranes. Bearing in mind the case of *Hooberman* v *Salter Rex* 1985, mentioned in the first section of this chapter, warnings must be included in the report based on what can be seen at the edges and the likely consequences of such decking being placed on top of old and worn-out or badly detailed construction.

Flat roofs are very often provided with lantern lights in different forms to light a part of the interior below. Badly designed and/or badly installed, they can be an endless source of trouble. Traditional types with properly detailed upstands and patent glazing generally perform well, within their limitations, if maintained satisfactorily, but the surveyor should seek out any deficiencies and any evidence from below of rain penetration. Older, single-glazed installations

90 The provision of paving to asphalt flat roofs, particularly in an even lighter colour than here, does much to combat the effects of hot sunshine.

91 Decking on a flat roof prevents the surveyor from seeing what is below, and therefore, in addition to describing the construction round the edges in the report, warnings must be included on what might be underneath and the possible consequences.

92 This purpose-built cover to a basement light well could act as a model for many roof lights. The lead detailing and toughened glass mean it will be watertight, durable and safe.

93 A series of proprietary domed roof lights which seem to have been fitted in accordance with the manufacturer's instructions. Modern lights such as these are double-glazed and secure from intruders. The surrounding flat roof has good detailing so it is disappointing to see indications that standing water accumulates on the left.

are likely to experience heat loss and condensation problems, but fittings installed since 1 October 2010 should comply with the requirements of the Building Regulations Part L 2010, or any later amendment if appropriate. Glazing should be toughened or laminated glass for safety, but this is not always found to be the case on older fittings.

Small, much cheaper alternatives to the traditional lantern lights are available in proprietary form with impact-resisting polycarbonate, one-piece, domed lights (**93**). They come with instructions for fitting, particularly with regard to ensuring that they cannot be lifted off by potential intruders from the outside. These are not always followed so that, as with traditional lantern lights, the surveyor needs to examine them carefully for flaws on installation giving rise to possible damp penetration.

Rigid sheet over-roofing

The fashion for building low- and medium-rise blocks of flats with mineral felt-covered flat roofs during the 1960s and 1970s left a legacy of damp penetration for occupiers and maintenance expenditure for the block owners – usually, though not exclusively, local authorities. During the 1990s and early years of the new century, many decided to bite the maintenance bullet and terminate the unremitting expenditure on roofing works by converting the roofs from flat to more durable and relatively maintenance-free pitched design. This can change the external appearance considerably, no doubt to the dismay of the original designers if still alive or causing those who have passed on to turn in their graves. Those buildings which have become listed icons of the modern movement cannot, of course, be subjected to such treatment, regardless of leaky flat roofs – possibly to the greater dismay of the owners and occupants. At least the newer RBMs now offer a more durable alternative to the original built-up felt covering.

As might be expected, there are many different ways in which over-roofing can be carried out and much will depend on the initial assessment which will have been carried out to determine how much extra load the structure will carry and what effect wind loading will have on any pitched roof structure if added. It is unlikely that a conventional timber-framed and tiled roof will be added, not least due to the additional load that would be imposed. Often the material deemed suitable for over-roofing will be some form of metal sheeting, due to its lightness in weight, supported on a light framing of either timber or steel. Sometimes such work is accompanied by over-cladding the elevations, and when the two are combined in this way, the components making up the new appearance fit together rather better than when the over-roofing is carried out on its own.

To avoid a repetition of previous failures due to wind suction of shallow, metal-sheathed pitched roofs, there is a necessity for such sheet metal coverings to be properly secured to the framing. The whole of the new roof structure framing must also be secured to sound construction below. Depending on the limitations of the inspection being undertaken, it may be possible for the surveyor to check these points if access if available.

If work has been carried out relatively recently, it is to be hoped that details will be available from some source; but if the roof was replaced some time prior to the inspection then information is less likely to be readily obtainable. The name of a well-known and reputable firm of structural engineers associated with the scheme will provide some reassurance. If access is not available or cannot be arranged, the advice given to the client is likely to be limited to what can be seen from ground level with binoculars along with what, if anything, can be seen from their flat and the common parts.

What is the surveyor to make of these sheet materials which are not usually associated with residential properties? Comment has already been made in the previous chapter (in the section 'Other forms of pitched roof coverings') about the steel panels made to give the impression of concrete tiles, which are often found in mobile home parks to give an impression of permanence but may also appear elsewhere (see **11** and **64** in Chapter 4). Others, of plastic-coated profiled steel or aluminium sheets in bright colours with contrasting coloured edge trims, are increasingly being seen on new blocks of flats even though some may think them more appropriate to warehouse or distribution centres. In the case of over-roofing, however, the decision to go forward with such materials has probably been taken on the basis that there is no practical alternative and that over-roofing is considered the best solution to long-standing problems. The outcome, nevertheless, applies to the roof covering but not necessarily the roof structure which supports it, which may have a limited life of around 20 to 25 years at most. The coated metal coverings that are generally used are usually subjected to cutting, and much other mechanical damage often occurs on installation. Though the manufacturers normally specify what is to be done to damaged edges in such circumstances, workmanship may be such that the procedure is not followed. Consequently, corrosion can set in at an early stage. The surveyor, given the opportunity of a close inspection, needs to look carefully for signs of corrosion not only on the outer surfaces but also, if possible, on the underside where undue levels of condensation have been known to add to the problem. Corrosion can also be accelerated by incorrect laying of the sheets. Sheets should be laid with the exposed edge of side laps facing away from the direction of the weather, but this precaution is often overlooked.

The noise of rain and even more of hail falling onto a metal-covered over-roof might be thought to be a problem for the occupants of dwellings immediately beneath. This should be less of an issue than it might once have been now there is a better understanding of acoustic insulation and there is a greater need for thermal insulation. However, it is not always possible for the surveyor to be sure what measures have been put in place, and if there are any uncertainties, it would be prudent to advise the clients to make appropriate enquiries to satisfy themselves that they will not be kept awake during nights when rain falls.

The reality, in the context of a flat in a large block, is that neither the seller nor the prospective purchaser can do very much to alter the situation as found. All the surveyor can do is to describe what has been seen, warn about what is likely to happen in the future and when this is likely to be, and indicate, if possible, how it might affect the purchaser's liability for a share of maintenance costs. It will then be up to the client to decide whether the risks associated with any uncertainties are worth taking on or whether the situation would be better elsewhere.

94 The historic problems with flat roofs mean that shallow-pitched metal profile coverings are now frequently used for modern blocks of flats. The first difficulty for surveyors is finding a vantage point from which the covering can be seen; the next is identifying the material, especially when viewing from a distance. This covering might be zinc or coated steel.

95 Here is another modern block where the surveyor can see nothing close at hand, unless inspecting one of the upper flats with a view of a lower roof. Inspecting with binoculars can be both time-consuming and frustrating. Information may be available from the management company or their agent, but not in every case. This roof is probably covered with coated steel.

96 Not all profiled steel sheet panels used for new roofs or over-cladding set out to imitate other materials in appearance. These are more distinctive in shape and colour and are provided with contrasting trims.

6

Chimney stacks and roof detailing

Introduction

Among the least troublesome of pitched roofs are those presenting plain, uninterrupted, sloping surfaces to the sky, which simply drain rainfall into gutters clear of the walls of the property below. Three separate factors coincided during the period between 1960 and 1980, and for a short time, such simplicity became the norm: the introduction of simple trussed rafter roof structures encouraged the construction of dwellings with a simple rectangular footprint; interlocking concrete tiles most economically covered roofs of the plainest design; and chimneys were no longer needed because open fires had fallen out of favour (**97**). Outside this period, however, few such straightforward roofs

97 Roofs don't get much more straightforward than this one from the late 1970s – no chimneys, projections, valleys, dormers or abutments, but it wasn't many years before complications started to appear again. A couple of hung tiles over the porch have slipped and need replacement.

exist. Most roofs have projections such as chimney stacks or ventilating pipes or both. Others may have dormer windows or, as do some flat roofs, projecting roof lights. Some pitched roofs change direction, necessitating a valley gutter between slopes, and all types of roof may abut walls rising above the slope. Such features introduce potential trouble spots. While there are long-standing ways and materials for overcoming the problems posed, there are also many less than ideal ways available both for new construction and for use when repairs are needed.

Chimney stacks

Chimney stacks can present the surveyor with some difficulty on account of their inaccessibility. It is generally impossible to look closely at back gutters on pitched roofs and often only side and apron flashings can be seen with the aid of binoculars. Back gutters are prone to acquiring debris and an accumulation of silt which can interrupt or even dam the flow of water, even assuming there is a good fall provided to either side in the hope of preventing rainwater from accumulating – not always the case by any means. If poorly designed and installed, rainwater can build up to the extent of overflowing. For this reason, chimney stacks spanning the ridge cause far less trouble and are preferred.

Metal flashings are much more durable than cement fillets and lead flashings are preferable to zinc, not only because the metal is more durable but also because it suggests a higher standard of installation as it is more expensive. Cement fillets are prone to cracking but any failure of the flashing or fillet around a stack can lead to damp penetration with the associated risk of deterioration to the internal fabric, whether staining to decorations or rot to adjacent timbers. Even when all seems sound from the external inspection, the interior of any projection through a roof should be carefully checked for signs of penetration.

Apart from back gutters and flashings, attention should be paid to the structure itself. In the past, chimney stacks were seldom provided with the damp-proof courses which would be or ought to be incorporated today. If the stack is unlined or if the lining has deteriorated, the mortar can be affected by sulphate attack from the condensed combustion residues, and this can result in characteristic cracking to rendering (**98**). Because of their position, they are subjected to the severest of weather conditions, and saturated brickwork can soon take on a sorry appearance, especially if the frost gets to it (**99**). Stacks are sometimes built unduly tall, particularly when rising from flank walls, so as to avoid the possibility of downdraught. The side of the stack most exposed to moisture from the prevailing winds will be affected more by any sulphate attack because the moisture will exacerbate the problem. This means that the mortar courses will expand more on that side and the stack will develop a lean away from the side most affected. High winds can subsequently bring about

98 The classic symptoms of sulphate attack are affecting the rendering on this stack. When the stack was demolished, the flue was found to be unlined and the mortar to the stack brickwork had become soft and crumbly.

99 Deterioration to this stack is so far advanced that it is beyond economical repair. Decisions will have to be made about the most suitable means of retaining the function of the single flue still in occasional use but there is no practical need for any external outlet to serve the other two, which are disused but properly ventilated.

a collapse so the surveyor needs to note all cases of cracked or leaning brick or stone chimney stacks and all cases of defective pointing or rendering. The flaunching around the pots at the tops of the stacks, if visible, should also be checked for signs of deterioration (**100**).

With the advent of central heating, many chimney flues are now redundant. The tops of many flues have been capped and many more stacks were completely demolished to below roof level. These demolitions are an understandable saving in maintenance expenditure from the individual owner's perspective, but they have often destroyed the attractive symmetry and repetitions originally present in the appearance of pairs and groups of similar dwellings and the consequent loss of character to many urban streetscapes is regrettable.

When old flues have been capped, whether above or below roof level, it has long been accepted practice to point out the necessity of maintaining a through current of air to all disused flues to avoid excessive condensation, which can cause staining to internal decorations. If vents are not present at the base and the top of disused flues, this needs to be noted and the possible consequences explained in the report. More recently, however, an alternative view has been put forward regarding disused flues which are wholly internal – that is to say with no exposure to external walls and where there is no longer an external stack. It is argued that for these flues only, there is more risk associated with constantly introducing potentially moist air at the base, while completely sealed internal flues will establish equilibrium with the surrounding areas, meaning that moisture levels will be constant and condensation will, therefore, not occur.

There has been a trend over recent years to resume using open fireplaces and particularly to install wood-burning appliances. This often necessitates

100 The flaunching to this stack clearly requires attention, and the shrubbery should be cleared as soon as possible to prevent greater deterioration. These are self-sown silver birch seedlings and are quite capable of thriving and causing significant damage if left undisturbed, despite their precarious perch. The stack rendering appears sound from this distance so repairs may not be too expensive if not delayed, subject to a thorough check when access is available.

101 Although the chimney stack on this eighteenth-century house has a rather unnerving appearance, it has clearly been comprehensively overhauled with the top part being rebuilt relatively recently. The surveyor may be reassured if details of the work are forthcoming from the owner, but full consideration of the internal supports is still required and periodic checks should be recommended to ensure the structure remains stable despite the eccentric loading.

102 Where there are clear signs lower down that disused fireplaces on external walls have been blocked up, it is important to check that through ventilation is still provided to the flue. This is usually done by fixing an unobtrusive ventilator grille in the room itself. At roof level, there are various ways it can be done. Here, most flues have been terminated with clay ware plugs with ventilator holes, bedded into the flaunching. Often terminals are simply fitted into the top of the pot similar to the remaining flues where purpose-made terminals, finished like half-round ridge tiles, have been fitted.

relining the original flue, which is 'notifiable work' under the Building Regulations. It can be difficult to see this from a normal inspection but there may be a new flue terminal visible externally. Perhaps of greater concern since open fires fell out of general use has been the tendency to remove internal chimney breasts, in many cases without providing adequate means of support to retained upper sections or chimney stacks. There are further comments on both these matters in Chapter 23.

Abutments to roofs

Arguably the simplest junction to make is that between a flat roof and a wall rising higher as part of a building, a chimney stack or a parapet wall. Usually this is formed by turning up whatever material has been used to cover the flat roof against the wall to a height of at least 150mm, considered the limit of splashing. Preferably the material so turned up is tucked into either a horizontal joint, if the wall is of coursed brick or stonework, or into a groove cut into uncoursed stonework. However, this is not always done and the material is often merely left turned up. Either way, the tucking in, or the top, needs protecting with a flashing. It is, moreover, essential that the top of the flashing is tucked into the wall – ideally for 25mm for a secure fixing – to make it watertight. To avoid any chemical reaction between different metals, it is important that the material for the flashing is the same as that for the roof covering; **81** in Chapter 5 is a good example – zinc for a zinc flat roof. If different metals have been used, the surveyor should look as closely as possible for any sign of degradation from chemical reaction; but even if none is seen, a warning of its possibility in the future should be included in the report. Degradation can also occur if lead and zinc are not coated with bitumen before being tucked in because jointing or subsequent pointing in Portland cement or lime mortar will result in alkali attack on the metal.

Given reasonable construction, this basic type of junction ought to be foolproof. However, the mere fact that it is necessary to bend material through 90°, or 45° if a tilting fillet is used, introduces a weakness. It is often at this point that a split or tear will occur, allowing dampness to penetrate. Such splits or tears are often difficult to see but a defect at this point is often the earliest sign that the material covering the roof is approaching the end of its life. Earlier patching with canvas or felt and blobs of bitumen will provide a clue that this is beginning to happen.

Flashings can become loose and require refixing because of inadequate tucking in, a lack of wedging, shrinkage of the mortar used for pointing over time, or physical damage. This can also happen if they are not held in place by tacks at regular intervals and are lifted up by the force of the wind.

With pitched roofs, the typical materials used as a covering, such as tiles or slates, obviously cannot simply be turned upwards where they abut a wall so the arrangement needed to cover the junction becomes more complicated. A traditional and cheap way of forming this joint is by providing a triangular fillet of cement and sand. The life of such fillets is very much determined by the quality of the mortar used for the purpose. Builders have a tendency to think that the more cement in the mix the stronger the mortar. However, such strong mortars shrink and crack far more than weaker mortars containing coarser aggregates and lime, and eventually, they part company with both the vertical wall surface and the roofing material itself. In an attempt to slow down shrinkage in the mortar, pieces of tile or slate may be embedded in the top of the fillet. Conceivably, this could lengthen the life of the fillet but only by a comparatively short period. If cracked, shrunk or partly missing fillets are encountered on a tiled or slated roof, the surveyor's advice should be to replace them with something better if that is a possibility. Inevitably, this costs more than merely replacing the old fillet with a new one so the advice frequently falls on deaf ears.

At the edge of a slope, the ideal way of forming this joint is to use lead soakers and stepped flashings (**107**). The soakers are short strips of lead, the length of each tile or slate turned up against the wall. They are either hooked over the top end of each tile or slate or nailed to boarding and laid in an overlapping manner so that if a defect occurs in the flashing or it is dislodged, any penetrating rainwater will follow a path down to discharge into the gutter. Soakers cannot be seen from above because they are covered by the flashings, but they might be seen from the roof space in the absence of boarding or roofing felt. The risk of damage from run-off or staining is too great for this form of detailing to be an appropriate use for copper. Using zinc instead of the more expensive lead is a cheaper alternative, but this is not so long-lasting. In any event, as mentioned in the previous section on chimney stacks, a combination of lead and zinc should be avoided in view of the possibility of electrolytic action between the two in damp conditions (**108**).

A sort of halfway house, to save on cost, is to provide soakers but, instead of a flashing, to use a triangular cement fillet to cover the joint. This is better than the same fillet without soakers provided that the soakers are well coated with bitumen before installation to protect them against alkali attack from the chemicals in the fillet or in the brickwork and pointing of the wall above. It is a method adopted in areas where the local stone is so hard that cutting a groove for tucking in the top edge of the flashing presents undue difficulty. However, if the top edge of the fillet shrinks away from the upstand, there is no protection against rainwater getting behind the soakers and perhaps causing stains internally as well as dampness in the timbers, with all the problems that could entail.

Although leadwork is regarded as the ideal, where it is possible for it to be used, on account of its greater durability, there are quite a few situations where there is no alternative or where a cement fillet might be preferred for aesthetic reasons. On the brighter-coloured profiled tiles, for example, the sight of strips of lead dressed over the tiles as flashings is not regarded by some as

103 The damage to this zinc flashing, above a zinc gutter, was probably caused when the old slates, bits of which remain in the gutter as potential obstructions, were removed and replaced with artificial slates. It leaves the top edge of the gutter, turned up against the brick parapet wall, exposed and providing a possible path for water penetration. The displacement may have been made easier by inadequate tucking in of the flashing to the brickwork of the parapet before pointing.

104 Shrinkage of a cement fillet to a chimney stack. Although not causing visible damp penetration at present, timbers are probably being affected and this, along with the condition of the slating, will cause more problems to develop in the future. Ideally it should be replaced with a lead apron.

105 Often the best the surveyor can hope to see is a neat cement fillet in reasonable condition to the joint between a chimney stack, as here between concrete tiles and brickwork. Encouraging too, that there is evidence of a damp-proof course to the party parapet wall. The cement fillet to the upper part of the parapet, however, seems to be cracking.

106 The newer pointing to the tucking in of this lead flashing is very crudely executed. It may be possible to check whether soakers have been provided from within the roof space. The crack and displacement of the brickwork at the end of the parapet requires further investigation in case there is a danger of collapse.

attractive, both at side abutments and even dressed down on to the tiles as an apron flashing. A coloured mortar in combination with matching tile insets, known as galleting, looks neater and less obtrusive. Against uncoursed stone-work, where stepping a flashing is not possible, the alternative of cutting a groove in the stone can be difficult and could lead to unnecessary damage depending on the type of stone; so even with ordinary tiles and slates, a neat cement fillet may provide a more suitable solution.

Occasionally, on older stone-constructed dwellings, the joint between the pitched roof covering and an abutment – usually a low parapet wall but sometimes a chimney stack – will be provided for when the stonework is laid by extending the coping or a section of stone outwards so that it overhangs the roof covering. This leaves the gap between the underside of the overhang and the top surface of the roofing material to be filled with mortar and pointed to provide the joint. This can be quite effective over a period of time. The method clearly required considerable forethought and is usually only found on the best quality of dwelling.

107 Stepped lead flashings to a tiled roof on a demonstration panel. At the top, a soaker is shown before being covered by the flashing and the top tile.

109 Junction of a pantiled roof, parapet wall and chimney stack formed with a cement fillet in which pieces of slate have been embedded to reduce shrinkage.

108 A zinc soaker, suffering severe deterioration, being removed from beneath a lead flashing around a 1930s chimney stack. This demonstrates the problem caused by having different metals adjacent to each other. A final soaker awaits removal.

Ventilating pipes

Although it is best to bring ventilating pipes on a swan neck out and away past eaves gutters, sometimes there is no alternative but for them to be taken through the roof slope. Providing a suitable flashing is, obviously, essential but not always achieved. In the absence of a purpose-made fitting, the preferred method is by way of a lead 'slate' dressed around the pipe. Securing the joint between the 'slate' and the pipe presents difficulty unless a special section of pipe with a shaped collar to receive the lead is used. If the flashing or joint are poor, there can be problems of damp penetration around the pipe (**111**).

110 Innocent-looking ventilating pipes projecting through pitched and flat roofs can be a source of damp problems if not properly detailed. On pitched roofs, the nearer they are to the eaves, as here on this 1920s house, the more likely they are to cause trouble. Though in this case, there is a lead flashing and any signs would first appear on the underside of the projecting eaves. Sometimes a swan neck detail takes the pipe outwards beyond the gutter, eliminating roof penetration, but this was not as easy with the original cast iron as it is with modern plastic pipework.

111 The cement fillet to this ventilating pipe is cracked and will not prevent dampness from penetrating, especially as tiles in the vicinity have cracked and slipped. A metal flashing is much more desirable. There should also be a cage or balloon fitted to the top of the pipe.

The surveyor should always try to see from below, ideally in the roof space or by examining the ceiling beneath the point where the pipe penetrates the covering; but frequently this will be located in an area, perhaps in a cupboard but very often within a casing, which unfortunately prevents the closest of examinations. Nevertheless, if the problem has persisted, it may be that staining will have extended beyond the casing. Staining is more likely to be apparent to ceilings where ventilating pipes have been taken through a flat roof because there is a shorter distance between outside and inside.

Dormer windows

Within the limitations of a visual inspection and the availability only of a 4m ladder, dormer windows – which are, in effect, mini structures in their own right along with roofs, side walls and windows – present similar difficulties for inspection as chimney stacks. However, as with other projections through the roof, the consequences of any defect may be apparent by way of damp staining on adjacent surfaces.

The presence of stains on new decorations suggests that perhaps a temporary repair has not been successful and something more needs to be done. In all cases, a moisture meter can be helpful in determining whether

112 Dormer windows can take on many different shapes. In the north of England and in Scotland they are often in the form of bay windows, which can make an examination of the sides and the junction to main roof a little easier. These seemingly overlarge versions from the early nineteenth century have the original small squares of glazing, which unfortunately have been substituted by large squares elsewhere in the dwelling. Old age at this stage can cause many problems and these certainly require attention. The apparent mixture of copper and lead on the parapet at the right is likely to cause problems with aggressive run-off.

113 The two large dormer windows to the dwelling on the left have angled sides and slate-hung and flat roofs; the smaller dormer in the roof of the adjacent house on the right is in the form of a bay window with a pitched roof, hipped at the front. Here, at least at first-floor level, the dwelling on the left has retained the glazing in small squares with the very slender bars appropriate to the dwelling's early nineteenth-century period.

114 The danger with dormer windows of this shape is that the shallower pitch of the dormer roof compared with the main slopes may not have been appropriate had the designer inadvertently chosen plain tiles for the covering. Here, with profiled tiles on this 1980s dwelling, there ought to be no problem. However, if there are damp stains from the entry of windblown rain, a check on this aspect should be made. The use of plain tiles would have looked better in any case and, if used on the sides of the dormers, would have avoided the need to show such a large expanse of lead. Being the 1980s, the open fire with chimney stack is back, though probably not originally with the rather quaint baffle.

115 It is instructive to compare the different treatments to the dormers on this Georgian terrace. The cheeks are variously clad with lead, tiles or boards, and they are roofed with either lead or mineral felt. The terrace is in a conservation area so there may be restrictions on future alterations. Other than the need for some redecoration, all seem to be in good condition.

penetration is continuing by checking whether dampness is still present. There may be assurances from a vendor that the problem has been solved but, like all such information, it either needs checking or should only be mentioned in the report with a warning that it has not been verified. Whether it is possible to check the information will depend on the circumstances and, in particular, the position of the damp staining. If it is on the ceiling, it is unlikely to be possible; but if it is at the sides or below the sill, opening the window and peering around could well provide useful information on construction and condition for inclusion in the report.

Parapet walls

As with chimney stacks, parapet walls are subjected to the severest of weather conditions. If old, they are seldom provided with features such as adequate copings or damp-proof courses of an appropriate sheet material type, all of which would help to keep them in good repair. If they are allowed to fall too far into disrepair, parapet walls can be in danger of collapse, which may not only have consequences for the owner or occupiers but may also injure a passer-by and perhaps result in a claim. As much as is possible should be seen of all parapets and all defects noted in the report along with any warnings it may be necessary to provide. The current requirements for parapet wall dimensions are set out in the Building Regulations but it is useful to bear in mind the rule of thumb that solid brick parapet walls should not exceed four times their thickness in height.

116 Parapets receive much battering from the weather and need to be examined carefully for any signs of sections becoming dangerous. These appear sturdy enough from a distance but it is not unknown for neglected sections to become dislodged and fall, to the possible danger of passers-by or occupants.

117 Many parapets will be found to be short of the ideal in regard to pointing of brickwork for weather protection, to have defective rendering, and to lack copings with an overhang; and few will have any damp-proof courses at all, let alone the two that are recommended. All these flaws can cause troublesome damp stains in the room below.

118 The sight of bituminous felt wrapped around a parapet in this fashion should immediately put a surveyor on guard. What is the felt hiding? This undoubtedly requires the insertion of appropriate warnings in the report.

7

Rainwater disposal

Arrangement

A quick and efficient system for transferring rainwater from the roof to underground drains or, at second best, surfaces at ground level is an essential feature for any dwelling. The system should be able to cope with the heaviest falls of rain without leaking or undue splashing of adjacent areas, even when accompanied by high winds. The normal method, external gutters and rainwater pipes, is safest where gutters are fixed onto generously overhanging eaves below a pitched slope because any overflowing or leakage from the gutter will probably fall free and away from the main structure. Additionally, the leak is usually clearly visible from below, and once noticed, arrangements can be made to have the defect repaired. There is more risk in the event of leakage when the gutters are fixed to a fascia at the top of the wall because leaks may not be noticed so readily. It follows that the most vulnerable arrangements in the event of leakage are those which are above internal accommodation, especially when concealed from view. When the design of the roof, such as a mansard roof (**379** and **380** in Chapter 18), means that gutters are located behind a parapet wall or arranged as a concealed valley gutter or in the form of a secret gutter, a sudden serious leak can cause extensive devastation to internal finishes, furnishings and personal possessions (**119**). More insidious, and often more expensive to remedy, are less obvious but long-standing leaks from such gutters, which can continue unnoticed for extended periods and can create conditions suitable for major structural problems like the development of dry rot.

The inspection of a dwelling during a period of high winds and heavy rainfall, although unpleasant for the surveyor, at least enables the system employed to provide a demonstration of its effectiveness. Water dripping down from holes in gutters or defective joints, water cascading over the gutters due to blockages or falls which are too steep, or water trickling down between the back of the gutter and the roof covering or fascia because the fall is too shallow are all indications that repairs or renewals are required. Split hopper heads or rainwater pipes, particularly at their backs, along with blocked and overflowing pipes are a frequent cause of other serious defects when water soaks into adjacent areas. It may be that a parapet or secret gutter is not actually leaking at the time of the inspection but, since internal leakage is obviously even more undesirable than leakage to the exterior, special pains must be taken during the examination of disposal facilities from the roof or in the related roof space areas to check for any evidence of leakage.

When conditions are dry, there will be no demonstration of effectiveness but there may well be evidence of flaws in the system, particularly if their consequences have been continuing over a period. The sky can sometimes be seen through holes or gaps in gutters when looking up from ground level, and the gutters can be viewed from a distance to check their alignment, which should be gentle rather than steep. Viewing from a distance also provides an opportunity to check whether there are any damp streaks on external walls below gutter level, a check which can be extended to the areas traversed by rainwater pipes. Continuing damp will be indicated by darker areas behind rainwater pipes (**120**) while white streaks will indicate where previous areas of dampness have dried out (**121**). This can give rise to the question of whether the repair was carried out in a temporary or permanent manner, needing investigation accordingly.

Gutters and pipes come in various sizes and should have been provided with a capability related to the area of the roof that they are serving. This is not always the case, and before examining the condition of the gutters and downpipes, it can be illuminating to spend a few moments thinking through exactly how the roof of a property is being drained. The locations of the downpipes should be noted – and incidentally also recorded on the site notes of the inspection. The source and potential volume of all the rainwater arriving at these downpipes should be considered in turn (**122**). Ideally on a modest two-elevation terrace, there will be a downpipe serving each elevation, and these serve only that one dwelling. Frequently, however, downpipes will be found to serve the gutters of more than one property in a terrace, so that rainwater from the roofs of several properties flows to a single downpipe at one end. In exceptional conditions, any overflowing is more likely to affect the dwelling nearest the downpipe, especially if there is any blockage there. If that is the property being inspected then the surveyor ought to make an appropriate comment, including checks by the legal adviser regarding rights and responsibilities for the shared drainage. On a detached property with a rectangular plan and a hipped roof with falls to each elevation, there will ideally be a downpipe at each corner; but frequently there will only be two downpipes, and those are unlikely to be located at opposite points of the building. Depending on the gutter falls, one may be draining very much more of the roof than the other. In most cases, experience will soon indicate whether or not the arrangement is likely to be adequate. But sometimes, with large roof areas or few downpipes, it might be necessary to do some quick calculations.

Detailed calculations for sizing gutters and downpipes, depending on the location in the country, are set out in BS EN 12056 and Approved Document

H of the Building Regulations. But for something more practical during the course of a site inspection, **Box 2** sets out a simplified way of assessing whether the gutters at a property are adequate to drain the roof areas.

A length of gutter will perform better if it has two outlets, a quarter of its length in from each end; twice that size of gutter will be needed if the only outlet is in the centre and four times the size, if the outlet is positioned at one end. This is because the whole of the run-off will be carried by the section of gutter nearest the outlet. Bends in gutters, often adopted in better-quality dwellings to avoid rainwater pipes appearing on the front elevation, reduce flow capacity by as much as 20 per cent if the bend is within 1.8m of an outlet but by only 10 per cent if the bend is further away. As to the size of rainwater pipes, provision is generally far in excess of strict requirement; for example, a

---

Box 2 Gutter adequacy calculations

This is a comparatively simple two- or three-step method of checking the adequacy of gutters on a dwelling, suitable for use during an inspection.

1   Calculate the effective area of the roof concerned in square metres:
    - For a flat roof, take the area of the roof on plan.
    - For a roof with a pitch of 50° or less, take the area of the slope on plan plus 25 per cent.
    - For a steep roof with a pitch of more than 50°, take the area of the slope on plan plus 75 per cent to allow for the greater speed of run-off.
    - For a roof that abuts a wall, add half the area of the wall.

2   Use the following table to check the resulting effective roof area against the maximum areas which should be served by particular sizes of half-round gutter. The table shows the largest effective roof area that should be drained into the gutter sizes most often used and the flow rates of the respective gutters.

Maximum effective roof area (m²)	Gutter size (mm diameter)	Half-round flow capacity (litres/sec)	Ogee flow capacity (litres/sec)
18	75	0.38	0.27
37	100	0.78	0.55
53	115	1.11	0.78
65	125	1.37	0.96
103	150	2.16	1.52

Half-round gutters have a better performance than ogee or other-shaped gutters even though they may be sold as the same size. So a third step is needed if ogee gutters are present or perhaps if the table indicates that the effective roof area is too large for the existing gutter size.

3   If appropriate, divide the effective roof area in square metres by 48 to find the run-off in litres per second which the gutter needs to serve. This assumes the heaviest of thunderstorms, which are estimated to unleash 75mm of rain per hour wherever the dwelling is situated. The gutter size that meets the flow requirement can then be determined from the table.

75mm downpipe, the most commonly found size on dwellings, has a flow capacity three times that of a 100mm gutter. It is not usually size that is the problem with downpipes but splits and blockages. It is best that joints are left open so that a blockage at a bend or a shoe at the base of a pipe will rapidly become apparent at the first joint above.

119 Secret gutters, whether within a floor or within a roof space as here on the left, are a frequent source of problems both from leakage and consequential rot.

120 A matter for urgent investigation. The extent and the degree of damp staining on the brickwork adjacent to this lead hopper head and rainwater pipe on an eighteenth-century mansion suggests that it has been going on for a long time. It might only be that the hopper head is blocked by leaves from the tree nearby, but it could be more serious. Such defects are a frequent cause of dry rot in timbers adjacent to the damp, and in this case, floors will need investigating as will the window frames and shutter boxes.

121 The white staining on this wall shows where a very long-standing leak, from a water storage tank overflow rather than a gutter, has thoroughly soaked the wall before being repaired and subsequently drying out. Internal timber in the affected area would have been at risk and a surveyor inspecting any of the flats adjacent to this location should ensure thorough checks are made.

122 The downpipe serving the rear main roof drains to the gutter serving the larger of the two rear addition roof slopes. Not an unusual arrangement but is the lower gutter large enough to cope with draining this total roof area? A quick calculation suggests not.

123 It would seem that the provision of guttering where none existed before on this late seventeenth-century house is relatively recent. However, it is a pity that plastic was used and that the arrangement and workmanship were not better. The appearance has been spoiled and a gap where the alignment is astray requires correcting urgently. The stone slating also needs some attention.

Materials

Gutters and pipes come in various materials. The earliest were lead, which gave long service provided they were not damaged. Often when early dwellings were provided with a wooden carved cornice, the top would be shaped to form a gutter and lined with lead; but when these cornices were found to contribute to the spread of fire, gutters in lead were formed behind parapets (**124**).

Later, cast iron was common for both gutters and rainwater pipes. They too give long service but do need regular painting to keep rust at bay. Downpipes, for this reason, should have been provided with distance pieces on the ears to enable them to be fixed at least 30mm from the wall; however, this is seldom done. The backs of ogee gutters can never be painted and this is a significant weakness of the design. The failure to paint the insides of gutters and the backs of rainwater pipes leads to earlier failure than might otherwise be the case. The small handheld mirror comes into its own here for examining the backs of rainwater pipes if cracks in the metal are suspected. Steel has been employed with the same profiles as cast iron – which makes the difference difficult to detect – but steel is reckoned to be less long-lasting as it is thinner, and it still needs regular painting in the same way as cast iron. These rigid types are less vulnerable to damage by ladders leant against them than gutters and pipes made of zinc, which can easily be distorted in shape. Gutters made of zinc also suffer from the disadvantage that they are manufactured with internal horizontal bars at the top for strength, which tend to provide an obstruction to the free flow of rainwater should there be an accumulation of leaves or silt.

Plastic is now the most widely used material for gutters and downpipes, both for new and replacement work. Its advantages are cheapness, ready availability in all shapes, sizes and colours, simplicity of installation and no maintenance requirement. Early versions of plastic suffered degradation from the effect of UV light, resulting in discolouration, brittleness and splitting. This is no longer a problem but plastic can be melted by undue heat and, therefore, needs to be kept clear of boiler flues. The lightness of the components is generally seen as beneficial and is particularly attractive to the handyman since assembly seems relatively easy, but this can be a disadvantage if fixings are given inadequate attention. The resultant sagging gutters and unstable down-pipes are often encountered (**125**), but at least they are unlikely to be dangerous, unlike cast iron sections which can pose a serious hazard if not properly secured.

One form of construction from the third quarter of the last century, found mainly on housing built by local authorities, utilised sectional concrete gutters which also acted as the lintels for top floor windows and a rebate for a wall plate. Known as Finlock gutters, these were lined with either asphalt or felt and bitumen, but they became discredited after differential movement tended to open the joints and cause damp penetration (**126**). They used to be condemned on sight but they can, in fact, be lined satisfactorily with metal or even, now, with lengths of the same reinforced bituminous membranes used to cover flat roofs (as described in Chapter 5), hot air welded on site to form one long length of sufficient weight that just needs to be spot bonded to the concrete, avoiding the effects of differential thermal movement.

Clearing gutters of sludge, grit and leaves once a year pays dividends if a dwelling is near deciduous trees and more frequently if the trees are coniferous as the needles can form a dense mass of obstruction. Such clearance provides the opportunity to check and, if necessary, coat the insides of metal gutters with bitumastic paint if such painting has been neglected in the past. The fitting of wire or plastic balloons at the gutter outlet can help to avoid blockages lower down at bends or at the base of the downpipe, which can be difficult to clear; but ironically, the balloons themselves tend to clog up with leaves and debris and can severely reduce the outflow from the gutter. Regular checks and clearances are, again, the answer (**127**).

Having traced the rainwater downwards, the surveyor should be alert for illegal connections to foul water drainage systems or septic tanks. The main problems of this sort tend to occur when extensions or alterations are undertaken. It may not be possible to be certain where underground pipes have been run but any clues should be sought and followed up as far as possible. If there is doubt, a caution should be sounded in the report.

Mention was made at the beginning of this chapter of the least safe type of gutter at the foot of a pitched roof slope, namely the gutter behind a parapet wall; and there is further consideration of this subject in the section dealing with 'Mansard roofs' in Chapter 18. In the absence of rooms in the roof space or trapdoors to the roof space and the exterior and if unable to gain access by a long ladder, little can be said other than to note whether or not there is evidence, old or new, of leakage on the ceiling below. Evidence of leakage necessitates the surveyor warning of the consequences to the timber supporting structure from rot and advising that the matter of opening up and repair is treated as urgent.

Even if there is no evidence of leakage, it should not be assumed that all is well. New decorations to the ceiling below any concealed area of guttering should arouse suspicion. A blob of bitumen could hold back rainwater

124 A simple timber gutter lined with lead on an outbuilding. Effective, but not something often encountered on a normal dwelling.

125 An additional support below this sagging gutter joint will remedy this problem, which will only get worse if ignored.

126 Finlock gutters – a post-war innovation which didn't stand the test of time – have concrete sections which also provide structural support over windows. Their propensity to leak can be overcome by providing durable linings – the edge of a metal lining is just visible here – but many have been replaced and, if encountered, these require careful consideration.

127 Fitting a wire mesh across the top of the box gutter serving this sheet metal roof in an attempt to prevent blockages has demonstrably failed, to judge by the staining and discoloration on the surrounding brickwork.

temporarily until the dwelling is sold and the burden of repair transferred. If significant parts of the dwelling are not accessible or visible, the client should be placed firmly on warning that, without an inspection, it is not possible for the surveyor to determine the condition of the concealed areas or to comment on whether or not substantial expenditure may be required. A parapet gutter, and presumably also parts of the main roof slopes as well, certainly comprises a significant part of the dwelling and is an area which carries a high level of risk. In these circumstances, the surveyor should recommend that the client commissions appropriate further investigations, with the use of a contractor's long ladder if necessary.

However, where there are rooms in the roof space or trapdoors, the surveyor should take every opportunity to examine the underside and exterior of parapet and valley gutters, noting the materials used, whether the designs are appropriate to those materials, and their condition. Indeed, if safe access is available to the exterior of the roof, the parapet gutter is not the only element which should be considered – all visible parts of the dwelling should, naturally, be included in the inspection.

8

Foundations

Subsoils, settlement, subsidence and heave

The foundations are never seen so it may seem perverse to include a chapter on them in a book dealing with the visual inspections of dwellings. Flaws in their design or construction or subsequent changes in the soil below, however, can sometimes have striking visible effects on the structure above ground level. By examining and analysing these effects on the visible structure, the nature of problems with the foundations can usually be deduced. Nevertheless, before any blame can be attributed to foundations or the subsoil, it is necessary to be reasonably sure that any defects seen are not due to some other cause. If other causes can be shown to be exclusively involved then there is little need to give the foundations of any dwelling much further thought as long as the dwelling is more than 10 years old because by this time, and very often sooner, it is generally thought that any fundamental flaws in design or construction will have become apparent. What does need further thought, however, is the subsoil on which the dwelling is founded and the surrounding topography – not that just within the boundary.

Catastrophic errors in the design or construction of foundations will usually reveal themselves within the first couple of years of a building's life, and in general, the longer a new building stands the more confidence the surveyor can have that the foundations are sound. If a dwelling is in the course of construction or has only recently been completed, the surveyor may be able to see details of the preliminary site investigation and examine how the results have been incorporated into the design adopted. This is certainly advisable since the foundations cannot be physically checked and the building has not been standing long enough for them to prove their worth. Having commissioned a private report, the prospective purchaser will hope to have some assurance of likely future performance, ideally beyond the fallback of a builder's warranty if problems should later develop.

The nature of the ground in which foundations are placed obviously varies considerably across the UK. Some is more suitable than others for building and some has been made more or less so by man's activities such as extraction of ballast or brick earth, underground mining, industrial processes and the waste tipping which produces landfill sites. It is an essential part of a surveyor's knowledge to be aware of the types of soil for the area across which the practice is exercised as well as the area's history. Accordingly, it is necessary for the office to have a set of the local 1:50,000 scale Geological Maps published for England, Wales, Scotland and N. Ireland by the British Geological Survey, now also available online and for mobile devices. The information on the Geological maps is a coloured overlay to the detail of roads, railways, rivers, etc. provided by the Ordnance Survey on its topographical maps to the same scale; this makes it possible to pinpoint with a reasonable degree of accuracy the location of any dwelling being inspected.

To a surveyor unfamiliar with Geological maps, it might be appropriate to arrange for a discussion with a local practising structural or civil engineer so that the meaning of the symbols can be explained along with how to determine, from the key provided on the maps, the type of soil on which properties are likely to bear and whether such soils are of types likely to cause problems. This information can be coupled with valuable local knowledge and perhaps a working relationship with the local Building Inspector or other professional colleagues. It is worth remembering that the Geological maps are only general guides. They do not always accurately record the shallow deposits within the top 4m to 5m of the surface where most foundation problems occur. Whether mention should be made in a report on the type of subsoil must be a matter of individual circumstances relating to each inspection, but in a Court of Appeal case, *Cormack* v *Washbourne* 1996, it was held that the surveyor was negligent in not consulting the geological map where there were diagonal cracks in the walls of a 1969 house and trees within 7m to 8m.

Methods of site investigation prior to building have developed very considerably over recent years, and where a large estate is proposed, this can be carried out comprehensively by taking deep samples from boreholes at many points and building up profiles. The end result should be a foundation design appropriate to the subsoil and the development. Special provision might need to be made where old features are to be removed, areas filled in or trees cut down. Notwithstanding, a sample cannot be taken at every point, and in many cases, it still boils down to a decision by the site team as to the appropriate level to place the foundation. While such a level should be approved for each dwelling by the Building or NHBC Inspector, it is inevitable that assumptions are made and that what is considered suitable for one dwelling will also be deemed suitable for others. The site investigation is an important part of the development process and there are clear expectations that correct decisions will be made. It is now only infrequently that cracks appear fairly soon after completion because a pocket of poor ground has been missed.

It is generally the level at which foundations have been placed that causes problems. Up to around 1900, most dwellings were provided either with cellars or substantial room-height basements. Accordingly, the foundations would be much lower than the normal ground level, and the removal of the volume of soil to form the basement and its replacement by the loading from the dwelling balanced the equation. It will be found that such dwellings experience few problems from subsidence.

While the loads in domestic construction are light, it is nevertheless necessary where foundations are placed at around ground level that they are at least below the level where the ground is influenced by climatic changes. Where areas are only grass covered, this is normally between 1m and 1.5m; but where vegetation, especially trees, are present, it can be much greater – perhaps as much as 5m – depending on the geology of the subsoil. This necessity has not always been followed with two- to three-storey suburban properties built since the early 1900s without cellars or basements. Even if an appropriate level is provided at the time of construction, the foundations can be upset by subsequent planting, or its removal, or other form of disturbance.

Where foundations are set around ground level, there is nearly always an initial overall settlement as the ground below is consolidated. It is usually quite small and, as long as it is of a uniform nature, often passes unnoticed (**Figure 1**). If the dwelling comprises a single structural unit but the initial settlement is not uniform, portions can settle at a different rate or by differing amounts and cracks can appear. These will almost certainly be diagonal in nature and the width of cracks will taper along their length.

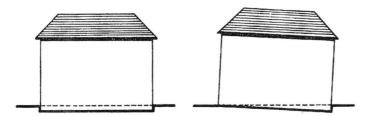

Figure 1 When a dwelling is provided with foundations at around ground level, there is a uniform settlement, often quite slight, as the ground below consolidates. Even if the whole dwelling tilts, as in the diagram on the right, this often passes unnoticed.

The pattern and tapering of cracks gives important clues to the location and direction of the movement. If the downward settlement takes place at the extremities of the dwelling, the cracks will be at their maximum width near roof level. If the settlement occurs at the centre and the ends are stable – a less likely occurrence – the maximum crack width will be at ground level, passing through the damp-proof course and probably extending down through the foundation itself; although this is not visible, of course, without opening up (**Figure 2**). The cracks will also appear in roughly the same place on both sides of a solid wall, but where there is a cavity, there could be a marked difference in where they appear, bearing in mind that it is not possible for a

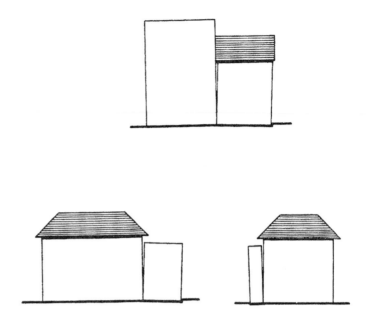

Figure 2 Different structural units that are attached can behave in contrasting ways and have a tendency to part company unless of very similar construction and with foundations at the same level. The top diagram shows a back addition attached to a taller main structure – very typical of dwellings up to the 1920s, with or without basements. The lower left diagram shows a garage alongside a two-storey dwelling with a room above while the lower diagram on the right shows a bay window, again, attached to a two-storey dwelling. All are parting company from their host structures.

subsidence crack to be restricted solely to one leaf. One instance where 'settlement' could appear to be taking place at the centre is where trees, which were formerly extracting moisture from a shrinkable clay soil, have been removed. On removal of the trees, the clay soil recovers the moisture formerly lost, swelling in the process. This is a phenomenon known as 'heave', and where the heave causes the extremities of a building to be forced upwards, this can cause similar cracking to that which develops when the centre has settled, creating the misleading impression of downward movement at the centre (**Figure 3**).

Of course, the walls affected by 'settlement' or heave are very often not blank walls but contain window and door openings, introducing areas of weakness which tend to focus the cracking at the corners of the openings. Accordingly, the cracks go round these, perhaps appearing as gaps between frame and wall (**132**). The pattern of any cracking between openings is a further clue to the direction and location of movement.

Movement causing cracks involving foundations and appearing within 10 years of construction is not covered by normal subsidence insurance. Insurers refer to such movement as 'settlement', restricting the term 'subsidence' to

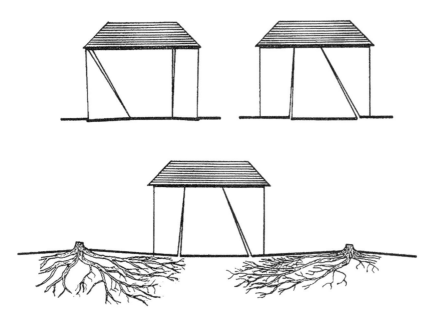

Figure 3 Differential settlement in a single structural unit invariably causes cracks, usually diagonal and tapering. Should the extremities settle and the centre remain stable (as in the top left diagram), crack widths are at their maximum at roof level. However, if the centre settles and the extremities remain stable, the maximum width of cracks are at ground level, continuing below ground through the foundations (as in the case of the top right diagram). In the example (at the bottom) of soil recovery known as 'heave' – where trees have been cut down before or subsequent to construction – the soil swells as moisture is recovered, forcing the extremities upwards. Heave cracks do not always display the diagonal tendency of subsidence cracks to the same extent but are always similarly tapered.

movement involving foundations occurring after the 10-year period. It would, however, be covered by the NHBC, or any other similar insurance warranty, provided the damage was notified in time. If the dwelling had no cover of this nature, a remedy might be sought from a builder or developer vendor but action would normally have to be taken within 6 years. If outside either the warranty or litigation claim period and a subsequent sale is involved, the surveyor could be in some difficulty should telltale signs be missed.

If the dwelling comprises a number of different structural units – for example, a projecting porch – or if an extension has been added at a later date, it is not uncommon for these to settle or be subject to the effects of ground movement in contrasting ways. Much will depend on how closely related are their forms of construction. More pronounced effects are often apparent where foundations are at a different level. One example is where a back addition, founded approximately at ground level, is attached to a main building with a basement; another is where there is a lightweight structure such as a garage at the side of a dwelling; but this occurs most often where there is a porch or bay window with minimal foundations, set out from a wall with foundations that extend deeper. Cracking often does not occur where the structures part company but will frequently appear a little way along the wall of the lighter structure at a plane of weakness such as a window or door opening (**133**).

The foregoing descriptions and the associated diagrams illustrate in outline the main types of settlement, subsidence and heave as well as their effect on masonry structures of brick, stone or concrete block – by far the majority forms of construction for dwellings in the UK. Of course, the evidence on site is seldom as straightforward as the descriptions, and outline diagrams suggest and require, at times, careful analysis to determine with any reasonable degree of certainty whether defects originate from below ground level or above. Movement in the structure above ground level from causes such as thermal effects and shrinkage produces crack patterns different to those caused by subsidence; other causes, such as failure in a lintel or beam, can produce diagonal cracks which also taper. However, they seldom extend and are traceable, even if interrupted along the way, throughout a wall. If such cracks extend below the lower windows down to ground level, however, they are highly likely to be caused by subsidence or heave. Careful viewing of whole elevations from the exterior will sometimes help the inspector to trace the entire length of a crack which may have been noted previously in sections internally from plaster cracks in rooms or the staircase.

The small but increasing proportion of dwellings which are of framed construction, timber, steel or concrete, if affected by subsidence, will develop a different pattern of cracking, more likely related to the separation of cladding from frame or the crushing of infilling panels. Those clad in brickwork may exhibit some similarities, but this may be modified by the effects of the movement on the underlying framed structure.

Any structure, framed or conventional, may be clad in timber or PVC boarding, tiles, steel or other forms of sheeting. Timber cladding, PVC boarding, or slate or tile hanging applied to masonry construction does not display the external evidence of subsidence so readily or to the same extent as masonry construction because the cladding components can adjust their position to accommodate the movement. In these cases, evidence may be restricted mainly to internal plaster cracks, although gaps in soffit or fascia boards are often an indication that movement has taken place. Fortunately, slate or tile hanging is seldom taken down to ground level because of the risk of physical damage, so cracks below the lower windows indicating likely subsidence or heave may still be visible.

Having decided that the visible defects are most likely to have been caused by foundation movement, the advice to be given to a client can only be formulated after a number of other aspects have been considered, relative, mainly, to the possible cause but also to the size of the cracks, whether they are likely to be progressive and the age of the dwelling.

Foundation movement occurs when there is a volumetric change in the ground supporting the foundation, causing the dwelling to move. Upward

128 Subsidence has severely affected this two-storey mid-Victorian house. The window openings of the right-hand half have been 'squared up', but many horizontal and vertical lines, especially the two corners on the right, are far from true. The cause is a well-recognised problem with the local ground conditions, but careful enquiries are required to ensure that all necessary works have been properly carried out. Even after all movement has ceased, the value will continue to be affected.

129 Subsidence in the pier between the two front entrances which is also, of course, the front end of the party wall of these three-storey, non-basement houses of the mid 1800s that have been converted two at a time horizontally into flats, hence the blocking off of one entrance. The effect is dramatic in that brick courses have dropped by as much as 75mm and cracks from ground level extend round the door arches to the sills of the windows above, disrupting, in the process, the adjacent windows and their arches and fracturing the artificial stone mullion, necessitating emergency restraint.

130 Close-up of the crack starting at ground level in the pier of brickwork between the door and windows of **129**, fracturing the bricks and increasing in width with height. Underpinning may be necessary in this instance if the movement is progressive.

131 Indications of movement to the two-storey bay window of an early 1900s terraced dwelling. The crack extends to ground level and increases in width with height, providing strong evidence that the cause is subsidence. The pattern is repeated on other dwellings in the terrace but, where carefully repaired, is difficult to detect. As plaster cracks have been repaired internally and show no sign of reopening, the movement is probably not progressive and careful repair here could restore appearance. It is regrettable that the original double-hung sash windows have been replaced by an out-of-style type. Surveyors should warn that this type of window is vulnerable in that the louvres can easily be removed from the outside.

movement is known as heave but the more frequent occurrence is downward, which is defined by engineers and the insurance industry as subsidence. Ground movement laterally down a slope or hillside, taking the property with it, is defined as landslip. Subsidence can be due to a shrinking or softening of clay soils, washing away of fine material in granular soils, further compaction of fill or compression of peat other than that due to the load of the dwelling itself. This latter cause, along with any movement within a structure due to the redistribution of loads and stresses within the various elements of its construction within the first 10 years, would be known as settlement within the insurance industry; and this is usually excluded from domestic policies covering subsidence, landslip and heave as it is considered the responsibility of the builder to put right.

132 Catastrophic subsidence along the flank wall has resulted in cracking which has exploited the weakness between the door and window openings on the front elevation. The increasing width of the crack as it gets higher reflects the downward and outward movement of the side elevation wall.

133 The foundations to the front elevation here have experienced greater movement than those supporting the party wall. The resulting differential movement has focussed on the line of weakness formed by the first set of windows in from the party wall, causing distortion to the window openings and unevenness to the brickwork at each level. Much of this is historic but there are sufficient areas of open cracking to raise questions about the amount of any continuing movement.

Volumetric changes to the ground only occur as a direct result of outside influences such as changes in groundwater levels, which may be caused by leaking drains, the extraction of water by trees or other vegetation, extremes of climate such as prolonged rain or persistent drought, decomposition of organic material and landslip, mining or other excavation. Mention has already been made of the need for surveyors to be aware of the subsoil in their area of practice. A very considerable proportion of dwellings have foundations placed in clay soil, some more prone to shrinkage than others but all with a propensity for volumetric change if the naturally occurring balance of composition is altered by external influences. Clays of highly shrinkable potential cover large areas, especially in the south and east of the country, and a surveyor's local knowledge should extend to an awareness of specific areas where such clay is likely to be present. In areas of highly shrinkable clay, a quick look around the garden of the dwelling being inspected, along with the grassed areas around trees in the locality, will sometimes reveal quite substantial fissures in the ground – as much as 20mm to 25mm wide and up to half a metre deep during extended periods of dry weather.

Trees

Of the various external influences which can cause volumetric change in clay soils, by far the most commonly occurring, giving rise to not far short of three-quarters of all claims under subsidence insurance in dry years, is the extraction of moisture by the roots of trees or other vegetation. This causes the clay to shrink and this can deprive the foundations of support. Knowledge that a dwelling is founded on clay necessitates the surveyor considering its proximity to trees and, if trees are present nearby, their type. The presence of such trees is not, of course, restricted to the gardens of the dwelling being inspected; they may be situated in neighbouring gardens or the public footpath.

Determining whether the trees are likely to be responsible for the damage can be a fairly complex matter and depends on a number of factors such as the species of the trees, the characteristics of their growing patterns and their ages as well as their proximity to the dwelling, the depth of the dwelling's foundations and recent climatic conditions. Each factor can be variable and the interaction between factors, even more variable. This results in the anomaly that damage can be caused, in some cases, by a tree at some distance away from a dwelling whereas, in other cases, a tree of the same species and size at a much closer distance causes no damage.

While it is probably true to say that any tree or large shrub with its roots in highly shrinkable clay and growing close to a dwelling can, in certain circumstances, cause damage, it is also true to say that many trees are unjustly blamed for damage that is really attributable to other causes. Trees, however, vary considerably in their size and characteristics, and some are more prone to causing damage than others. The trees which are known to cause damage are more likely to do so when long, hot summers are separated by relatively dry winters, 1975–1976 and 1989–1990 being prime examples when subsidence insurance claims reached record levels. There was a further peak in claims in 2003 and some more minor blips in 2006 and 2010.

Fast- and vigorous-growing broad-leafed trees with shallow roots that have reached maturity and are producing large crowns and a substantial area of leaf surface are the most likely to cause damage. Such trees spread a myriad of fine roots over a considerable area of ground, much nearer to the surface than is commonly believed, and extract moisture at a rate often well in excess of what can be recovered in the winter months. As such, there is a continual deficit of water in the ground when compared with its natural state. This deficit of moisture can only be made up if the tree dies or is cut down; but this can produce a substantial problem from heave as the soil can take many years to recover all the moisture that has been lost, continuing to swell as it does so. Rather than cut the tree down, reducing the size of the crown to lessen the summer requirement for moisture could lead to a less damaging, gradual recovery instead.

BRE and the Institution of Structural Engineers (IStructE) have both published lists of trees, indicating the types which are most likely to be implicated in causing problems to buildings; and the Association of British Insurers (ABI) has produced a guide for tree types with recommended distances from the property, which can be accessed through their website (www.abi.org.uk). **Table 1** is based on ABI data and represents maximum distances. On non-shrinkable soils, the distances can be reduced. Perhaps its greatest value for the surveyor, however, is as a guide to indicate the higher-risk tree species of which the surveyor should be aware.

The risk of damage from trees increases substantially the closer the tree is to the dwelling, and those dwellings with shallow foundations will always be more vulnerable. Two-storey dwellings without basements from the 1920s until the 1980s are the most likely types to be affected while extensions, porches, garages, conservatories and bay windows of any era are even more at risk.

It is not always an easy matter to tell one type of tree from another. The expert advice is that for identification, the tree should match with its descrip-tion in all respects, not just the leaf shape, since different families of tree have leaves of comparatively similar shape. Nevertheless, it is important for surveyors to ensure they are familiar with the most common types of tree and that they are able to identify with a fair degree of confidence most of the trees in the top half of **Table 1**, especially those nearest the top which pose the greatest risk to buildings on shrinkable clay soils. The judge in the case of *Daisley* v *BS Hall & Co.* (1972) 225 EG 1553 said: 'it really is essential to make sure of your tree recognition because Poplars are not the only trees of which there is more than one species'. The case concerned an 8-year-old house where the surveyor had attributed cracks to 'minor settlement and bedding down', failing to mention either the subsoil or a line of trees 8m to 9m away which were a close relation of the black poplar.

Table 1 Relative subsidence hazard of full-grown trees– i.e. suggested safe distance (metres) from tree to building

Risk group	Tree type	Distance for individual type (m)
40 metres	Willow	40
35 metres	Poplar	35
30 metres	Oak	30
	Elm	30
25 metres	Horse chestnut	23
	Plane	22
	Ash	21
	Cypress	20
	Lime	20
	Maple	20
	Sycamore	17
	Beech	15
	Walnut	14
	Hawthorn	12
	Cherry	11
	Plum	11
	Rowan	11
10 metres	Apple	10
	Pear	10
	Birch	10
	Laburnum	9
	Pine	8
	Spruce	7
	Holly	5
	Laurel	5
	Magnolia	5
	Yew	5

Source: based on ABI data

There are numerous publications to assist with tree identification but a good pocket version should be regarded as an indispensable part of the surveyor's equipment and a copy should always be on hand in the car to assist with the inevitable uncertainties which will crop up. It is not always appreciated that trees are often easier to identify during the winter because the differing bud shapes and arrangements on the twig can frequently be more readily distinguished than the summer leaves. The chosen tree identification guide should, therefore, include details of bud patterns as well as the more obvious leaf shapes. As a first step, taking the time and trouble to examine and become familiar with the identities of the common trees encountered on a daily basis around the home and office, and in all seasons, will soon establish the basis for the necessary expertise. Differentiating between the various members of the willow and poplar families is always liable to be difficult, but merely recognising that a tree *probably is* a willow or a poplar is already winning more than half of an important battle. Most identification is straightforward and there should certainly be no reason for a competent surveyor to make the embarrassing error, revealed during one complaint investigation, of describing an innocuous silver birch as 'a willow-type tree'!

Ground movement

In the absence of trees but still with the evidence of subsidence damage to a dwelling, other possible causes may need to be considered. These can fall into two categories: those which are within sight; and those which can only be a matter of pure speculation but, nevertheless, are put forward with a distinct degree of possibility, reinforced, if possible, by the surveyor's local knowledge.

The most obvious possible cause within sight could be the effect, if any, on the dwelling of its proximity to sloping ground. Surveyors should be aware of the tendency of sloping ground to slip. Dwellings on or at the top of slopes can accordingly be affected by subsidence or, more accurately, landslip. Steep slopes are more inclined to move than shallow, and looser soils more so than those that are stiffer or more cohesive, although even rock is not immune if cracks or fissures are present. Clay interspersed with layers of sand is much more sensitive to ground movement. Indications of unstable slopes include out-of-plumb fence posts and trees, trees which have grown with a curve upwards from the slope at their base, fissures on the surface of the ground and local 'slumped' areas of ground. Where clay is known from local knowledge to be present, a slope of 8° is the maximum that can be thought of as free from risk. The surveyor needs to note where there is evidence of sloping ground and any signs of downhill movement in the dwelling. If excavations have been carried out nearby then their influence or possible effect needs to be considered.

Leaking drains can be the cause of subsidence. The external area in the region of cracks thought to have been caused by subsidence should be inspected for the evidence of drainage runs, inspection chambers, gullies, soil and rain-water pipes. Leaking drains can cause a softening of ground below the foundations, rendering the ground incapable of supporting the load of the dwelling; or they can cause ground to be washed away, depriving the foundation of support – a particular problem with sandy soils. This can be a smaller-scale version of what can happen when there is movement of groundwater caused by, for example, a fractured culvert, pumping of groundwater out of excavations, discharge of rainwater pipes on to the ground or leaching of water into backfilled material in excavations. The likelihood of a dwelling being founded on subsidence-sensitive soil other than shrinkable clay needs to be borne in mind, although probably only a surveyor's local knowledge will lead towards such a conclusion.

The action of both ground and surface water on some soluble rocks such as chalk and limestone can cause cavities to form in the ground, and these may be infilled with soil, water or just air. If the infill material destabilises then subsidence known as a sinkhole can occur, often with little warning, which is more than a little disconcerting for those in the vicinity. The exceptional rains during of the winter of 2013–2014 resulted in a rash of sinkholes appearing. Fortunately, those in this country are usually not very large compared with those in some parts of the world, but they can still pose a severe threat to the stability of nearby structures and urgent remedial measures will be needed.

Unnatural cavities can be produced by the mining of below-surface deposits, the collapse of the roofs of which can cause subsidence. Those mined in the past include coal, slate, limestone, tin and iron among others, and local archives may need to be drawn upon if such a cause is suspected. Some mining techniques, such as longwall mining, may deliberately allow the roof of worked out sections to collapse (**134**). Others, such as the pillar and stall method, utilise intermediate supports but deterioration of the pillars can cause collapse and subsidence many years after mining has ceased. Land reinstated after opencast mining becomes another backfilled site.

Backfilled sites can continue consolidating for some time, depending on the material, the placement and subsequent compaction. Decomposition can continue for many years as well and particular dangers from subsidence occur where a development straddles the boundary between a filled site and virgin land. Vibration due to piledriving on nearby sites can cause yet further consolidation.

Peat causes particular problems because it is highly compressible and liable to be displaced under load. Non-uniformity is a problem and significant variations in depth can occur. IStructE describes peat as 'reinforced water', giving a good idea of its load-bearing capability. Lowering groundwater in peats, silts and loose granular soils can reduce volume levels and cause subsidence. Freezing of the water content of soils and its consequential expansion can cause uplift or 'frost heave'. This is an unlikely source of damage to heated dwellings but could affect unheated garages or outbuildings if the foundations are less than half a metre deep in parts of the country where extended freezing conditions may occur.

To conclude the description of less likely, but possible, causes of subsidence, mention should be made of chemical attack on foundations by sulphates in the soil, only feasible in the presence of water. Sulphates occur naturally in some soils: for example, Kimmeridge, Oxford and London clays, Keuper Marl and Lower Lias clays. Solutions of the sulphates in such soils attack the hardened concrete, causing it to expand. Sulphate-resisting cement should, of course, have been used for the foundations if the presence of sulphates in the soil had been known about or ascertained, but tests were not always carried out in the past and may not always be on smaller developments today. Even if sulphates were not present in the soil, the use of sulphate-laden backfill materials, such as colliery shale, brick rubble with adhering plaster and industrial

134 A clear example of subsidence affecting this terrace of late 1800s, two-storey, stone-fronted cottages, situated in a mining area. The whole of the terrace beyond the rainwater pipe – some ten dwellings – is tilted outwards in relation to the section on the left. It is just possible to see, by the rainwater pipe, the brickwork which has had to be used to make up the gap in the party wall created by the subsidence.

wastes, have caused problems in this regard, lifting both ground-floor concrete slabs and the walls they support. Trench foundations, piers and piles are, however, less liable to damage from chemical attack than slabs.

Assessing movement

As to the investigation into the causes of subsidence and the remedies available, IStructE produced a booklet entitled *Subsidence of Low-Rise Buildings* (second edition, 2000) for buildings up to four storeys. It is the work of a task force brought together by IStructE, representing owners, insurers, loss adjusters, lenders, contractors, local authorities, surveyors, arboriculturists and, of course, engineers.

Although some have criticised the repetition that originates from its multiple authorship, the booklet in its entirety contains much of value, especially the first six chapters. This will be of considerable interest to surveyors who are on the receiving end of a certain amount of implied, although not direct, criticism in their capacity as advisers to owners, buyers and lenders and, sometimes, insurance companies.

Prior to preparing the first edition of the booklet, published in 1994, IStructE had recognised that a combination of increasing home ownership, changes in building design and construction over the years which seemed to make dwellings more brittle, two hot summers separated by a dry winter, and the perceptions of occupiers had led to a vast increase in the number of claims against insurance companies. These were made under the insured risk of subsidence and landslip, first introduced generally in the early 1970s at the request of the Building Societies and extended to include heave in the 1980s following publicity from a number of cases concerning the removal of large, mature trees from public footpaths.

Whereas occupants of rented accommodation would tolerate cracks as not being their concern, homeowners could see the value of their investment, and lenders, the value of their security, being put at risk. In particular, after the very dry summers of 1975 and 1976, a substantial industry had grown up devoted to underpinning dwellings at insurers' expense. Further dry summers in the 1980s and 1990s led to annual expenditure on such work often amounting to between £300 million and £500 million with an average of 40,000 claims a year. Some 20 years later, the insurance industry continues to quote such figures as averages subject to annual variations.

Originally, much of the expenditure on underpinning came about because the prospective buyer of a dwelling up for sale could not obtain finance due to the presence of cracks, thought to be due to subsidence. If confirmed as subsidence and the insurer accepted the claim, pressure would be brought to bear for something to be done quickly; and whereas in other circumstances, a much fuller investigation over time as to causes and possibly less drastic remedial measures might have been acceptable, underpinning would be carried out with a view to getting the sale back on track. Even so, some insurers and lenders continued to be cautious about dwellings that had been underpinned. Research demonstrated that 65 per cent of all subsidence damage to dwellings can be attributed to trees and large shrubs on clay subsoils – and a greater proportion in years of drought – so it is not surprising that if a tree was nearby, it became the culprit and the object of removal with other possible causes given scant attention.

It was perhaps the concern of arboriculturists, as well as some engineers, which led to the preparation of the second edition of the booklet, a mere 6 years after the first, containing as it does much more information on tree management and recommending the felling of trees only in exceptional circumstances. With a wider and better understanding of the genuine concerns of all parties at times when subsidence is suspected but, particularly, when dwellings are in the process of being sold and bought, it is to be hoped that with better knowledge of causes and a greater understanding of the variety of remedial measures available, there can be more cooperation and less confrontation between the parties involved. This is perhaps a pious hope but it is, of course, the surveyor inspecting a dwelling and perhaps owing a legal duty to both a lender and potential purchaser who brings such questions into the open. Too much caution on the surveyor's part can perhaps hinder a sale but too casual an attitude, or recklessness, can expose the surveyor to a charge of negligence.

Categorising cracks by their size and dealing with their structural significance is dealt with in the IStructE booklet, which in turn followed a 1990 revision of a BRE Digest with a table dealing with 'The Assessment of

Table 2 Crack width classification and repair considerations

Damage category	Approximate crack width (mm)	Definition of crack and description of damage	Definition of crack and repair type/considerations
0	Up to 0.1	**Hairline:** • Negligible structural implications and expected to occur on almost all buildings at any location • Not generally related to subsidence	• Can be filled or covered by wall covering and redecorated
1	0.2 to 1	**Fine:** • Generally has some structural significance • Often more visible inside the building than outside • Generally located at points of structural weakness (doors/windows) • Indicates slight foundation movement	• Can be filled or covered by wall covering and redecorated
2	2 to 5	**Moderate:** • Most likely will have some structural significance • Almost always occurs at points of weakness or hinge points • Generally visible internally and externally • Indicates foundation movement • Enough to distort door and window frames and make doors and windows stick	• Internal cracks are likely to be raked out and repaired to a recognised specification • May need to be chopped back and repaired with expanded metal/plaster, then redecorated – if not, cracking is likely to return • External walls require raking out and repointing – any cracked bricks should be replaced
3	6 to 15	**Serious:** • There will be some severe compromise of the structural integrity • Weathertightness will be impaired	• Internal cracks repaired as per **Moderate**, with reconstruction if required • Rebonding is required • External cracks will also require reconstruction, and some sections of brickwork may require replacement • Specialist resin bonding techniques may be required
4	16 to 25	**Severe:** • Structural integrity severely compromised – floors sloping; walls leaning or bulging; lintels suspect	• Major reconstruction works to internal and external wall skins • Realignment of windows and doors

Table 2 Continued

Damage category	Approximate crack width (mm)	Definition of crack and description of damage	Definition of crack and repair type/considerations
		• Pipe fracture inevitable • Windows broken	
5	Greater than 25	**Very severe:** • Potential danger from failed or fractured structural elements and from instability • Safety issues must be considered	• Major reconstruction works, plus possible structural lifting or sectional demolition and rebuild • Window and door replacement • Construction (Design and Management) Regulations will probably apply

Damage in Low-Rise Buildings with Particular Reference to Progressive Foundation Movement'. The BRE classification categorises the likely seriousness of cracks into six groups depending on their width and indicates the level of repair which might be required for each, although the subjective descriptions of each category may not reassure a worried homeowner (see **Table 2**).

For note-taking and recording purposes, the width of cracks should be noted and if they are tapering, the width at each end given. There is no objection to a general collective adjective being used in the report to describe cracks in a room or dwelling – for example, 'the cracks are serious for a house of this age' or 'considering the age of this house, the cracks are slight'; but the site notes should be factual and avoid, as far as possible, elements of subjectivity. Using a crack gauge rather than a tape measure or rule will help greatly when recording details of cracks (**7**).

Describing and classifying cracks by their width and giving a number to a category of damage, as in the BRE Digest and the IStructE booklet, goes only a small way towards assessing their real significance. Their location, the pattern they form and whether cracks that taper are widest at the top or the bottom will indicate whether they are likely to be attributable to subsidence; but what clients – whether owners, buyers or lenders – want to know is whether they are progressive or not (**135**). If a reasonably definite answer to that question can be given then a much better picture of the significance of the cracks will be obtained – and even more so when they are related to other aspects of the dwelling like its age, its site characteristics and the surrounding features.

All the aspects of the cracks listed in the previous paragraph, bar one, are matters of fact which can be recorded on a single inspection. What cannot be recorded on one visit is whether the cracks are progressive or not. The only way to be able to express a positive view on that is to monitor any changes over a period of time sufficient to determine whether movement is in excess of that which can be considered acceptable (**136**). An opinion, but only an

136 The amount and direction of movement can be accurately measured over time with a crack monitor. The two ends are firmly fixed to the wall on either side of the crack and regular readings are taken from the visible grid in the centre.

135 Detailed measurements are not required to confirm that there is a V-shaped vertical crack at the junction between these two structures. The older brickwork on the left has dropped dramatically away from the more recent structure on the right. The cause of the movement needs to be identified and the nature of the remedial works must be determined.

opinion, might be expressed as to whether cracks are likely to be progressive or not, but that opinion would usually be based mainly on evidence other than the cracks themselves.

As already discussed, dwellings will probably suffer an initial settlement when first built, which may or may not be noticeable by way of cracks. Because most new dwellings involve a mixture of wet and dry trades in their construction, they may also exhibit cracks from shrinkage as the water used in those constructional processes dries out. These cracks will show up on the initial decoration provided by the builder. It is generally considered that any cracks arising from these two causes should never exceed 2mm in width and will generally be very much less. Repair will, accordingly, be comparatively simple, involving filling in at the time of the first proper redecoration.

In addition to initial settlement and shrinkage, the surfaces of a near new dwelling, and indeed a dwelling of any age, have to cope with seasonal variations between hot and cold and wet and dry periods. These variations can cause slight flexing of the fabric of a structure, particularly when it is founded in a clay soil; and this flexing will often manifest itself at a point where any previous cracking has developed. When a wall has cracked, for whatever reason, filling in and redecorating does not mechanically rejoin the two parts together; it merely fills the gap with a material which, in itself, may be subject to shrinkage as it dries out and hardens. This shrinkage and any slight flexing between seasons can be sufficient to reopen cracks to an extent of 1mm or so. The traditional way to combat this is to line the repaired area though the slight movement may still be sufficient to tear the lining material.

It is not the purpose here to deal with any concerns that the purchaser of a new dwelling might reasonably have about the scope of a warranty for a newly built property. A surveyor's advice on that and on the new dwelling itself would take on a greatly different form to that derived solely from a visual inspection. However, in the majority of cases where a surveyor is instructed to carry out an inspection on a near new dwelling being sold to a second or subsequent purchaser within 10 years of the issue of a new-build warranty, there may be cracks of no more than 1mm to 2mm in width present or evidence of making good to cracks. This degree of disturbance would be consistent with the effects of initial settlement and shrinkage coupled with any subsequent seasonal movement in a dwelling on a reasonably flat site with foundations in a clay soil which is not unduly shrinkable and where the dwelling is well away from any trees. Provided the cracks or making good did not form an array – i.e. a series of cracks fanning out from a single location – the sum of the width of which should be taken as the equivalent of the width of a single crack for assessment purposes, or a pattern which could be attributable to subsidence, and furthermore a period of 5 years had elapsed since construction had been completed, it is not considered that there would need to be any cause for alarm. Nevertheless in the apt phraseology of the IStructE booklet, relatively new dwellings are 'untested' and it is, therefore, necessary to adopt a particularly cautious attitude to such dwellings if cracks or other evidence of structural movement are found to exist or if there are indications that such cracks have been made good. For this reason, any inspection of such a dwelling needs to be even more exhaustive than usual to ensure that differentiation is made, as far as is reasonably possible, between cracks which may have different causes. Having done this, it is important to demonstrate in the report how conclusions have been formed and to set out the evidence to show that they were not reached lightly but only after due diligence and consideration of all the factors.

This care and diligence is particularly important in the circumstances set out above in view of the different considerations which apply to a second or subsequent purchaser in relation to a new-build warranty. While the warranty will be carried forward, it will only be honoured in respect of defects which develop after completion of the sale and not to the repair of those which existed, or which could have been ascertained to exist by inspection, at the time the sale took place. For this reason, if a favourable view is to be given in a report then it may be that furniture has to be temporarily moved in order to obtain an adequate sight of all parts of the dwelling.

It is unlikely with a near new dwelling that such a favourable view could be expressed in any other circumstances than those described above. Cracks or evidence of making good to cracks larger than 2mm, a shorter period of time, a pattern of cracks pointing to the view that subsidence could be the cause, a steeply sloping site, or clay soil with trees nearby are all possible reasons, either singly or collectively, for the surveyor to indicate a distinct degree of caution – the reasons for such wariness being fully set out in the report.

Only 1 per cent of valid claims against NHBC now relate to foundations, with a further 4 per cent relating to 'substructure and ground floors'. There is some reassurance in knowing that these proportions are relatively low, but if NHBC house registrations are typically around 100,000 a year, this means there are about 1,000 cases annually of foundation problems. Of greater significance is that these are inevitably expensive problems to remedy; the 1 per cent of claims eats up 16 per cent of the annual settlement costs, a further 19 per cent of annual costs dealing with the claims for substructure and ground floors. Perhaps even more surprising is that 8 per cent of claims costs in 2013, nearly £4 million, related to buildings near trees. With remedial costs at these levels, an aggrieved second purchaser of a nearly new home, finding that a new-build warranty does not provide the expected cover, is unlikely to be deterred from pursuing an action against a surveyor who may have missed possible evidence of foundation failure.

There is, therefore, a possibility that a surveyor will be inspecting a dwelling where it is purported not only that initial settlement and shrinkage have been dealt with but also that a degree of subsidence due to unsatisfactory original construction work to foundations has been stabilised. If such stabilisation has been carried out within a period of 5 years before a sale to a second or subsequent purchaser then the only way to demonstrate that the work was successful is for it to be monitored professionally for at least that period, irrespective of claims about the high standards of the earlier investigation and the work of stabilisation. Even if the situation had been monitored for at least a period of 5 years and there was no sign of further movement, it would still be necessary for the surveyor to consider critically what had been done or, if unable to do so personally, to advise employment of an engineer for the purpose.

Such considerations apply equally to any dwelling of an age beyond where it can be considered new or near new; although in the absence of a new-build warranty, there may be less temptation for a prospective buyer to proceed since there is no other party who can be expected to bear the cost of repairs if the surveyor has been sufficiently careful to stress the risk. The most dangerous situation from the surveyor's point of view is where a dwelling has been completely and newly redecorated and there is evidence of cracks having been made good. Here, again, the surveyor is in no position on a single inspection to say whether cracks are likely to reopen or not and if they did, whether they would be progressive. The redecoration may have been innocently carried out to improve the marketability, as is sometimes advised, but as the judge in the case of *Morgan* v *Perry* 1973 noted, human nature being what it is, vendors will from time to time take steps to cover up defects. The surveyor is not to know which of these circumstances apply and questioning the owner will not always elicit a satisfactory reply. The case mentioned concerned the inspection of a 4-year-old house built on sloping ground, the surveyor mentioning only 'hairline' cracking but no location given. He made no comment on making good to a cornice or the redecoration which had been carried out, the sloping floors or cupboard doors which were sticking. More cracks appeared after purchase and damages of £11,400 were awarded.

When dealing with a newly redecorated or renovated property, on finding evidence of making good in places where subsidence cracks might have occurred as well as the presence of any of the higher-risk factors – such as clay soil, presence of trees or sloping ground, for example – there is no alternative but for the surveyor to advise that proceeding to purchase would involve risk – perhaps more in some cases than others, but that would only be a matter of degree. The risk could only be discounted by a period of monitoring to ensure that cracks did not reopen to an extent of more than 0.5mm over at least 2 years. If evidence such as photographs or professional certification could be produced to show that such monitoring had been carried out professionally for at least that period before the entire redecoration, the surveyor would have no valid reason to stress the risk of proceeding unduly; but he or she should be very careful to explain the reasons for being less concerned so that, if challenged at a later date, it would be possible to demonstrate the execution of all the reasonable care that any prudent surveyor would take in similar circumstances.

Where the decorations in a dwelling are clearly of some age with wall coverings faded, paintwork yellowing, worn and chipped, and ceilings and walls perhaps discoloured by the occupants' smoking habits or marked by centrally heated air currents, cracks can usually be discerned as either old or fairly new. Cracks that are as grubby as the rest of the interior and that contain dust are clearly not of recent origin. If they look sharp-sided and have a fresh appearance and are clean on the inside, possibly with clean plaster or brick dust visible, then these are good indications that the cracks are recent. Caution is still needed because older cracks seen on a single inspection show the total amount of movement since the decoration was carried out, but not conclusively whether it was comparatively sudden at some more distant time or whether it has developed slowly over time. Advice here will depend very much on individual circumstances. Local knowledge as well as the type and age of the dwelling, the soil type and the presence of trees and sloping ground will all need to be taken into account for an opinion to be formed. When none of the circumstances likely to induce subsidence are present or if they are but they are too remote and can be discounted, it may be possible to conclude that ground-related movement in the past has ceased, is of long standing, is non-progressive and is acceptable. It could be that the older the dwelling the more likely is this to be the case. That may not be the end of the surveyor's movement-related investigation, as the cause of cracking, especially if of recent origin, may be related to some failure in the above-ground structure, but as far as the foundations are concerned, the surveyor's mind can be at rest.

9

Main walls – ascertaining construction

The inspection

The purpose of inspecting the external walls is to ascertain their construction, so far as is possible, and their suitability for fulfilling the basic functions of supporting loads without deforming, keeping the elements at bay, and insulating the dwelling's occupants against undue cold, heat and noise – all at a minimum cost as far as maintenance is concerned. Faults in design and defects in performance can best be described in a report if the construction is known or an assumption can be made to a reasonable degree of certainty. Consequently, the inspection of the external walls needs to be devoted initially to ascertaining construction.

Construction is likely to be governed to a very large extent by the age of the dwelling, which will either be known from information given or have been assessed on the initial reconnaissance. For example, there is generally a very fundamental difference in constructional form between the nearly 6 million dwellings built before 1920 and the 18 million built since. The former will almost exclusively have external walls of solid load-bearing construction even though the components within the wall's thickness may differ, as in the case of expensive stone backed by cheaper brickwork.

By way of contrast, from 1920 onwards, the majority of external walls of dwellings will be found to be of load-bearing cavity construction. Although some external walls were built with cavities before 1920, these were relatively few in number. In some areas, up to about 1940, load-bearing cavity construction vied with solid construction, but it became the predominant form thereafter and continues to be so. However, BRE has identified that over the same period of 1920–1940, about half a million low-rise dwellings were built of framed construction. The frames were of steel, timber or prefabricated reinforced concrete in about equal proportions – the first and last dropping out of the running to a large extent by about the late 1970s but timber continuing strongly in favour. Many of the framed dwellings can be considered to be of cavity construction since the outer leaf is tied to the frame across a gap, although the constructional principle is entirely different from that of load-bearing construction, both in its solid form and when a cavity is included. About another half a million dwellings have been identified by BRE as being constructed of no-fines *in situ* concrete over this same period, and many of these types also include a cavity.

In addition, there is also the relatively small number of dwellings provided in high-rise blocks. Although a number of high-rise blocks have been demolished in recent years, quite a few still remain, often constructed of reinforced concrete frames or an assembly of storey-height prefabricated panels bolted together. Commenting on the structure of such blocks is beyond the scope of this book, but the walls enclosing the flat being reported upon need to receive the same consideration by the inspector as the walls of houses and those of flats in low-rise blocks as to construction and performance.

In response to the need to comply with government codes on sustainable construction – not least the requirements for greater levels of insulation – along with an increasing willingness to embrace sustainable forms of construction, there is a growing interest in and use of alternative forms of construction. During the first decade of this century, providers of social housing were targeted by the government to increase the proportion of dwellings built using so-called 'Modern Methods of Construction' or MMC; and although this overt pressure has been since removed, surveyors must reconcile themselves to an increasing number of modern properties which are not built with conventional forms of construction. Identifying different methods of 'whole-dwelling' construction, in addition to that of the walls, is likely to become a significant challenge in the future. Until recently, wall thickness has been a reliable indicator of construction type; but with newer houses, more caution will need to be exercised. For properties built since 2010, wall thickness should be used primarily only as one indicator of possible construction type rather than a defining characteristic, as is usually possible with older dwellings. Some of the nontraditional construction forms in current use are considered in Chapter 15. Nevertheless, for decades to come, an overwhelming majority of the nation's housing stock will continue to comprise conventional forms of construction and it is these, along with some earlier and not always successful digressions into nontraditional forms, which will be considered in this and the following chapters dealing with main walls.

Even when comprising a mixture of materials, the total thickness of most load-bearing walls of solid construction in the domestic field is made up of small units: bricks, stones, pieces of timber. Such small units are able to adjust their position through their jointing material and thus accommodate movement. This is particularly so if the jointing material is in the softer lime mortar used in former times rather than the harder cement mortar used today. Even when such movement is of some consequence, the external evidence may be difficult to see. On the other hand, the internal face of solid external walls, when covered with an expanse of homogeneous plaster, will crack relatively easily as there are no joints to take up movement. In many instances, these cracks will be strikingly visible.

On an internal inspection, plasterwork around windows, in corners and at the junction of ceiling and external walls as well as general wall surfaces are all possible locations where cracks might indicate movement of some sort; and, accordingly, these need to be noted in each room when they are found to be present. The note should extend to taking down details of the size of the crack, using a crack gauge at both ends if tapering, and whether its direction is horizontal, vertical or diagonal since it will be necessary at a later stage to decide whether it can be attributed with reasonable certainty to problems above ground level – settlement – or to foundation and ground movement – subsidence. Sketches may be considered advisable at times to form part of this notation. It is best to use specific measurements and avoid trying to standardise subjective descriptions of cracks, such as fine, moderate or serious, since perceptions vary as to the implications of such words.

At the same time, other defects such as damp penetration perhaps due to inadequate thickness or defective pointing, rising damp perhaps because of the bridging of a damp-proof course, defects in horizontal features such as sills producing damp staining below windows, can be seen more readily and noted. In many places, when it is safe to do so, the opportunity can be taken to open windows to look out at the sides and head to see if there are any gaps or disturbances to the frame and to see at close hand what might be wrong.

Quite often, solidly constructed load-bearing walls will be rendered externally as well as plastered internally; sometimes the rendering will extend to the whole of the elevations and sometimes only to part. In the latter case, the material used for constructing the wall can usually be ascertained, but if the elevations are totally covered then means of identification other than by appearance need to be used. Whatever the situation, the pattern of defects seen from the interior may be repeated on the exterior rendering whereas they might not have been so apparent on an unrendered surface.

Initially, it is important to ascertain the type of construction used for the external walls of the dwelling, so far as it is possible to do so from a visual inspection, together with the materials adopted for the purpose. For this to be achieved, both internal and external faces need a thorough inspection, and this should extend from top to bottom externally and through all rooms internally at each floor level. Walls should be related to each other and to other structural features such as the roof, floors and partitions, which could be providing essential restraint to a length and height of wall which would be far too weak to stand on its own.

Ascertaining construction

Ascertaining the construction of the external walls of a dwelling involves collecting a number of items of information. There is the appearance, which might already have contributed towards an assessment of the dwelling's age – in itself, an important item of information – there is measurement of thickness, there is the sound made when tapped sharply with knuckles and there is also evidence which may have been derived from elsewhere. There is also the question of quality which may need to be taken into account on an overall view.

As to appearance, the initial reconnaissance will have revealed, among other information, what materials the dwelling presents to the outside world. Brickwork will be the most frequently encountered but there will also be stonework and occasionally concrete block – sometimes in imitation of stone-work – as well as applied surfaces such as rendering, slate and tile hanging or weatherboarding, which tend to hide the true construction underneath. The elevations may present the same material on all faces or they may be different according to aspect. Different materials may be used on parts of the same elevation. All need to be noted.

Mention has already been made of opening windows if possible and looking out and around to see if there is any evidence of factors contributing to defects noted internally. This should be combined with measurement of wall thickness and ideally should be carried out at each floor level and through a window or door opening on each elevation. Measurement of the thickness will throw up a number of possibilities as to construction which, combined with appearance, tapping both faces and knowledge of or an assessment of the age of the dwelling, should narrow down the likely form.

Tapping both sides of external walls is an essential requirement on all inspections. A 10p coin carried specifically for the task will save skin on the knuckles, especially when checking sharper materials such as pebble-dash; but the denomination of coin – or the individual's preferred substitute – should always be the same because the surveyor is listening for differences in sound so the striking object needs to be constant. The sound made will help to determine whether rendering or plasterwork is on a solid background of bricks, blocks or concrete or on a hollow background of wood or metal lath or, internally, comprising a layer of plasterboard on studding.

Evidence from elsewhere in the dwelling could help to narrow down the possibilities even further. The bare appearance of the inside face of gable or separating walls may provide a clue to construction of the walls lower down – useful where the interior face will normally be covered by plaster and decorations and even the exterior may be totally covered by rendering. Sometimes the only way to tell that an apparently brick-built dwelling is steel framed is from the presence of steel sections in the roof space. The presence of a cellar might also be very helpful in providing information from surfaces not normally decorated; and even if they have been covered, the cellar might still provide a glimpse of the inner face of the outside walls from beneath the ground floor.

As to quality, it is to be expected that dwellings for the top end of the market will be built better than those built at the lower end, and this is as true today as in the past. While the law requires a certain minimum standard from plans submitted for new work and most of the really awful poor-quality housing of the past was demolished in the post-war slum clearances of the 1950s and 1960s, the refurbishment of older dwellings is not always directed to matters which a surveyor would consider a priority. At times, it can be more cosmetic, leaving basic design faults as they were.

Brick faced

For ascertaining construction, it is probably fortunate from the inspector's point of view that the majority of dwellings present a face of brickwork to the outside world. Coupled with other information, an elevation of brickwork can reveal factors about construction in a more immediately straightforward manner than many other materials. For example, to achieve effective strength for carrying loads, a solidly constructed brick wall needs to be built so that the bricks bond together. This bonding can be seen on the outside face as a pattern of headers and stretchers (**137**). If there are no headers to be seen but only stretchers then the wall is only a half-brick's length in thickness. Such a wall has its uses in light structures such as garages and bin enclosures and also to define boundaries, particularly when strengthened along its length by piers one brick thick. As part of a dwelling, however, the sight of stretchers only will be a very strong indication of load-bearing cavity construction (**138**). The half-brick outer leaf will be tied to an inner leaf carrying most of the load but some of that load will inevitably be transferred to the outer leaf by the wall ties.

Confirmation of load bearing cavity wall construction can be obtained by measurement and soundings. Both leaves consisting of brickwork at half-brick thickness with the usual 50mm cavity in between and plaster internally will measure between 290mm and 300mm compared with only 240mm to 250mm thickness of a solid one-brick wall with plaster. Both walls will sound solid when tapped and both are perfectly satisfactory for a purely load bearing point of view for the average two- or three-storey building. Larger dwellings can, of course be constructed in both load-bearing cavity and solid brickwork and measurement in excess of each of the above figures by about 115mm would indicate a more substantial wall of another half-brick thickness appropriate for dwellings taller or of larger area.

In the case of cavity construction a measurement of some 10mm to 15mm less with a house built since the 1950s could indicate an inner leaf of concrete block rather than brick and this would be a reasonable assumption for a dwelling built in the last 50 years, provided that tapping the inner leaf still produced a solid sound.

If the sound is hollow when the inner face of the wall is tapped, then with any measurement of thickness around the 240mm to 300mm mark the assumption must be of framed construction. For a dwelling built between 1980 and 2010 the framing would almost certainly be of timber but from before 1980 it could be of either timber, steel, or, less likely, prefabricated reinforced concrete. Framed construction since 2010 is still most likely to be timber framed

137 Dwellings built with solid load-bearing brick walls require the bricks to be properly bonded for strength, showing a pattern of both headers and stretchers on the elevations – as here, known as Flemish bond. There are other types of bond. Dwellings with bonding such as this can be from all periods.

138 On a brick-faced elevation, stretcher bond, as shown here, is a good indication of cavity construction, but what lies behind may require further investigation. Many dwellings built since the 1920s will be of cavity construction, usually load bearing over the thickness of two leaves but sometimes as a non-load-bearing outer leaf attached across the cavity to a load-bearing frame. Since the 1970s, framed structures are usually of timber construction but the use of light gauge steel frames, while still relatively rare, is now increasing.

139 Stretcher bond on this 1980s brick-clad house might initially suggest load-bearing cavity construction, but the hung tile detail to the upper part of the party wall would not be expected and indicates timber-frame construction with a brick outer skin. Further support for this conclusion is provided by the age of the dwelling, the allowance left for differential movement around openings between the timber frame and the brick outer leaf and a hollow sound produced by tapping the inside face with the knuckles. Hopefully, this will all be confirmed by sight of separating walls in the roof space that are not masonry but boarded.

but other forms of construction are being used, including light gauge steel frames. In the absence of any clear positive indications in the roof space of steel or timber framing, evidence of which type would have to be sought elsewhere internally. One method of seeking confirmation of whether the dwelling is timber framed, is to examine externally around window and door openings where either a gap of 6mm to 12mm or a compressible filling of about that thickness should have been left to allow for differential movement between the external brickwork, or alternative material used for the outer leaf, and the timber frame.

Render faced

Presented with the sight of rendering to all of the elevations from ground to roof level on a dwelling built before 1940, the initial assumption will be that the wall is of solid load-bearing construction. Rendering was usually necessary in many parts of the UK for weatherproofing purposes, particularly on walls one brick thick. Cavity walling had been specifically developed to avoid that need and, therefore, rendering would not normally be required on one of the increasing number of cavity wall properties being built prior to 1940. There is a little more to it than that, however; but combined with knowledge of or an assessment of age, measurement and tapping could provide the inspector with a much more reasoned assumption of what lies below the rendering. Knowledge of the locality could be of help too.

Rendering of any type will usually add to the order of 15mm to 25mm to any of the previously mentioned figures for both brick-faced solid and cavity construction. A solid sound when tapped on the inside face will merely confirm that the wall comprises masonry of some sort, but a hollow sound could indicate a framed structure or, in comparatively rare instances, a dry lining on internal studding. Bringing age into the equation should narrow the assumptions down. Totally rendered dwellings from before 1940 are far more likely to have been provided with solid load-bearing walls, the rendering to keep the weather at bay or to cover a material not particularly attractive in appearance – cheap bricks or ugly stone perhaps. In the latter case, the wall would of course be a great deal thicker.

If there is render on a dwelling built since 1940 where measurement indicated load-bearing cavity construction and there was a solid sound when tapped, probably the only reason for the presence of rendering would be to obscure a material of unprepossessing character or appearance. If measurement and a hollow sound when tapped suggested a cavity and framed construction then, again, the rendering is probably presenting a face to the outside world more attractive than would have been the case if it had been left off. Insulating concrete blocks provide an example of this when they are provided as the outer leaf for some of the framing systems in use about this time; these certainly needed to be covered over.

Where rendering has parted company with its substrate, tapping will produce a hollow sound; but this should only occur in localised positions, in most cases, and checking elsewhere should prevent any confusion with framed structures. However, insulation cladding can be used externally, as well as for interiors, of solid external walls and this could produce a hollow sound when tapped (**142–144**). External insulation eliminates the grumbles of occupiers about the provision of secure fixings for wall cupboards and shelves and it is becoming increasingly popular, but serious problems can develop if it is not properly installed.

As to the assumptions about the construction of totally rendered external walls, in the absence of any firm evidence from elsewhere in the dwelling, these must remain as assumptions and should be reported as such. It may be that the inspector will know from the location of the dwelling that it is one of a particular privately developed estate that is known for its shortcomings or that it is one of a local authority's, built using a particular system (**145** and **146**). The inspector might know that an estate had been developed by the local authority using a no-fines *in situ* mass concrete system. Such systems account for about 2 per cent – roughly half a million dwellings – of the total stock, but as examples are unlikely to be encountered very often, they could confuse if the inspector lacks the benefit of local knowledge. Walls of such construction tended to be built thinner as time went on. Those built before 1951 were commonly 300mm thick, those built in the 1950s and early 1960s about 250mm, with those built later down to 200mm thick.

Another example where local knowledge may help to ascertain the construction behind rendered elevations is where the surveyor is practising in an area known for building in the local unbaked earth. The chocolate-box white-painted cottage with its thatched roof, the retirement dream of many, will have the appearance of a dwelling with walls rendered in lime plaster and limewashed, but the rendering may cover load-bearing structural walls of cob – i.e. made from clay mixed with straw and water. These cottages are common in the West Country and in South Wales but also exist elsewhere. For example, in areas where chalk occurs in southern England and in East Anglia, this too becomes part of the mix – the material being known as witchert in the former area and clay lump in the latter. Dwellings constructed of witchert are found in a band across Buckinghamshire and Oxfordshire and south into Hampshire and Dorset (**147** and **148**). The construction is similar to cob but because of the chalk content of the subsoil, the wall thickness need only be in the order of 350mm to 450mm compared with between 600mm and 900mm for cob walling. In East Anglia and up into Lincolnshire, the clay lump is produced from the chalky subsoil mixed with straw and water in moulds and air dried to produce mud bricks (**248** in Chapter 12), again to be built off a stone plinth as in the case of walls of witchert.

Another form of building with unbaked soil is *pisé de terre* or *pisé*, where the mix minus the straw is rammed between temporary shuttering which is struck later to allow for drying. This produces similar plain-finished walls, again raised off a stone plinth, about 350mm to 450mm thick but much quicker to construct than by the other methods. This method found much favour in Scotland, particularly along the south coast of the Moray Firth east of Inverness and in pockets of coastal areas in the east between Aberdeen and Dundee, having been introduced from France in the 1790s.

Knowing or assuming the age of totally rendered dwellings is a help towards making a reasonable assumption of what lies below since some types of construction are less likely to be found and can, therefore, be eliminated as possibilities. Most of the existing examples of construction in unbaked earth are thought to have been built in the eighteenth and nineteenth centuries, spurred on by a tax on bricks between 1794 and 1850, though some may have been built much earlier. The twentieth century has produced relatively few examples because of the more widespread adoption and enforcement of bylaws in the late nineteenth and early twentieth centuries, which deterred the use of these methods since they were not covered by the regulations. Although numbers are small, some are said to have been built during the 1920s; and the increasing emphasis on conservation since the later years of the twentieth century has led to a reassessment of the use of unbaked earth for repairs and extensions to existing dwellings and in the construction of new dwellings.

The functional requirements imposed by the Building Regulations are much more flexible than the specific requirements of the old Model Bylaws and the earlier Building Regulations; so it is possible, just as with more modern innovative forms of construction, to demonstrate that walls constructed of unbaked earth are capable of performing the functions for which they are intended. Consequently, approvals under all parts of the Building Regulations are potentially obtainable, subject to meeting all of the relevant requirements. Accordingly, it is conceivable that individual dwellings recognisably built within the last couple of decades might be found to be constructed of unbaked earth. Since the building material is usually sourced from the very site on which the dwelling is constructed and virtually the only energy required is the physical effort of digging, mixing and lifting the earth onto the walls, this method is as 'carbon free' and sustainable as building construction can possibly be.

As well as the possibility of encountering dwellings of unbaked earth, surveyors in the far south-west of England need to appreciate the likelihood that the presence of external rendering may indicate a dwelling built with an *unsatisfactory* local material that was all too readily available at one time. Cornwall, from Greek and Roman times, was for long the world's major producer of tin; and in the late 1700s and early 1800s, Cornish underground mines also produced half of the world's copper supply. After years of decline, the last mine closed in 1998; but the waste material left behind from underground mining was very considerable and was there for the taking by small local plant operators and local contractors. Along with the waste material from china clay working – that other long-standing but still-operating Cornish industry – the waste material from mining was initially thought to be highly suitable as an aggregate in the making of concrete blocks, which were intended to replace the use of expensive natural stone for building or to save on the cost of transporting bricks for this purpose. This followed on from the general use of Portland cement in the early 1900s, particularly between 1920 and 1950.

While the waste from china clay working and some sources of waste from underground mining have proved suitable as aggregate for concrete block making, there are some sources of mining waste that have proved to be highly unsuitable. Substantial deterioration was found in the 1960s and 1970s in a number of dwellings identifiable as being constructed with blocks made with aggregates specifically from some of the underground copper and tin mines. The deterioration was so severe that a number of dwellings had to be demolished. The problem was diagnosed as being attributable to the presence of mundic – the old Cornish Celtic word for pyrites, or iron sulphide – in the aggregate, and it was rightly concluded that a significant number of dwellings built mainly between 1920 and 1950 were affected. Sulphate attack is the cause of the degradation when sulphides oxidise, but this was on a scale far in excess of what happens to mortar in normal brickwork. The chemical process has proved to be complex but, in essence, if concrete blocks made of Portland cement contain enough of the susceptible aggregates, it only needs the presence of dampness to initiate a sulphate attack. The resulting large expansion and loss of particle adhesion causes catastrophic failure of the affected components, with potentially disastrous implications for the whole of the building structure (**149**).

The Council of Mortgage Lenders became alarmed and, in 1992, banned lending on suspect dwellings until a reliable screening test could be established. This was found possible through detailed petrographic examination of core samples to identify the sources of many of the aggregates and to link them with consequential degradation. Unsurprisingly, it was found that areas having the highest proportion of dwellings with defective concrete corresponded with the districts that had had the greatest mining activity, especially around Camborne and Redruth, areas in East Cornwall and around the Tavistock area in south-west Devon. An acceptable test, based on petrographic study of concrete specimens, was introduced by the RICS Mundic Group, with BRE input, in 1994; this was refined in 1997, since when a standard testing process has been applied to all properties in Cornwall and south-west Devon which may be affected by the problem.

The introduction of British Standards for aggregates meant that it was much less likely that unsuitable material would be used after 1950, but there is evidence of occasional use of unsuitable aggregates in some localities in the years immediately after 1950, and occasionally even later; so it remains necessary for practitioners in the areas affected to exercise vigilance. Lenders do not normally require tests on post-1950 concrete-constructed properties unless the surveyor has determined that there are visual or other signs to suggest the possibility of mundic-related deterioration.

Before being considered suitable for mortgage lending, every property in this age range and located within the defined area must be tested unless the construction can be positively identified as not containing any potentially mundic-bearing material. Where properties are visibly wholly of brick or natural stone construction, matters are straightforward, but the mortgageability requirements mean that every property where any part of the main wall construction is concealed – for example, by external rendering – is considered at risk and requires a mundic test to be completed.

The testing procedure is invasive and involves drilling core samples (**150**) so it is beyond the scope of a visual inspection, which has to be limited, in the first instance, to identifying the need for testing. The petrographic examination allows the aggregate in the samples to be categorised and the classification determines whether or not the property is suitable for mortgage lending. More than 80 per cent of properties tested are completely sound and only a small proportion of the remainder are found to have mundic deterioration present, necessitating remedial work. Refinements to the original tests and procedures have been periodically introduced – most recently in 2016 – with the most significant change being the introduction of an exhaustive test to determine the long-term likelihood of deterioration where the initial

140 The totally rendered dwelling, as here, presents difficulty for ascertaining construction. Measurement of wall thickness and assessment of age and even type of rendering can be pointers. This early twentieth-century house with its roughcast rendering is almost certainly of solid load-bearing construction, probably of cheapish brick throughout. It should be noted that chimney stacks are no guide to the basic construction of a dwelling as even an old timber-framed wattle and daub cottage with a thatch roof usually has a brick or stone stack in an attempt to keep fire at bay.

141 A typical 1930s private sector semi-detached house, totally rendered. Normally, only the first floor would be rendered and the visible bond of the ground-floor brickwork would confirm the basic construction. The use of cavity or solid construction at this time varied across the country; knowledge of local practice should assist identification but it would not be wise to make an assumption without measuring the thickness of the walls and sounding the inner face.

142 Hung tile cladding of a 1970s house being removed prior to fitting external insulation. The original single-leaf inner skin and tile hanging on properties of this era provides relatively little insulation, but cavity wall insulation self-evidently cannot be installed.

143 The external insulation is fitted to prior to being coated with external rendering. This local authority-financed improvement to its own housing stock did not include privately owned dwellings such as the one to the left.

144 The finished face of the external rendered section can be distinguished from the unimproved properties on either side, previously purchased under Right to Buy. Identifying external insulation can be more difficult when whole terraces are upgraded.

145 Totally rendered walls of a 1970s three-storey block of flats. Measurement indicates a wall thickness much less than cavity construction and even slightly less than solid load-bearing construction. The elevation is very plain, untypical of private sector flats at this time, but the windows are sheltered slightly by the bell mouth formed in the rendering. All this evidence points to no-fines *in situ* mass concrete construction, much favoured by local authorities. The rendering has probably been renewed. Tapping the inner face would, of course, have elicited a solid sound.

146 These dwellings from the 1930s, totally rendered externally but sounding solid when tapped from the inside, are in fact steel-framed Dorlonco system houses. Although all Dorlonco houses are characterised by being double fronted and have steel roof trusses which can simplify recognition, enquiries would need to be made about their history and how they have performed over the years. The conclusion by BRE is that most steel-framed houses even of this age, while subject to some corrosion, are still likely to provide long service if well maintained. Nevertheless, a prospective buyer should be advised to have the structure opened up at an exposed point to examine the steel for corrosion, and it may well be that the metal lathing used as a base for the rendering has almost totally corroded.

147 The base of the walls of this limewashed and roughcast-rendered cottage are of rubble stonework. To an inspector from the area where it is situated, this usually indicates construction to the upper part of chalky earth and straw mixed with water and left to dry before being rendered to protect it from the weather. Solid sounding on both exterior and interior faces, overall thickness will be somewhere in the region of 350mm to 450mm, and the construction is known locally as witchert.

test results show that, although a property is sound at the time of testing, it contains aggregate with the potential for deterioration in the future.

The mundic problem, as currently defined, is an issue which is specific to the far south-west of England, and the necessary testing of properties which are potentially at risk is now an accepted part of the house-buying process in the affected area. There are, however, well-founded suspicions that serious structural failure caused by pyrites following the use of unsuitable aggregates has also occurred on occasions elsewhere, so practitioners should take nothing for granted.

148 In the same area where the cottage at 147 is situated, portions of the rendering to the upper parts of other cottages are left off where owners seem to be saying, 'look my house is not built of earth and straw', and providing visible proof. Overall wall thickness would probably have helped to differentiate anyway as rubble walling is usually 600mm or more in thickness.

149 A classic example of a property affected by mundic, which is beyond economical repair. The effects of the problem mean that cracking doesn't follow the normal patterns which might be associated, for example, with structural movement or cavity wall tie failure.

150 The first part of undertaking a mundic test is to extract a series of core samples; here, one is ready to be bagged and labelled, having been taken from the wall of a rendered bungalow. The samples are then subjected to a petrographic examination so the aggregate can be classified in accordance with the agreed procedure.

Stone faced

An elevation of stone, real or artificial, presents the inspector with more difficulties for ascertaining construction than does brickwork but, on the other hand, less than is the case where walls are totally rendered. Once again, as with both brick and rendered surfaces, there is no reason why the external walls should not be of either load-bearing solid or cavity construction with a further alternative of a casing over a frame as yet another possibility. Much the same procedure should be followed for ascertaining the construction. Unlike an elevation of brickwork, however, appearance is not much help in ascertaining the type of construction. The shapes visible on the face of the stonework representing, as they usually do, one or other of the traditional types of walling are no guide as to how the wall was put together. Age is an important factor, on the other hand, since there is also the matter of distinguishing between the real thing and the artificial variety, which is in effect a concrete block although the aggregate used in manufacture includes stone particles.

Both the natural and the artificial can be produced in a thickness of around 115mm to 120mm, appropriate for inclusion in the outer leaf of a cavity wall, and in varying heights and lengths so as to enable variety to be maintained on the elevations, which is so much a feature of most types of stone walling. Since the 1960s, building in the areas where stone was the traditional material has been required, or at least strongly encouraged, to match in with the surroundings. For a special one-off, privately built dwelling, it may be that the owner could rise to the expense of real stone. For estate development, however – even when relatively top of the market – the expense of the natural product would usually be such as to turn the developer towards the artificial. The products of the reconstructed prefabricated stone industry have improved immeasurably over the years and are acceptable to most planning authorities. The inspector is, therefore, more likely to encounter the artificial than the real on dwellings built since the 1960s.

Measurement of thickness to the same order as with a facing of brick – i.e. around 300mm – and a solid sound when the inside face is tapped will suggest load-bearing cavity construction. Hollow sounds from the inside face are more problematic. It might simply be a dry-lined plasterboard finish to conventional construction, but it could alternatively suggest framed construction which is, of course, the other alternative to load-bearing construction. For a dwelling built since the 1960s, any framing is far more likely to be of timber since, by this time, the use of prefabricated reinforced concrete and steel for domestic low-rise construction had fallen out of favour.

Having ascertained the type of construction of the post-1960s dwelling with an apparent stone facing, a close examination of the facing should reveal whether the stone is real or not; although this information is not vital from a structural point of view. Reconstructed artificial stone can be made to look remarkably real; but although panels may be made from a range of masters to avoid undue repetition, a thorough examination may reveal some panels which are identical, perhaps along with joints moulded within the blocks as distinct from genuine joints between blocks, duly pointed. One telltale indication of artificial stones can occur when minute bubbles of air are trapped as the concrete mix is poured into their stone-shaped moulds. As the blocks dry out, the bubbles leave small but characteristic hemispherical indentations on the surface (**151**). Bedding courses can usually be detected on natural stones of sedimentary origin, and in localities where natural stone is widely used, practitioners should quickly become familiar with its appearance.

For dwellings built in the periods 1920–1940 and 1940–1960, the use of artificial stone can effectively be discounted. In the latter period, the concentration was on producing dwellings in large numbers as quickly as possible to make up for wartime losses and a growing population. The appearance of stone on a dwelling would have seemed an unnecessary luxury and there would hardly have been any demand for the real, let alone the artificial, leading to a marked, indeed near fatal, decline within the industry. In the 1920s and 1930s, particularly the latter, the typical suburban house could be produced in a range of finishes according to location and what was perceived as likely to sell. The increasing use of cavity construction was more suited to mass production in brick rather than stonework, so the bay-windowed three-up, two-down brick or rendered semi-detached or its detached equivalent tended to pre-

151 Reconstituted artificial stone is more realistic than plain concrete 'stones', but sharp edges from the moulds can be seen on many of these blocks and, even more of a giveaway, is the shape of small air bubbles sometimes captured when the mix was poured. Most of these blocks, especially the two headers on the right, have examples.

dominate. At the top end of the market, however, dwellings with natural stone-faced load-bearing cavity construction could make an appearance. The precast concrete industry was not geared to producing artificial stone in the form of realistic blocks or walling panels at this time, although it could produce smooth, well-formed sections for sills, copings and door surrounds. Occasionally, stone-shaped concrete blocks with textured surfaces will be encountered on individual interwar-built properties. In many cases, these have weathered relatively well but a close examination will reveal that their origins are simply concrete and not artificial stone, let alone natural stone (e.g. **257** in Chapter 13).

Between about 1840 to 1860, when the railways began to transport mass-produced common bricks to more or less all parts of the country, and the widespread adoption of cavity construction in the 1920s, walls could, in many cases, be of composite construction with stone facings, having squared blocks when required for better appearance. This consisted of stone with a backing of brick, which not only saves on the cost of the stone but also produces a better surface for internal plastering. The stone would be reduced to blocks in sizes of multiples of a brick to enable a satisfactory bond to be maintained without showing brickwork on the face.

Wall thicknesses would depend on the suitability of the local stone for reduction to blocks of small size with the

152 This 1970s development was built in an area where there are many buildings constructed with a local stone which does not lend itself to such regular coursing (see the property to the left). The use of artificial stonework, presumably intended to make the houses blend in, has the opposite effect as the colour and coursing are completely at odds with the surroundings and not helped by the rendering. The presence of a natural stone plinth at the base of the wall is a curiosity.

153 This pair of 1930s semi-detached houses presents a distinct contrast, perhaps through neighbourly rivalry. The one on the left has had unnecessary rendering applied to the load-bearing brick cavity wall construction. The one on the right has had its entire front elevation covered with an imitation stone veneer to represent squared rubble built to courses. Because of its location on an estate, the cladding cannot be mistaken for a wall constructed of either real or artificial stone. Even when applied to all elevations on a detached house, an inspector should not be confused. The waney edge boarding on the gables gives a rustic feel to the suburban setting.

154 This small 1900s terrace house is wider than its neighbours at first-floor level as it extends over the 'gennel' or 'ginnel', giving access to the gardens at the rear. Nevertheless, in an area where much of the housing of the same age and of similar size is built of brick, this house would be considered a bit upmarket with its genuine stone, regular, coursed rubble facing backed by brick, which is in evidence on the rear elevation and on the flank walls at the end of the terrace.

155 Stone-faced dwellings before the early nineteenth century, as here, in areas where bricks were not produced will traditionally be constructed in two leaves with through stones for transverse bonding across a hearting of rubble and mortar. The thickness will seldom be less than 600mm and often more, particularly in dwellings of more than two storeys.

equipment available at the time. With some types of stone, blocks with the thickness of a brick – around 115mm to 120mm – were possible. Accordingly, walls could be built of a comparable size, the thinness of the blocks being balanced by lengths and heights greater than a brick providing a variety of shapes on the face of the stonework. A British Standard of the 1950s, although withdrawn for new construction, still has relevance and can be borne in mind when considering existing properties; it states that solid stone walls, even with a backing of brick, should generally not be less than 400mm thick since the width on bed of most building stones varies from 150mm to 300mm and difficulties with bonding could be encountered with any lesser thickness – a point to be remembered if a stone-faced wall of unsteady characteristics is encountered. Where a rustic or rural finish was acceptable – for example, uncoursed random rubble or flint – the backing might still be of brick but the overall thickness of the wall is more likely to be to the order of 400mm to 500mm with a fair amount of small stones and mortar making up space around the through stones that provide the transverse bond, though these, of course, will not be seen.

Where stone-faced dwellings were built before around 1840, local custom will have prevailed so that, in a brick-producing area, the aspirational owner could have what appeared to be a stone house – so as to be thought superior to his neighbours – but the stone would be backed by brick. In areas where bricks were not produced locally, the long-standing traditional method of constructing a stone wall would be employed: two stone leaves with a hearting of rubble and mortar in between. Wall thicknesses, in these cases, tend to be at least twice those of the comparable requirements for brick though will often be found to be far more, depending to a large extent on local custom and the materials available.

Information on age, the appearance, thickness, the sound on tapping and, perhaps, evidence from elsewhere should provide the inspector with what is needed to ascertain or make a reasonable assumption of the construction of the external walls of dwellings with elevations mainly of brickwork, rendering or stone, whether real or artificial.

Faced with tile, slate, shingle, timber, steel or other material

Other materials that dwellings present to the outside world are somewhat less common and are generally, apart perhaps from weatherboarding, on only part of the elevation. Slate or tile hanging, for example, is usually confined to upper floors in view of the danger of accidental damage or damage by vandals at low level. While both can be applied to solid walls, they are often found applied to timber framing so that tapping the inside face is always a necessity, even if the dwelling's basic construction has been established from elsewhere. There might have been a change in the form of construction from one floor level to another or between parts of the dwelling.

Weatherboarding, whether of timber or plastic, can be considered in much the same way as slate and tile hanging. Although not so subject to low-level damage and, therefore, occasionally found as a material covering the whole of a dwelling, it generally appears on part only so that the basic construction can usually be ascertained from elsewhere. However, it too can be applied to both solid and cavity load-bearing walls as well as framed walls so that tapping the inside face is always necessary. Knowing or assessing the age of a dwelling is still a useful guide to ascertaining what might lie below slate, shingle or tile hanging, or weatherboarding even though these finishes have been applied to all types of construction in all periods. Where it can be a really significant help is in identifying newer materials or materials and systems which were used for particular reasons at specific times. As already stated, the period 1940 to 1960 saw the use of various systems of framing to speed the process of housebuilding in a time of shortage of both materials and skills. Different systems employed different external treatments to cover the frame, with prefabricated reinforced concrete panels, arranged in a variety of ways with different finishes, used in many cases; but there were also steel and asbestos sheets, timber boarding and proprietary panels of insulating material with factory-produced external finishes. Some framing systems offered alternative types of finish and, all told, there were up to two hundred different types of frame and finish available.

There are also dwellings where the walls are covered with coloured vitreous enamelled steel panels, profiled aluminium sheets or glazed tiles. It is more likely to be a matter of speculation as to what lies below these finishes, but the same procedures should be followed to arrive, hopefully, at a reasonable assumption, having regard to all the possibilities which were available at the time of the original construction and also, perhaps, at the time of any refurbishment or upgrading that may have since been carried out. Needless to say, during the course of the inspection, there may be evidence available from the performance of materials, such as deformation or delamination, which might alter a preconceived view or help to reinforce it. Terraces built with cross wall construction, popular during the 1960s and 1970s, used the party walls for structural support, so the non-load-bearing front and rear elevations could be, and were, clad in a whole variety of materials, usually on a timber frame. More traditional finishes came back into fashion during the latter decades of the twentieth century as a reaction against some of the poor-quality post-war construction; but increasingly, innovative forms of construction are being introduced in response to the demand for greater levels of sustainable construction. The ingenuity of the suppliers to the building industry is limitless and new 'wonder' finishes are continuously being developed – all the more reason for the surveyor to keep abreast of current developments.

It is necessary to warn against being deceived by appearances. Certain building characteristics suggestive of an earlier age may lull an owner into thinking a property is older than it really is and can also mislead the unwary surveyor. Some features are fairly easy to detect even from the exterior. Mock-Tudor framing, for example, consisting as it does of planks stuck on the outside,

156 Tile hanging at first-floor level on a late Victorian dwelling. The inspector would need to ascertain that there is not a change in the form of construction at first-floor level from load-bearing, rendered, solid construction to timber framing. Could the tile hanging and the rendering have been a subsequent application to bare-faced brick walls to cure damp penetration and if so, why was this and are they working?

157 With not a brick in sight, there is little doubt that these two-storey, fully boarded houses are of timber-framed construction. Tapping inner surfaces will confirm or disprove, and an examination of the inside face of the gable wall from the roof space will provide the answer to the type of framing or the material used for the walls in the unlikely event that they are load bearing.

158 Knowing the age of a dwelling is a help towards ascertaining what lies beneath weatherboarding or other forms of timber or plastic cladding. On this 1980s two-storey house, the inspector would need to ascertain whether load-bearing cavity brickwork is the form of construction, as might appear, or whether both brickwork and the timber are carried on a frame. Tapping and an examination of the inside face of the gable wall could provide the answer.

159 This cottage has clearly been subject to much alteration over the years and it is not clear at first glance whether the boarding on the first-floor gable end is suggesting timber-framed construction above the ground-floor brickwork or is simply providing weather protection. As the boarding extends across the rear at first-floor level and is mirrored on the gable at the opposite end, the probability is that timber-framed construction is present.

160 This two-storey semi-detached house of the 1950s is a British Iron and Steel Federation (BISF) steel-framed dwelling with a rendered ground floor, profiled steel sheeting to the first floor and either steel sheeting or corrugated asbestos to the roof. It has the distinctive large ground-floor windows; these are usually retained in shape if the dwellings are upgraded, which is a help towards recognition when the overall appearance may well be substantially altered (see **286** in Chapter 14).

161 The undulating appearance of the cladding to these 1970s three-storey terrace houses with integral garages may cause puzzlement. Tapping will reveal that the panels are of off-white vitreous enamelled steel to the exterior and there is solid-sounding construction internally. Construction could be of brickwork but the wall thickness is less than that required for load-bearing cavity work. Enquiry would elicit that it is in fact a load-bearing no-fines mass concrete *in situ* form. The replacement white-painted metal windows spoil the symmetry but may be more practical with their top-hung opening vents.

162 It is not clear why painted metal panels have been used on this local authority-built block of flats, other than as a means of varying the external appearance. Enquiry is needed to know what lies behind the panels since tapping will reveal only whether the construction is solid or hollow. The amount of dampness on the brickwork at high level does not speak well of the external weather detailing. If any remedial work should become necessary, an owner may share some responsibility for the cost even if their own flat is not directly affected.

even including simulated dowels, lacks the drunken air brought about by the shrinkage of the green oak timber of which the genuine would be constructed. Those early timber frames which have been entirely rendered are likely not be fakes in the true sense but treated thus to keep out the draughts. On the other hand, there were many timber-framed dwellings which were given a

163 Although there are dwellings of much the same height and appearance from the 1500s, the inspector ought not to be mistaken into believing that this is a genuine timber-framed example. Even without knowledge of its age, it will be revealed from its too-precise appearance along with measurement and sounding that it is a block of three-storey, load-bearing, brick-constructed flats with applied decoration – the 'mock Tudor' for which there was a vogue in the period 1920 to 1940.

164 The appearance of these buildings – a certain impreciseness coupled with the jettied construction at first-floor level – along with measurement and tapping would suggest that these are genuine timber-framed structures perhaps of mixed dates, the dwelling on the left earlier than that on the right, certainly above first-floor level. A check on the listing on the Historic England website (http://historicengland.org.uk/listing/the-list/) would confirm a mixture of sixteenth- and seventeenth-century construction.

165 The true construction of this totally rendered two-storey dwelling is given away by the projecting jetty supporting the front wall above first-floor level. Such construction was a feature of the fifteenth and sixteenth centuries. The rendering was probably applied for no other reason than to keep out the draughts.

166 On arrival here, the less than wary surveyor might be anticipating an inspection of a fine example of an early eighteenth-century dwelling. Once inside, it becomes apparent that this is, instead, a timber-framed structure from the sixteenth century, given a makeover so that its owner could keep up with the latest fashion.

makeover in the eighteenth century to become, from the exterior at least, up-to-the-minute examples of fashionable building. The interior will probably tell a different story, but the surveyor needs to be careful all the same. To be wary must indeed be the advice for any surveyor inspecting dwellings, for whatever reason. It is best to anticipate the likelihood of meeting many different types of construction, some more frequently than others. With this attitude and the avoidance of hasty assumptions, meeting the unexpected will not come as too much of a shock.

If the comparatively rare is encountered then the surveyor must be prepared to make enquiries from the current owner, or the local authority if necessary, and to use resourcefulness to track down information if it should be available. The surveyor's own library and file of local information should often be of sufficient scope to help, websites can be searched, and enquiries made of BRE or Historic England, local historical societies and perhaps, in appropriate areas, organisations of local enthusiasts dealing with an unusual type of dwelling that is common to a particular area.

Of the roughly 24 million dwellings in England, Scotland and Wales, it is believed that about 9 million are enclosed by solid mass load-bearing walling, brick, stone or concrete or a combination of these materials with a relatively few built of unbaked clay or chalk. The majority of the dwelling stock, however – something like 15 million units or about 60 per cent – is enclosed by walls of cavity construction, mostly load bearing but with perhaps around a million or so hiding a frame.

The external walls of all dwellings have a number of basic functions, one of the most important of which is making sure that rain driven against the outside face does not penetrate to the interior. Having ascertained the construction, it is necessary in the first instance to consider whether that construction is adequate to cope with the vagaries of the local weather.

Assessing exposure

The climate is a never-ending presence and the UK's dwellings should be, but are not always, built to withstand that climate. The questions arise: were the external walls of the dwelling under consideration designed and built to withstand the local climate and, in particular, is the wall facing the prevailing wind suitably built for the purpose? It is the prevailing wind which, at its strongest, can drive rainfall at its heaviest through the external walls to the interior.

Local custom and tradition often used to govern the form of construction of the external walls. What was successful in keeping out the rain in a particular area would be repeated and what was not would be abandoned. Accordingly, in the wetter and windier parts of the country, thick walls of harled stone are seen as successfully keeping the rain at bay; but in the drier and calmer parts, unrendered one-brick walls perform the same function equally well. However, such custom and tradition would be ignored by, for example, landowners and builders putting up dwellings at the least cost possible for renting out during the Industrial Revolution. Such practices continued even though, by the start of the twentieth century, local laws often imposed some minimum standards. In the decades after the 1950s, cost limits on public housing did not always mean that the best decisions were taken as far as construction to cope with exposure was concerned. In the private sector, the volume builders are governed by the bottom line; and on their estates, the best case scenarios rather than the worst case ones often governed the type of external wall construction adopted throughout.

Rainfall varies considerably across the UK but it is largely unaffected by local features. On the other hand, the general wind speed does not change very considerably but it is much affected by local features such as trees and buildings and whether the ground is flat or steeply rising. Consequently, it is the degree of exposure to the wind, especially the prevailing wind, which determines how much a building is likely to be affected by driving rain. Considerable research was undertaken into the varying levels of exposure across the country during the second half of the last century, and information is now available to allow detailed calculations to be made for the exposure of walls to local wind-driven rain. The results will depend on orientation and such variable factors as topography, surrounding obstructions and the roughness of terrain. While such calculations are not something that the inspector will expect to undertake at all frequently, an awareness of the general exposure levels within the inspector's working area is certainly something which ought to be second nature in order to recognise the vulnerability of particular forms of wall construction in certain situations.

Approved Document C of the Building Regulations includes a national map, derived from the detailed ones in British Standard 8104: 1992, which shows four exposure zones across the UK. Surveyors should know which overall zone is applicable for the prevailing wind in the locations where they work as that will have a bearing on the types of construction which are appropriate. This information can be gleaned from an accompanying chart in Approved Document C that shows the maximum recommended exposure zones for a variety of insulated wall constructions. In practice, the zones will usually be modified downwards by one zone value in urban situations where protection is provided by surrounding housing or trees. Conversely, a property in a particularly exposed position – facing down a valley or on the crest of a hill, for example – or one standing above the surrounding buildings might have more than average exposure to the prevailing wind and warrant an increase of one zone value.

The site research carried out to produce BS 8104: 1992 confirmed what had been previously assumed from hearsay evidence about certain aspects of exposure conditions and enabled the factors recommended to be given appropriate relative weight. For example, the walls of dwellings on the edge of an estate facing open countryside, a park or playing fields are at a far greater risk of rain penetration than the walls of dwellings within the estate – even one row in – except where walls have a clear view down a road or are exposed above their surroundings on a knoll or a gradual slope. It also confirmed the value of overhangs, projecting sills, string courses and the like as protective features for walls, determining that water running off falls vertically to the ground rather than being blown back on to the wall lower down. As an example, the success of rendering or cladding the top half of an exposed two-storey gable wall as a repair depends on the lower edge of the rendering being finished with a bell mouth or any cladding being tilted outwards at that point.

However, the research also cast doubt on the effectiveness of often-used features such as copings at the top of parapet walls and overhangs at the top of gable walls. Eaves overhangs to sloping roofs where the pitch exceeds about 25° are effective because wind flow patterns at the wall face tend to act as though the pitched roof extends the height of the wall and, to an extent, the eaves overhang acts like a long sill. However, with copings to parapets and gable projections, the high wind uplift tends, to a degree, to increase the volume of rainfall which collects below the projection – in many cases to a greater extent than if there had been no projection at all. This effect could be a contributory cause of the problems of excessive dampness appearing in the parapets of the dwellings shown in **176** (Chapter 10).

10

Main walls – cavity construction

Introduction

As demonstrated by **137** and **138** in Chapter 9, the external appearance of brickwork can be a guide as to whether the wall is of load-bearing cavity construction or load-bearing solid construction. Stretchers alone indicate cavity work whereas a mixture of headers and stretchers signify solid construction. Snapped headers could, of course, be used in a wall of cavity construction to imitate Flemish bond or some other bond where headers are visible but would be expensive and laborious as the half bricks would have to be cut flush; this is justifiable only rarely, perhaps to match other existing work. Unfortunately, the appearance of other facing materials, as stated previously, provides no ready guide in a similar way as to type of construction. Rendering is just rendering and natural stone could be in any form. Artificial stone, on the other hand, might provide a clue if a difference between real and moulded joints can be detected and the dwelling is of an appropriate age and in a traditional stone building area.

Although by no means a new idea, cavity construction – i.e. two leaves of brick or block, of whatever type, separated by but also held together by ties across a gap ideally of a 50mm minimum – began to be much favoured in the late nineteenth century as a way of preventing damp penetration through walls of one brick thickness. This was all that was required structurally for the two-storey dwellings then much more in vogue. It had been found that a wall of this thickness was not really proof against penetrating damp, particularly where exposed to driving rain at first-floor level. To overcome the problem, walls had either to be built one and a half bricks all the way up, costly and wasteful in the circumstances, or more attractively from the cost point of view, rendered.

As might be expected, the latter option was usually favoured and indeed this continued to be the practice of builders of new dwellings for many years into the twentieth century. In the period 1920 to 1940, much of the development of two-storey dwellings by private builders, particularly in the south-east, was carried out in solid brickwork, albeit with the top storey rendered and the bay window tile hung. This practice continued even into the 1950s when private development was allowed to start up again after the Second World War. A number of developers in the south-east, catering for the upper end of the market and perhaps employing a designer, turned to cavity wall construction earlier; outside the south-east, cavity wall construction was more typically adopted from about the 1920s. In some localities, most especially in the north and west and in places along the south coast where the weather is more severe, cavity wall construction was widely used even before that. Where adopted, cavity walls were generally provided with brick outer leaves, although natural stone was also available in some localities in blocks of varying lengths and heights and thin enough for the purpose.

In the public sector, however, cavity construction was thought of as best practice by designers either employed or engaged by local authorities, and its use will be found in many of their developments from the 1900s onwards. Sometimes the decision would be strongly influenced by location. For example, local authority housing in the north of Scotland built in the late 1920s was provided with cavity walls. Most public sector development of low-rise housing up to three storeys has been built in cavity work, mainly with a brick outer skin but for cheapness, to meet cost yardsticks, alternatives such as concrete blocks and rendering could have been used. In the private sector from the 1950s onwards when cavity work became almost universal, brick remained the favoured material for the outer skin except in areas where stone was the traditional building material. Here, town planners and the conservation movement pressed strongly for the use of either natural or artificial stone, the appearance of the latter being much improved from the 1960s onwards.

When soundly built, cavity wall construction solved the problem of damp penetration through one-brick walls and could be cheaper than a one-brick rendered wall. While it also improved the insulation qualities of the dwelling – always provided ventilation to the cavity was limited to weep holes at the base and over window and door openings – its detailing was far more compli-cated and required more careful workmanship than that involved with a solid wall. Such demands are all too easy to skimp on at construction sites, and poor workmanship on sites where standards and supervision are lax can lead to problems, most of which, if left uncorrected, are all too readily covered up by the time of completion. Typical of these include mortar left on wall ties and collecting at the base of the cavity, damp-proof courses laid incorrectly, weep holes omitted and wall ties sloping downwards towards the inner leaf. Of these – and there are many others of varying degrees in importance – only the omission of weep holes will be visible to the inspector once the dwelling has been completed. On the other hand, the consequential patches of damp penetration may well be all too obvious on walls not recently decorated or perhaps hidden behind items of furniture strategically placed. Suspicions might well be aroused if walls have, in fact, been recently redecorated, and detection of damp might involve touching with the palm of the hand and judicious use of the damp meter, particularly on a wall exposed to the prevailing wind.

167 Cavity construction was well known by the 1920s and 1930s although many speculative builders, particularly in the south-east, continued with solid construction. Whether built with solid or cavity walls, as here, rendering the first floor and providing bay windows helped to distinguish private developments from those of the local authority – a vital selling point.

169 Stretcher bond on this two-storey detached house dating from about 1890 indicates load-bearing cavity construction. While early for some parts of the country, the construction is normal for this particular location. What advice should the inspector give about the wall ties? There is no evidence of the use of aggressive or soft mortar, there is no pollution in the area, the dwelling is not unduly exposed and there are no signs of cracks in the mortar joints suggesting wall tie failure. However, the wall ties are over 100 years old and only by exposing a tie can the actual condition be checked. If local knowledge raises any concerns, a check should be recommended; but in any event, at the very least, a warning should be given.

168 Some private builders continued to build in load-bearing solid construction during the 1930s but others, especially in more exposed areas such as here in the north, turned to building in load-bearing cavity construction, thus avoiding the need to render at first-floor level, except on the retained bay window where the construction would usually be framed up in timber with a panel of rendering.

The surveyor needs to be attentive to detailing around windows and doors and at the base of walls where faults common to cavity construction – in both design and workmanship – are prone to allowing damp to penetrate. General surfaces of walls need to be scanned for the same reason. Other faults to walls which also cause damp penetration – such as inadequate projection of sills, copings, eaves and verges and, for example, defective pointing – are not restricted to cavity wall construction; if present, these should obviously not be overlooked but neither should they be confused with those typical of that form of construction. Other faults such as those related to the performance of materials or defects due to ground movement, both probably causing cracks, also apply to all forms of construction and will be dealt with under their appropriate headings. However, there are two defects, wall tie failure and sulphate attack, which are more properly dealt with here as the first is more or less unique to cavity work and the second usually affects cavity construction in a more pronounced way.

Wall ties

It is obvious that the more exposed an outer leaf of a cavity wall is to the weather in the form of driving rain the more rainwater will pass through. All being well, the bulk will drop down inside the cavity so that no dampness appears internally. However, a proportion will remain on that part of the wall tie embedded in the outer leaf. That proportion will be greater and remain longer where a cavity has been provided with insulation, whether injected some time after construction or provided at the time of construction. The outer leaf of such a wall will remain colder, being deprived of a proportion of the heat escaping through the cavity, thus reducing the rate of evaporation. The life of a metal wall tie depends on its ability to resist rust in the presence of that dampness along with the quality of the mortar in which it is embedded and the shape of the tie, and in this, lies a potential vulnerability.

In the late 1800s, comparatively heavy ties in shapes not too dissimilar from those used today might be made of cast or wrought iron, the former tarred and sanded and the latter, in some cases, galvanised. Both lasted quite well, often giving 100 years of service. Hollow clay blocks were also available in the early 1900s, some glazed to shed any moisture and cranked down in shape from the inner to the outer leaf. In the absence of movement in the structure, which would break them, these ties are practically indestructible, and it is conceivable that these were used in some of the early local authority developments of cavity construction between 1900 and 1940.

By the 1920s, cheaper mild steel ties were being manufactured in twisted, flat, bar, fishtail and butterfly shapes. Sometimes these were left unprotected and sometimes they were galvanised or alternatively coated with black bitumen paint. These types became subject to a British Standard BS 1243 in 1945, which curiously required differing levels of galvanising for the two types. Even more curious, the specified levels of galvanising were reduced in 1964.

It was only when wall tie failures attributed to the omission of a galvanised coating or the corrosion of an inferior protective layer came to light in the 1970s that the Standard was again revised in 1978 to bring the galvanising to a much higher level than before. For buildings constructed since then, the quality of the wall ties has no longer been seen as an issue.

When defective wall ties were being investigated by BRE, it was found that corrosion could be accelerated in certain circumstances. Ties embedded in black ash mortar corroded much more rapidly because of the high sulphur content in the ash, particularly where dwellings were situated in areas where exposure to driving rain was the norm, in areas where there was a high level of industrial atmospheric pollution and in coastal areas where the atmosphere contained a high level of salt. It also found that ties positioned in gable walls were much more susceptible, as were those set in the weaker lime mortars.

Where the thicker vertical twist bars are present, any rusting will result in a substantial increase in bulk (**171**), and the combined effect of all ties being affected along a single mortar course is sufficient to cause a horizontal crack along that course. When this is repeated up the whole height of a wall, a regular pattern of horizontal cracks can be discerned, typically every five to seven courses of brickwork (**172**). The cracking will be more pronounced at upper levels, partly because the greater exposure causes more rapid deterioration but also because there is a reducing mass of brickwork to constrain the effects of the rust expansion. The expansion means that affected walls actually increase slightly in height and can cause an uplift to sections of roofing which bear on them. This is most likely to be noticeable when it causes a so-called 'pagoda effect' where a gable causes the end of the ridge of a roof to lift. Often, however, the weight of the roof prevents the wall from moving upwards; then, it may bow outwards, especially around window openings, resulting in cracking to the reveals. When deterioration becomes very advanced, the ties break off and no longer restrain the wall so the bowing becomes ever more exaggerated (**Figure 4**). Unless remedial work is undertaken, this will increase until such time as the detached section of the outer leaf collapses outwards, with obvious results.

The horizontal pattern cracking – which may only be visible during a careful check of the upper sections of brickwork with binoculars – outward bowing and cracking along window reveals are all classic indicators of wall tie failure for the inspector to look out for. Unfortunately, wire ties – the type most often used in post-war housing – do not have the same volume as the strap type and the effects on the adjacent brickwork are unlikely to be visible (**173**). During the exceptional gales of 1987 and 1990, a significant number of failures occurred where outer leaves were sucked out by the winds; and the BRE investigation recorded that many of these were due to corroded wall ties.

As a result of all their investigations, BRE concluded that any dwelling with cavity wall construction built before 1981, the year when the standard of galvanising for mild steel ties was raised – estimated to be some ten million properties – is at some degree of risk unless it is known that stainless steel ties were used. It was said at the time that most collapses of this nature involved

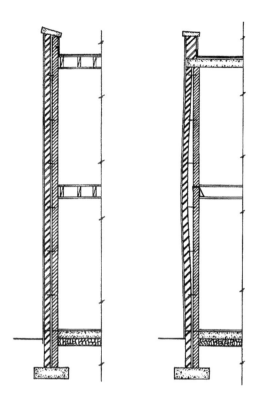

Figure 4 Typical box-like dwellings of the period 1960–1980 with their flat roofs frequently have inadequate damp-proof courses in the walls at roof level and, if rendered, allow yet more damp penetration through shrinkage cracks. Wall tie failure and sulphate attack cause different problems depending on whether the roof load is transferred to the foundations through both leaves or merely via the inner leaf. This applies irrespective of whether roofs are flat or pitched. On the left where there is no roof restraint to the outer leaf, the expansion in the joints tends to be progressive towards the top, tilting the wall ties towards the inner leaf and probably causing damp penetration to appear on plastered walls internally. On the right, the roof restraint to the outer leaf causes it to bulge away from the secure fixing provided by the ties, eventual collapse being a possibility. Horizontal cracks to all brick joints, or on the rendering on the line of the joints, indicates sulphate attack. If the horizontal cracks are intermittent, often at about 450mm vertical intervals, then wall tie failure is the defect. There could, of course, be both types of defect happening simultaneously.

inferior ties, the use of permeable lime or aggressive types of mortar, such as black ash, or exposure to severely polluted or marine environments. Reassuringly, there has not been a rash of failures since, but the surveyor should be aware that this is likely to be an increasing area of vulnerability.

Clearly it is not appropriate to recommend exposure works for every property in the 'at risk' group, but without actually seeing any wall ties, how is the surveyor to build up an understanding of their likely condition? Part of the answer is to take every opportunity to increase local knowledge of wall ties by looking at their condition whenever it *is* possible – as they are exposed during alterations or demolitions, for example. It also may be possible to obtain useful information about wall ties across the working patch from builders or, ideally, local wall tie replacement contractors. In any event, careful checks for evidence of deterioration should be made during every inspection and, when appropriate, suitable warnings about the long-term risks should be given. In high-risk situations – where black ash mortar is present or in marine environments, for example – it may be necessary to recommend precautionary investigations so the actual type and condition of representative ties at the property can be determined.

When wall tie problems are identified or suspicions of problems are confirmed, it will be necessary to commission a thorough investigation and to replace ties where necessary. There are now specialist companies which will undertake all of the necessary work, and a number of different techniques are

170 A pair of semi-detached houses in a coastal town built around 1880 with walls of load-bearing cavity wall construction. The rendering on the right-hand house shows signs of horizontal cracking in several locations. Some of this seems to be due to normal lintel movement but elsewhere may be an indication of wall tie expansion. Despite being a couple of miles inland, there will still be regular exposure to a salt-contaminated atmosphere and the wall ties are over 130 years old so it is imperative to advise the seeking out and exposure of a tie to establish type and condition.

171 Progressive rusting of this metal bar above the window of a small outbuilding has caused it to increase in thickness by at least three times its original size, disrupting the brickwork above. The same effect on a series of individual wall ties spread along the same course of mortar in a wall will cause horizontal cracking, despite the weight of the brickwork above.

173 The galvanised protective coating on the end of this 1960s wire tie, exposed during alterations, has broken down, allowing it to becoming lightly rusted. There is still plenty of life remaining but deterioration will continue and this tie is not in the most exposed location at this property so it would be worthwhile checking others in more vulnerable areas. The metal tie does not have enough volume for rust expansion to disrupt the brickwork so any failures will not be obvious. The cavity has been filled with urea formaldehyde foam insulation.

172 A repeated pattern of horizontal cracks in a cavity wall is a strong indication of cavity wall tie deterioration. Sometimes, as here, owners will repoint the cracks, either because they do not appreciate the cause or in the vain hope of preventing damp penetration causing further rusting. The result is unsightly but very helpful to the surveyor.

174 Gable walls in load-bearing cavity wall construction are at greatest risk from wall tie failure, particularly as here where the dwelling is from the early 1920s and the ties must be around 90 years old. To be assured that collapse is not likely, even though conditions are favourable as regards mortar and exposure and there is no evidence of defects, advice to expose a tie needs to be given to ascertain type and condition. It is surprising that the stack, 3.5m tall, is not showing more signs of distortion than it does.

175 In extreme cases, expansion of rusting cavity wall ties can cause bulging and cracking to walls, especially at higher levels where greater exposure tends to cause more rapid deterioration. The results are shown graphically here where displacement of the outer leaf has caused a dramatic opening to develop around the window frame. The brickwork of this entire gable required rebuilding.

available whereby ties can be replaced from the exterior without the need to demolish the outer leaf, either leaving the old ties in place or having them removed, depending on which type was originally used.

Frequently, there will be little visible evidence that wall tie replacement work has already been carried out at a property. Sometimes it may be possible to see the locations of drill holes where replacement ties have been inserted and original ones have been removed or physically isolated from the outer wall to prevent further damage. At first glance, these might be mistaken for the holes drilled when cavity wall insulation has been installed, but insulation requires fewer holes and they are much more widely spaced. Replacement wall ties should be installed at regular intervals along equally spaced courses of brickwork with additional ties around openings in the wall. When completed to a good standard, the holes should be made good to blend well with the original masonry and should not be at all obvious. If the work has been completed on a rendered wall, it may be impossible to make out.

Whether from personal observation or because a vendor or estate agent has thoughtfully provided the information, whenever the surveyor becomes aware of prior wall tie replacement work then it is helpful to see the associated specification and documentation if they are available to help form a judgement about the standard and extent of work. Similarly, the client's legal adviser should check the standing of the company which undertook the work and the value of the warranty or guarantee which hopefully was provided upon completion. If any of this seems unsatisfactory, it may be necessary to recommend exposure works to check what has been done.

176 There are clearly problems with the parapets of these two-storey flat-roofed terraced houses from the period 1960 to 1980, built in load-bearing cavity construction. Brick-on-edge copings are inadequate on their own and need something more to keep damp at bay. If the bricks used have high sulphate content, there is a considerable danger of sulphate attack developing. Extreme care would need to be taken when inspecting these dwellings – and others nearby in the same form of construction – to check what has gone wrong and whether any remedial measures taken are, or are likely to be, effective.

177 The more innovative designs which began to appear during the later years of the last century can produce their own problems, and the inspector needs to check that sound principles have been followed. On this three-storey terrace, the parapet and semi-circular feature to the front wall needs a sound coping and good bricks and pointing as, like all parapets, it is exposed to the weather on both sides. Although there may be no defects apparent internally, it is clear that damp at least is beginning to affect the parapet, staining the brickwork and eroding the pointing; and if the bricks have high sulphate content, there can be a danger of sulphate attack. It may be possible to gain access to the parapet gutter and peer over the top for a closer look.

Sulphate attack

The indications of sulphate attack on the jointing and pointing of the mortar of brickwork and any applied rendering could be confused with those of wall tie failure. Although sulphate attack can affect brick walls of solid as well as cavity construction, it only occurs on those dwellings built in the last 100 years or so where Portland cement has been used in the mortar mix. Since there are fewer dwellings of solid construction from this period, sulphate attack is more likely to be encountered on dwellings of cavity construction. Against this, however, is the fact that much rendering with Portland cement in the mix was applied to dwellings at first-floor level in the 1920s and 1930s and, when applied overall, constituted a feature of those purporting to be in the 'modern style'. Prior to the 1920s, most jointing, pointing and rendering was based on lime or older forms of patent cement that were not subject to sulphate attack.

Sulphate attack is caused by the presence in the bricks of soluble salts. When brickwork remains wet over long periods as it can do where dwellings are situated in exposed positions, particularly in gable walls and where there is little protection from overhanging eaves, rainwater dissolves the salts and carries them into the mortar of jointing and pointing. The salt solution attacks the tricalcium aluminate in the Portland cement causing it initially to expand but eventually reducing it to mush. Signs that this is happening are cracks in all the horizontal joints of the pointing in the damp areas of the brickwork, a whitening of the mortar and often signs of a whitish staining around the edges of the bricks. In contrast, the visible signs of wall tie failure are limited to cracking along the horizontal joints where the wall ties themselves are located. If neither of these defects can be seen at low level or from windows, an examination of walls at high level through the inspector's powerful binoculars may reveal such defects in the brick joints and enable the two types of defect to be distinguished.

Where an external wall has been rendered with a mix containing Portland cement over bricks with a high soluble salt content, the fine shrinkage cracks which often accompany the drying out of such a rich coating allow for moisture to penetrate through to the brickwork and for it to be in contact with the bricks for far longer periods than would have been the case if the wall had been left uncovered. Damage can then be even more severe than on un-rendered walls and a pattern of frequently occurring cracks in rendering along the line of the majority of brick joints, both vertical and horizontal, will appear. These clearly defined cracks should not be confused with the crazy pattern of cracks running in all directions caused by the shrinkage of a rendering mix too rich in Portland cement, although these sometimes precede sulphate attack by allowing dampness through to the brickwork. The problem can then become compounded as yet more rainwater is allowed through to the brick-work and eventually areas of the rendering become loosened and bulge. If the defect is confined to wall tie failure then the rendering will only be cracked along the horizontal joints where wall ties are located – still, of course, indicating the need for attention (**Figure 5**).

Both wall tie failure and sulphate attack can lead to expansion of the mass of brickwork and, in a cavity wall, can cause disruption at the position of the ties and possible bulging of the outer leaf when there is sufficient loading at the top to prevent an upward movement in the brickwork.

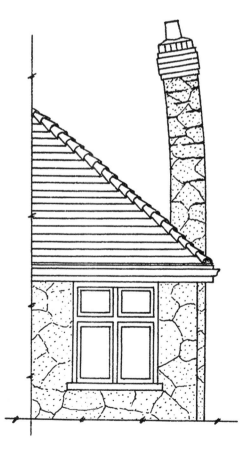

Figure 5 At first-floor level on the walls of this typical semi-detached house of the period 1920 to 1940, there is the characteristic crazy pattern of shrinkage cracks caused by the use of a rendering mix too rich in Portland cement. The more pronounced horizontal cracks on the leaning chimney stack – too tall for its thickness – are caused, in this case, by the heavy deposits within the stack from the use of solid fuel, inducing sulphate attack on the brick joints and rendering. There is obviously a danger of collapse. Should the bricks in the main wall contain sulphates, damp penetration through the shrinkage cracks will bring about a more extensive sulphate attack on the mortar joints and the rendering and more pronounced horizontal cracks will appear. If the walls are of cavity construction, wall tie failure will cause the same type of horizontal crack to appear even if the bricks are free of sulphates but the extent of the cracks will be limited to the joints where wall ties are situated, but probably only if the vertical twisted bar type of tie has been used.

Remedies for sulphate attack can range from renewing individual defective bricks to, in serious cases, renewing all the bricks on an elevation. Reducing exposure to dampness can sometimes be achieved by perhaps adding a protective flashing to a band course or a canopy over a porch. Other methods, all of which are primarily concerned with preventing rainwater reaching the brickwork, can extend to slate or tile hanging or covering elevations with weatherboarding.

Insulation

A cavity wall, in its basic original form of two leaves and a ventilated cavity, was better at preventing damp penetration than a one-brick solid wall and also had the added benefit of providing a slightly better level of insulation. This situation was thought satisfactory when domestic central heating was practically non-existent. However, when all-embracing warmth became desirable and was achievable from the 1950s, it was found that much of the heat went out through the walls, so steps started being taken to improve the insulation properties of the wall. For new building this, initially, consisted of sealing off the cavity from the outside air, but retaining, of course, enclosed trunking for the apertures to provide underfloor and other necessary ventilation, and constructing the inner leaf of load-bearing insulation blocks. These measures only produced a marginal improvement and attention then turned to providing insulation within the cavity itself even though this compromised the basic principle of that form of construction.

For new work to meet the insulation requirements of the Building Regulations, first introduced in 1966 but progressively increased since, the provision of insulation to the cavity may be in the form of bats or slabs or various forms of insulating material. This might also involve the use of longer wall ties as it was deemed essential to maintain the recommended 50mm clear cavity. However, incorrect or a lack of proper fixing could cause subsequent damp penetration to the interior when, for example, the cavity became blocked by slipped or loose sections.

For existing cavity wall construction, in many cases now well in excess of 50 years in age, a variety of materials have been employed, each with their advantages and disadvantages. Urea-formaldehyde foam was often utilised following the 1970s oil crisis but adverse publicity originating in the US and Canada, later declared unfounded, caused it to be banned for a while and, despite being given a clean bill of health, it still retains something of a stigma. A number of other foams may be encountered, including polyurethane foam. Some of these shrink as they dry, causing fissures which can channel moisture across the cavity. Loose fill materials may be used but, whereas foams fill the cavity void as they expand and set, there is a risk that if installation is not properly undertaken then loose fill materials may migrate unevenly within the cavity and leave areas uninsulated. Expanded polystyrene beads or granules react chemically with the PVC coating on electrical cables so the risk of this occurring should be taken into account during installation. Additionally, expanded polystyrene is flammable and gives off toxic fumes if it catches fire so care is needed in proximity to hot areas such as flues. Man-made blown mineral fibre is chemically inert and has become one of the most popular loose fill materials. An advantage of loose fill materials compared with foams is that removing them from the cavity is usually much more practicable if that should be required for some reason at a later date.

Before any form of insulation is injected into the walls of an existing dwelling a preliminary survey should be carried out by the contractor to assess its suitability. Framed forms of construction, such as timber-framed properties, should not have insulation injected. If this does happen, it is likely to render the property unacceptable to lenders for mortgage purposes, with a consequentially serious impact on value and saleability.

The surveyor may see external clues in the form of filled injection holes, but often the only way to tell whether a cavity wall has been insulated is by catching sight of hardened foam or some other insulating material at the top of the cavity in the roof space and, not infrequently, spilling into the roof space from the top of the wall. Failing visual evidence, it is necessary to rely, in general, on statements made or documents produced by the seller. For a new or near new dwelling, the specification approved under the Building Regulations will hopefully be available and the surveyor may be able to form an opinion on the way in which the insulation requirements have been met. Builders are finding that achieving the progressively more demanding requirements for insulation in new-build properties can no longer be met simply by providing insulation within the conventional 50mm cavity, and increasingly a variety of alternatives are likely to be encountered. These typically include increasing the size of the cavity to incorporate additional insulation or maintenance of the normal 50mm cavity sealed from the outside air but with insulation fixed to the inside face of the inner leaf, raising the overall dimension to between 300mm and 330mm. Also typical are cavities filled with mineral wool or fibreglass, both of which are thought unlikely to provide a passage for the transmission of water to the inside face of what is, in effect, a solid wall about 280mm in overall thickness. Insulation by the use of foam – either urea-formaldehyde or polyurethane – is not likely to be encountered for new work, and other regulations apply in Scotland where the effects of exposure to weather are more severe.

Dwellings in exposed positions are unsuitable for cavity wall insulation. If there are problems of damp penetration in a dwelling situated on a hillside or in a coastal region, unscreened by neighbouring dwellings and in an area subject to heavy rain and gales, it is fairly elementary to conclude that undue exposure is at least one of the causes. As mentioned previously, the inspector needs to be aware of the direction within the area of practice from which rain is driven at its maximum. While for most of England, Wales and Scotland it is from the west, there are, in fact, areas in both England and Scotland where it is from the east.

Wall tie failure, sulphate attack and problems of damp penetration to the inner leaf either due to faults in design or workmanship – particularly around openings for windows and doors – or from the subsequent addition of insulation to the cavity where none existed before, are defects particularly associated with load-bearing cavity wall construction. Of course, such construction is not immune from other defects due to causes originating above ground level, such as roof thrust, overloading, movement due to temperature changes, expansion and shrinkage, which also affect load-bearing, solidly constructed walls (considered in the next chapter). They are also not immune from defects originating from below ground level such as inadequacy of foundation design in relation to the subsoil, leaking drains and tree root action, among others, which have already been considered in Chapter 8. Cavity construction tends not to accommodate movement in the same way as earlier solid brick construction, which was usually more substantial and generally more forgiving because of the weaker lime mortar. The mismatch of materials normally used in cavity work and the fact that ties are set a mere 50mm in the outer and inner leaves, together with the use over the last 100 years of the stronger cement-based mortars as against the weaker lime-based mortars, often makes the effect of movement more dramatic and serious when it does occur.

11

Main walls – solid brick construction

Bricks

Well-selected clay dug out of the ground, weathered and suitably fired in a kiln can last an astonishingly long time, and bricks made of such material are no exception; witness those of Roman origin incorporated in buildings constructed much later but still in sound condition. On the other hand, soft under-burnt bricks made from unsuitable clay much more recently have long since crumbled away. Most familiar to the surveyor are the poor-quality red bricks which seem to have been made in great quantities from the 1880s to the 1920s. They are vulnerable to weathering and their disintegrating faces are a depressing sight in urban and country districts alike.

The satisfactory performance and the length of life of clay bricks in a wall – whether of cavity or solid construction – is governed by the bricks selected to withstand the degree of saturation likely to be experienced, not only in the general area of walling between damp-proof course and eaves but also at other positions in the dwelling. In addition, the composition of the material used and the standard of workmanship adopted for jointing the bricks and the profile of the completed joint along with its subsequent maintenance have a strong bearing on the durability of the bricks.

Bricks used in parapets and chimney stacks, as copings, string courses or in cornices but, particularly, those below ground and up to the level of the damp-proof course – within the 150mm splashing distance of impervious paving – are most at danger of becoming totally saturated with water. If this happens, it can lead to breakdown by frost action and, perhaps, sulphate attack. An examination, through binoculars if necessary, of the dwelling's vulnerable positions where bricks have been used plus any other positions, such as in free-standing or retaining walls, where they may have also been used, can provide crucial information. This applies for a dwelling of any age. The current state of the bricks can be assessed and, perhaps, advance warning obtained of possible future problems even if the condition of the walling between the level of damp-proof course and eaves is generally satisfactory.

As to the bricks themselves, where the surveyor is inspecting an older dwelling or where there is no specification available to provide information on their origin, the evidence of the eyes must be relied upon. Those locations mentioned which are especially vulnerable to becoming saturated need to be viewed from as close a distance as possible or through binoculars for any signs, however slight, of failure in the form of spalling or erosion. Areas where the brickwork is a darker colour than elsewhere are often an indication of damper conditions and repay even more detailed examination and the judicious use of the penknife to scrape the surface if that is at all possible.

On a new development, a copy of the brick specification should ideally be available for examination and consideration. With regard to modern existing properties, a surveyor who is a bit longer in the tooth might have noticed new developments in the area during years in practice and taken note of the make and type of brick being used. Such local knowledge is invaluable. Where

179 A terrace of four-storey houses built around 1800 and categorised as second-rate under the Building Act 1774. The use of different, variously coloured bricks in the rebuilding of the upper part of the top storey is unfortunate for appearance; the surveyor will want to know when it was done and why it was necessary.

178 The Building Act of 1774 categorised proposed terraced dwellings into one of four different rates according to a number of factors. These included proposed floor area, anticipated construction cost, the number of storeys and the height above ground level, prescribing the thickness for external and party walls of each. This photograph shows a first-rate house of five storeys built around 1790 where localised repairs have been carried out in many areas, not always with a pleasing appearance.

there are no defects visible and the dwelling is on an estate with an age assessed at up to 60 to 70 years, it would be sensible to view the exterior of a dwelling of the same age in a more exposed position if at all possible. This might show defects likely to become apparent on the subject dwelling in time. If the dwelling itself has an assessed age of 60 to 70 years or more and there are no visible defects on either it or similar dwellings nearby then it could, within reason, be said that it is unlikely to develop faults in the brickwork unless the ground is disturbed or maintenance is neglected. This comment could be reinforced if the surveyor's local knowledge allows a reasonable certainty that the bricks are of a variety commonly used in the area with a reputation for long life. There might be evidence from brickwork of the same type on older dwellings in the area to justify the comment.

181 Rows and rows of three-storey fourth-rate houses as categorised under the Building Act 1774 were built, often very cheaply, in the latter part of the eighteenth century and the early part of the nineteenth century. Many have long since gone by way of redevelopment, with those remaining often very flimsy. Tie bars have been fitted to the front of two here to prevent outward movement of the unrestrained front walls.

180 A terrace of four-storey houses built in the late eighteenth century and categorised as third-rate under the Building Act 1774. As with the first- and second-rate houses shown in **178** and **179**, the need to rebuild at top-floor level in the past, either wholly or in part, should remind the inspector that it is not only adequate thickness that has to be considered but also essential restraint. Usually, floor joists span from party wall to party wall, the shortest distance, leaving the front wall unrestrained for much of its height. The outcome of the practice, at the time, of building the party walls before the front walls of terraced dwellings often necessitated the provision of restraining discs at mid height to secure the front wall at its centre, as in **181**.

Facing bricks

Facing bricks are selected for their appearance. Historically, they contrasted with 'common' bricks, which were used internally or as backing before concrete blocks became ubiquitous. Clay suitable for making facing bricks is now obtained from many areas in the UK but this was not always the case. When the only way to make bricks was by hand, only the softer clays with good plasticity from certain areas were suitable. This was the situation up to about 1850, and it had been the case for 600 years since the re-introduction of brick making from the Netherlands after a lapse of 800 years following the departure of the Romans.

The early processes to produce hand-moulded bricks were very time consuming, usually carried out near where the bricks were destined for use and involving at least one winter when the clay was weathered. Even so, the results were variable since the firing first by wood, up to about 1700, then by coal was by rule of thumb and relied on the brick maker's expertise. Initially, shades of red towards brown, depending on the proportions of iron and lime in the clay, were produced, and these remained in favour until about 1750 when paler colours of grey, yellow or even off-white began to be considered much more fashionable. Under-burnt bricks came out softer and paler-coloured than properly burnt bricks and even more so than those which had been over-burnt, which came out a very dark colour and were often misshapen and brittle. Some, depending on their position in the kiln, might be vitrified either wholly or in part by the heat, turning a very dark purple in the process. The under-burnt should, of course, have been rejected, but the colour of the over-burnt might be exploited in the creation of walls with the perennially popular diaper pattern.

It is the irregularities in shape, texture and colour due to clamp burning and often the presence of small cinders which make old, hand-moulded bricks attractive to many and, for which some prospective owners are prepared to pay at least three times as much as for machine-moulded bricks. These characteristics are not entirely achievable by processes relying on the machinery developed since the late 1850s and early 1860s. This included grinding machines, which could process the harder and less plastic clays of the Midlands and the North, drying and moulding machines, and continuously burning kilns. These developments all made uniformity, much prized at the time, easier to achieve.

Perhaps the greatest undermining of local brick making came later, towards the end of the nineteenth century, with the extraction of the shale-like Knotts clay with its higher percentage of lime than iron from the vast Oxford clay bed formation running through Bedfordshire and into Lincolnshire. This has properties which, without the need for weathering or drying, enabled it to be pressed into shape by the new machines very quickly and allowed subsequent kiln firing, which only required a third of the fuel needed for processing other clays. Even with comparatively high transport costs, local brick makers could be undercut, particularly when – from the late 1920s – with additives, texturing the surface and controlling the amount of air in the kiln to determine colour, passable facing bricks could be produced by the big manufacturers, along with the basic, sharp, arrised, pasty-looking, pinky yellow, smooth-surfaced Fletton brick, which is not helped in appearance by the striped effect produced by kiln burning. For a while, only those areas inaccessible by rail remained free of the products from Bedford, Peterborough and North Buckinghamshire; but cheaper and more reliable motor transport and now shrink-wrapping means that bricks are available wherever they are required, not always with advantage to the landscape if insensitively used.

As far as age is concerned, it is therefore quite safe for the surveyor to say that the original brickwork of any dwelling assessed as having been built before 1850 is hand moulded. For dwellings built subsequent to that date, both hand- and machine-made brickwork is possible and other evidence will be needed to suggest which is the most likely. Size and quality of dwelling can be an indication since the larger and better-quality dwellings tend to have hand-moulded bricks while at the other end of the scale, cheapness will have directed attention towards the machine-made product.

Using a higher proportion of fine sand, increasing the time for weathering and specially washing loamy clay produces the denser-textured and harder brick suitable for cutting with a bricklayer's saw and rubbing down to produce the fine, precise gauged brickwork with its very thin joints used in arches (**187**) and quoins and also for carving capitals and the bases of pilasters. Clay can also be pressed into specially prepared moulds to produce ornamental panels with patterns in relief either of geometrical shapes or of leaves and flowers, popular from the middle to the end of the nineteenth century (**188**).

While many varieties of red facing brick have excellent qualities of durability, sometimes the less well burnt but better-shaped bright red bricks would be used in arches and around window openings to contrast with the bricks of more sober colour used for general walling. Such bricks will not generally be confused with the much harder bricks used in gauged work and will often be found to have become eroded by the weather much sooner than the brickwork of general walling, requiring replacement.

The muted but variously coloured, matt, slightly creased, wrinkled, open-textured surface of the hand-moulded brick represents the ideal; and while the wooden moulds of machine-moulded bricks can be sanded before the clay is squeezed in by machine, the resulting effect is never quite the same as achieved by throwing the clay into the mould by hand. When built into a wall, the slightly open texture allows a certain amount of rainwater to be absorbed into the brick and, provided the pointing is of a similar strength and texture to the brickwork, the 'overcoat' effect is achieved. When the rain ceases, evaporation enables the wall to dry out.

The direct opposite to the ideal open-textured surface for a brick is the dense, close-textured, smooth, shiny-surfaced brick, often very fiercely or very darkly coloured and appearing almost glazed in the firing of ordinary facing bricks but, sometimes, with a glaze deliberately applied before a second firing.

182 Bricks from Roman times were more in the form of thick tiles than the sizes of bricks we know today. These are incorporated in a much later stone wall but have lasted very well.

184 White bricks of gault clay, containing lime but no iron, and red bricks where the clay contains much iron.

183 Bricks from the hottest part of the kiln usually came out harder and darker than the remainder but these were often smaller and brittle, easily broken if handled carelessly. They were often incorporated as headers in a diaper pattern, as here in a wall from the 1500s. The over-burnt bricks are clearly wearing better than the red bricks but even most of these have already lasted over 400 years. More effort should have been made to obtain a better match for the replacement bricks, especially where the overall effect has been diminished.

186 The end house of the terrace of which **185** forms part adjoins a narrow road entrance where the corner is curved. The headers used in the wall to turn the bend were those partially vitrified in the kiln with a hard, dark blue surface – not quite hard enough, however, to resist scraping by carts and motor vehicles over 170 years.

185 A little judicious scraping reveals that these silver-grey bricks, very fashionable in the first half of the nineteenth century when this house was built, are in fact the same red bricks used as dressings around the windows. The colour change was produced by a coating of sand to which salt had been added before firing. Natural weathering has had very little effect over 170 years, though the position is comparatively sheltered.

187 Fine-sawn and rubbed brickwork from the early eighteenth century.

188 Moulded panels of red clay used for decorative effect on the bay of an unremarkable late nineteenth-century house, beautifully cleaned and in near pristine condition after renovation works. Such attractive details are often sadly hidden by years of grime or obscured by successive coats of paint, so their delicacy is usually unappreciated.

189 Salt-glazed brickwork has a surface highly resistant to rain penetration, but this has the unintended consequence of concentrating rainwater run-off into the joints between the bricks where any weakness in the pointing will inevitably be exploited. Any damage to the face of the bricks is likely to remain an unsightly blemish.

190 Excellent bricks of an attractive red shade and sand-faced texture from the mid eighteenth century, of regular shape, rubbed to sharp arrises and meticulously pointed with fine, near matching joints, now slightly eroded in places.

Sometimes such a surface was introduced in the mistaken belief that it prevents rain penetration. Raincoats are better without joints, viz. a cape; and it is the joints in a wall of such bricks that are its cardinal weakness since rainwater streams down the face of the brick to seek out any minor cracks and open pores in the mortar of the joints, of which, inevitably, there are many. It is rather like water running down the glass of a window and seeking out weaknesses in the lower section of the frame and sill. In other circumstances, the firing of the hard, stiff local clays produced bricks with hard, dense surfaces.

The problems these bricks created in the Midlands and the North – where they are more common than elsewhere – from rain penetration probably led to the more ready adoption of the hollow cavity wall in those areas.

Another reason for glazing bricks was the belief that they would better resist the salty atmosphere of coastal regions. In fact, bricks are not generally affected by salts in the atmosphere, unlike stonework which readily succumbs to acids in any atmosphere, be it from the sea or industrial processes in whatever the locality. A further use for white glazed bricks was as facings for the enclosed areas of the tall blocks of flats popular in the late nineteenth and early twentieth centuries to reflect light, as well as in other positions where natural light was limited.

Brickwork bonds

Defects such as leans and bulges in walls can be caused by poor workmanship in the form of inadequate bonding, even where the thickness of the wall is adequate. Up to the eighteenth century, English bond was commonly used but thereafter Flemish bond became much more common. It is thought to have been first used at Kew Palace, Surrey in 1631. It was thought at one time that English bond was the stronger of the two types but it would appear that there is no evidence for this.

It is possible, of course, to build a wall entirely of headers, the very opposite of a wall built entirely in stretchers; but it is very rare to find a dwelling built throughout in header bond – although some do exist (**192**). However, headers alone are useful for taking brickwork around fairly sharp curves – on bow windows, for example – and around the bends in walls, as was popular in the 1930s. Using stretchers for this purpose would produce too many sharp angles, which would look ugly if not covered over by a rendering (**193**).

Less strong than either English or Flemish bond are some of the other bonds which may be encountered. For instance, English Garden Wall bond – one course of headers to three or five of stretchers – is less satisfactory, having a long, continuous vertical joint between each three, or five, courses of successive stretchers that causes a deficiency in transverse strength. Rat Trap bond has too many voids, looks awful and, as its name implies, can provide a refuge for vermin. It is reckoned, however, to provide a saving in bricks of about 25 per cent over a solid wall (**195**).

Jointing of the brickwork should, of course, be properly carried out with the mortar applied to all surfaces of the brick for the full thickness of the wall; and if the bricks have frogs, these should have been positioned uppermost. To an extent, all this has to be taken on trust once the wall has been completed. However, the sight of brickwork in the roof space on the internal face of gables and on party or separating walls might provide a clue, if it is of poor construction, as to the situation found elsewhere, should this be thought of as the possible cause of defects (**Figure 6**). Nevertheless this also has to be

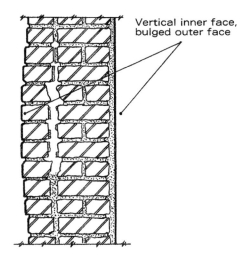

Vertical inner face, bulged outer face

Figure 6 There are usually consequences to building solid one-and-a-half-brick-thick walls with a facing skin of good-quality bricks and a backing of those of poorer quality, a practice common in the eighteenth and nineteenth centuries. Often headers would be snapped to save more money and only a few facing headers carried through to provide a bond. Poor-quality bricks and mortar in the backing compounded the problem, resulting in sections bulging forward and causing even those headers intended to provide the bond to snap.

somewhat in the realm of speculation as the bricklayer could have adapted a much more relaxed attitude to construction at this level compared with elsewhere.

Also speculative, to some extent, must be a suspicion that the wall in question, although supposed to be of solid construction, might perhaps be in two skins. Much brickwork in terrace construction built speculatively presents a fine face of quality brick to the street. This could, however, have been applied to far less satisfactory bricks behind and, to save yet more money, the headers seen on the face could in fact be half bricks. Such practices were the subject of much complaint even in the seventeenth century. It took a long time before legislation, but more importantly enforcement, curbed speculative builders from adopting them. The strength of a wall would depend very much on the backing brickwork and its workmanship, but very often, the bricks used for backing would consist of misshapen and under-burnt examples despite an Act of 1764 requiring bricks to be 'good, sound and well burnt'. Furthermore, even the mortar might be of poor strength and thinly applied, leaving many voids. Bulges when the two skins separate are not uncommon, but hefty and hasty, ill-considered tapping to see if there is a hollow sound is not to be recommended. The inspector might disappear in the subsequent clatter as the two skins part company completely.

The practice of building solid walls in two skins for the sake of appearance in speculative building is of long standing, and this continued to a greater or lesser extent up to around 1900. To keep the rougher-quality backing brickwork more or less in line with the courses of the finer-quality facing bricks and to achieve at least a modicum of bond at intervals between the two, the use of bonding timbers was common (**Figure 7**). Because these are usually within a half-brick's length from the face of the wall, they are inevitably affected by damp over the years which, coupled with the absence of ventilation, causes

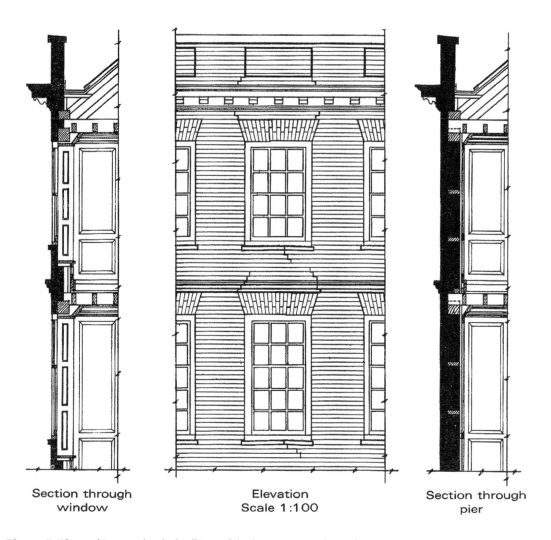

Section through   Elevation   Section through
window   Scale 1:100   pier

Figure 7 The multi-storey brick dwellings of the late seventeenth century and eighteenth century usually incorporated a considerable amount of timber within the walls, originally to hold the structure together; but this was to cause problems later when damp penetrated – perhaps due to lack of maintenance – to the unventilated areas, causing the timber to rot. As the section through the window shows, there was little brickwork at this point, but a continuous large ring beam at each level would be connected to floor and roof structures and there would be panelling below the windows and shutter boxes at the sides. The section through a pier shows the ring beams and the bonding timbers which were used to maintain the levels for the rough backing brickwork and to provide a fixing for wall panelling, as here, or later for lath and plaster. When the bond timbers rot, the brickwork in the piers settles, disturbing the arches over the windows and producing cracks in sills and brickwork. The house illustrated was perhaps built between the two Acts of 1707 and 1709. The first banished wooden cornices so that this dwelling was provided with one either in stone or built up in tiles and mortar, along with a parapet. The window frames, however, are still set level with the face of the brickwork; although the Act of 1709 required them to be set back a distance of half a brick's length, enforcement took a while and it also took a length of time to filter down outside the centre of London.

them to rot and lose their strength to hold the brickwork above in position. The differential movement as the inner skin drops down leaving the outer skin at its original level causes what headers there are to snap and the outer skin to bulge outwards (**Figure 8**).

Scale 1 : 20

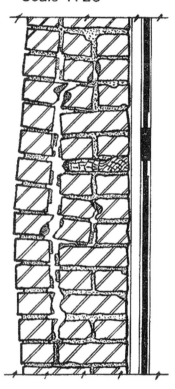

Figure 8 The outer edge of the unventilated bonding timbers found in the walls of many brick-built dwellings of the eighteenth and nineteenth centuries is usually within a mere half-brick's length of the outside air and, therefore, subject to penetrating damp, causing rot and loss of strength. In consequence, the bricks above drop down, fracturing any through bricks and causing the outer face to bulge.

191 An example of brickwork in English bond.

192 It is unusual to find a dwelling built almost entirely in header bond. This photograph shows part of the front wall of a cottage where the person who built it clearly wanted to make a statement by using this expensive form of construction.

193 Headers alone are useful for taking brickwork around fairly sharp bends, such as on these brick bow windows to a 1930s detached house. On a curve of this radius, stretcher bond would introduce many projecting edges, although this can be overcome by rendering. At first-floor level, the bay is usually timber framed and either finished with tile hanging or, as here, rendered.

194 English Garden Wall bond comprises three, four or five courses of stretchers to one of headers. This example has three courses of stretchers. As this bond has fewer headers than either English or Flemish bond, it is easier to obtain a fair face on both sides of a one-brick wall, hence its primary use in the past as a division wall between gardens. Although it is considered deficient in transverse strength because of the continuous longitudinal vertical joints running through each of the successive stretcher courses, it is normally thought suitable for dwellings not exceeding two storeys and is cheaper than using either English or Flemish bonds. Nevertheless, a surveyor needs to look carefully to see whether the front 'layer' of stretchers is parting company from those behind and bulging outwards, particularly, as here, where the pointing has been drastically neglected.

195 Walls of one-brick thickness built in Rat Trap bond, where all bricks are laid on edge and the stretchers have a 75mm gap between, are about 25 per cent cheaper to construct than walls in English or Flemish bond. The problem with this bond, as the name implies, is that the gap in the centre provides a haven for vermin – less of a problem when used for a garden wall, as here, but less suitable for the walls of a dwelling.

Jointing and pointing

As important as the masonry units themselves in keeping the weather at bay is the quality of the mortar in which they are bedded and jointed, coupled with the workmanship employed. With its microscopic labyrinth of voids, the mortar in the joints is the most vulnerable part. This is particularly so where the masonry units lack a degree of absorption and rain is inclined to run down the face of the unit, finding minute cracks in the jointing and allowing water to be carried through, by capillary action, to the interior.

It is axiomatic that all joints should be fully filled with mortar, but as this is a matter of workmanship, it may not be achieved unless there is the closest of supervision on site. Bed joints are usually filled as the bricklayer will recognise that not to do so could affect strength. It is the perpends that are often neglected to the extent that only the visible ends are filled – 'tipping and tailing'. In the case of a new or near new dwelling where the specification is available, the surveyor will be able to check that all matters of mortar mix and workmanship have been covered; but once the dwelling has been completed, it will be impossible to see such detail unless the bricklayer has been so careless as to leave visible joints only partly filled.

What can be seen and checked is the profile that the joints present to the outside. This can have a pronounced effect on keeping the rain at bay. With new construction, it is better to finish the joint as the work proceeds since this avoids the need for any raking out of the bedding mortar. If pointing is carried out after bedding and jointing, however, it should be in a mix similar to that of the jointing material which, in turn, should have matched the strength of the bricks. A tooled finish is better than a flush joint, which leaves the surface more open; this is because with the former, mortar is pressed into the joint, assisting adhesion to the bricks, while the tooling tends to compact the surface. Thus bucket handle or struck weathered joints are preferable to merely striking off the mortar with a trowel to produce a flush joint (**Figure 9**). Recessed joints, providing a ledge for water, possible rain penetration and a risk of frost damage, are to be avoided. While many consider that brickwork can be made to look more attractive with recessed joints, their use is best confined to interior fair-faced work.

Attention should be given to what can be seen of the pointing, preferably at the highest level on the most exposed wall, perhaps at the side of a window. BRE suggests that pointing should last 30–40 years but, depending on the quality of the original mortar, the pointing profile adopted and the exposure conditions, much will last a lot longer. Nevertheless, the age of the dwelling will determine, in broad terms, when repointing is required. However, this is seldom carried out when it should be. Loose, crumbling or soft mortar high up on the most exposed wall is an indication that it really is required. The trouble is, of course, that lower down on the most exposed wall and on other less exposed walls, pointing is usually in better condition; accordingly, it all gets left unattended for many more years. This can be a false economy should

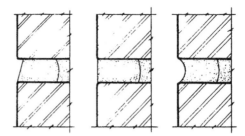

Figure 9 Profiles of three types of pointing: on the left, struck weathered; in the centre, flush; and on the right, bucket handle. Those to the left and right are much preferred as the tooling compresses the mortar to provide better adhesion to the bricks and closes up the pores on the surface.

rain penetrate solid brickwork and cause damage internally or, in the case of both solid and hollow walls, allow the bricks to become saturated with the danger of frost action and sulphate attack.

Repointing of brickwork on dwellings built before the early 1900s often ignores current advice to match the strength of the bricks and the strength of the lime mortar with which they were originally jointed. Many examples will be found of pointing carried out in a mortar based on too much Portland cement, based on the theory that it must be 'stronger'. In practice, it usually shrinks considerably, not only producing fine cracks on the surface which can allow water to penetrate but also shrinking away from the original jointing mortar and the bricks themselves. After a few years, it is often easy to lift sections of the pointing out of the joints, and if the joints have not been prepared properly – i.e. not raked out sufficiently and wetted – the pointing will often fall out of its own accord.

In careless or unskilled hands, the raking out of joints in preparation for repointing can cause much damage to the edges of the bricks. It should be done carefully by hand to a depth of 19mm to 20mm but all too often a mechanical disc grinder is used for speed, with dire results. In consequence, a much wider band of mortar has to be used for the repointing than would otherwise be the case with a detrimental effect on the appearance and poor adhesion to the bricks. Even when raking out is done by hand, carelessness can cause damage if the bricks are soft or unduly brittle. Brickwork that has been repointed many times in the past can often be seen to be affected in this way to a lesser or greater extent.

If bricks do become damaged in this way, tuck pointing might sometimes be adopted. This type of pointing was commonly used for new work in the eighteenth century and less frequently in Victorian dwellings when window openings and quoins were dressed with a softer, smoother red brick to contrast in colour with the rougher, less well shaped bricks used for the general walling. To match the finer jointing of the dressings, the rougher bricks would be flush jointed in a mortar to match the colour of the brick used and a groove left to be subsequently filled with a narrow projecting band of contrastingly

196 Recessed jointing to brickwork on a modern three-storey block of flats. The designer has taken a chance in that such jointing is not recommended except where the exposure category is sheltered; but here it is moderate. The use of a cavity wall will probably help to reduce the likelihood of rain penetration to the interior but the bricks in the outer leaf are likely to become saturated and subject to repeated freezing and thawing. Despite the known problems with this finish, it was a popular detail on developments and estates built towards the end of the last century. Already, at the base of the wall, there is evidence of the brickwork disintegrating because of saturation and consequential frost action due to external paving being taken above the damp-proof course level.

197 Disintegrating brickwork to a boundary wall. The usual cause of such severe deterioration is under-firing when the bricks were made. Initially, the soft inner material is hidden by the external surface, but if pointing is defective then dampness will penetrate at the top and failure of the brick face follows. Once the protective outer skin has gone, the problem is compounded by persistent dampness, and a boundary wall such as this will always be particularly vulnerable. If only a few bricks in a wall are deteriorating, individual replacements may be satisfactory; but if a significant number are affected, complete renewal will be required.

198 What happens to brickwork after many years of frequent saturation and subsequent freezing and thawing. Mortar stronger than the bricks often gets left behind.

199 Inadequately filled perpend joints to a 1960s cavity wall of sand-faced Fletton bricks, the surface of which can be scraped off. Bad workmanship will allow more water to penetrate to the cavity – perhaps seeking out flaws in the drainage channels around windows and doors – than would otherwise be the case if the joints had been properly filled. The effect would be much more serious if the wall had been solid.

201 Tuck pointing on yellow brickwork of the mid nineteenth century, used to disguise the damaged edges of the bricks and to produce a fine, precise appearance. The white 'tuck' contrasts with the base mortar, which matches and is flush with the face of the yellow bricks.

200 It is often surprising how much neglect a building will tolerate. This wall probably dates from the 1770s and it is unlikely that it has been repointed since it was built. The degree of erosion is clear at the corner and many of the bricks have lost their faces. But fortunately, most of the wall is in a somewhat better condition so it is still stable and essentially sound. Long-overdue repointing with a suitable mortar will prevent further deterioration and restore it for many more years.

203 Tuck pointing used to good effect on contrasting red and blue–black bricks of a Georgian house. The two colours of the base mortar, matching the respective bricks, can be picked out in places, especially where the white tuck pointing has worn and fallen away. Poorly finished repairs at the top of each window highlight the delicacy of the original work.

202 Three varieties of brick on an early eighteenth-century wall. On a pilaster: smooth, red, rubbed bricks with fine jointing on the right; ordinary, greyish brown stock bricks for general walling in the centre; and on the left, dressings of red bricks around a window (not visible in the photograph). The technique of tuck pointing for the dressings and general walling would have produced a more precise finish to match the rubbed bricks on the pilaster, but the slight setting back of the repointing is a reasonable alternative and more durable than a thick flush joint or the ribbon effect which a struck weathered joint would have produced.

coloured mortar – usually white but occasionally black (**201** and **203**). In repair work, this type of repointing can successfully disguise damage to the bricks caused by careless raking out. It is, however, not normally very durable and costs at least five to six times as much as ordinary repointing.

Wall thickness

When ascertaining the construction of the external walls, the inspector will ideally have taken a measurement of the thickness at each floor level and, if possible, on each elevation. Determining at the design stage what thickness of brickwork ought to be provided for an external wall of solid construction has always been governed by the ultimate height and, in more recent years, by the length of the wall. A series of additional design considerations, including wind loads, construction materials, the provision of lateral support and end restraints are now required by Approved Document A of the Building Regulations 2010 for England and Wales, and the Small Buildings Structural Guide 2010 for Scotland. For practical purposes during an inspection of an existing dwelling up of to three storeys, however, the minimum thicknesses outlined in **Table 3** can be used as a rule of thumb to see whether the thicknesses seem adequate should there be a doubt in the inspector's mind as to whether defects in the form of leans and bulges can be attributed to inadequate thickness for the wall's height or length or alternatively whether the defects seen are due to some other cause. Experience suggests the latter to be far more likely in the case of low-built dwellings.

In the absence of any visible defects in the form of leans and bulges – observable by looking along the wall from the ends, upwards and along both ways from the midpoint at the base, over the top of a parapet or by looking out of windows – there is usually no need to take other measurements in addition to thickness, which is checked against **Table 3**. Most often the thickness will be adequate; but on occasion, in times when the enforcement of regulations was less stringent, builders would cut corners for the sake of cheapness and walls will be found to be half a brick less in thickness than they should have been. The margin of safety incorporated in the regulations for the bricklayer's practice of the time was such that the gamble often paid off, but not always if other factors in the construction such as poor and/or insufficient mortar or soft and misshapen bricks came into play.

The regulations governing the thickness of solid brick walls have been refined from experience down the years. For example, the Building Act of 1774 categorised buildings into seven rates, four of which applied to dwellings, and prescribed the thickness of party and external walls for each. The categorisation took into account various factors including floor area, the number of

204 Areas of darker-coloured brickwork usually indicate where damp is concentrated more than elsewhere, and these are likely to be the areas where future problems from frost damage or sulphate attack will occur.

Table 3 Suitable thicknesses for domestic solid load-bearing brick walls of various heights and lengths

Height of wall	Length of wall	Minimum thickness of wall in brick lengths
Not exceeding 3.5m	Not exceeding 12m	1 brick (190mm) for the whole height
Exceeding 3.5m but not exceeding 9m	Not exceeding 9m	1 brick (190mm) for the whole height
	Exceeding 9m but not exceeding 12m	1½ brick (290mm) from the base for the height of one storey, 1 brick (190mm) for the rest of its height
Exceeding 9m but not exceeding 12m	Not exceeding 9m	1½ brick (290mm) from the base for the height of one storey, 1 brick (190mm) for the rest of its height
	Exceeding 9m but not exceeding 12m	1½ brick (290mm) from the base for the height of two storeys, 1 brick (190mm) for the rest of its height

Source: derived from the Building Regulations 2010, Approved Document A: Structure, incorporating 2013 amendments (Table 3).

storeys, the anticipated construction cost and the height above ground level. In some ways, the definitions were contradictory and open to various interpretations, resulting in some proposed dwellings being tagged with what could be considered an incorrect rate. It is said that a fourth-rate house could comply by its anticipated cost and floor area yet still be built higher and have more storeys than it should. As a result, some builders took advantage and built with walls of a lesser thickness than required. Some dwellings had to be demolished on becoming a danger but many still remain, often with parts rebuilt.

Contemporary pattern book illustrations show houses of various rates, but even when built in accordance with the Act, a third-rate house built after 1774 of basement and three upper storeys could be built with walls of one-and-a-half-brick thickness in the basement and a mere one-brick thickness for the three upper floors. Lacking restraint from floors but carrying a roof load, it is not surprising that the front walls of upper floors will often be seen to have been rebuilt; and this will be found to have been even more of a requirement for dwellings built in the hundred years or so before the 1774 Act was passed. If not actually rebuilt, the inadequacy of thickness and lack of suitable restraint also accounts for the numerous metal ties seen on many a dwelling of the eighteenth and nineteenth centuries.

As time went on, the proposed height and length became the sole governing factor for determining the thickness required for a dwelling's external and party walls, and reference to **Table 3** should be made if there are any doubts about adequacy. The third-rate house mentioned above should have had the first storey above basement level of one-and-a-half-brick thickness in addition to the basement but, even so, would also need to have been provided with sufficient restraint by attachment to roof and floors.

206 The exterior of one of the windows of a very early eighteenth-century four-storey dwelling. The brickwork is typical of the dark shade popular at the time but relieved by bright red dressings, and window frames are set flush with the face of the wall. Subsidence at the corner has caused the lower brick courses to be out of alignment and the sill to fracture.

205 The contrast between the thickness of the external envelope at windows and elsewhere can clearly be seen in this photograph taken in a Georgian dwelling where window frames are slightly recessed behind the outer face of the wall. The walls at lower levels of the building are even thicker and may be fully panelled internally. The inside of shutter boxes and the panels above, below and at the sides of windows must be inspected for any signs of rot.

207 There is clear evidence of substantial differential movement in the front wall of this early 1700s terraced dwelling. There is no obvious recent cracking externally and the upper windows have been 'squared up'. If internal floors and fittings have been properly made good and there are no signs of recent cracks in plaster or any other signs of continuing movement, it would be reasonable to conclude that the movement is historic. If the surveyor is not entirely satisfied in every respect then it would be necessary to advise further investigation to ascertain the cause of the movement and whether it is progressive.

Restraints and ties

A feature of speculative building of the eighteenth and nineteenth centuries for terrace housing was the practice of building all the party walls before the main front and rear walls. The time lag did not make a great deal of difference since most terraced dwellings of these periods were provided with basements dug out of the ground and foundations, such as they might be, were at a low level, which avoided many of the problems of differential settlement. Bonding between party and other walls was often rudimentary and, at times, a straight joint would be left relying on perhaps just strips of iron to hold the front and rear walls upright. This practice can be a further cause of bowing of brickwork as the strips fail to hold and pull out. Even if not unduly serious, the movement can produce vertical internal cracks to the plasterwork in the corners of rooms and is probably the reason why it was found necessary to provide restraint to the front wall on the party wall line of the terrace in **209**.

Current requirements for dwellings lay considerable stress on the need for restraint to be provided for external walls from both the roof and floors, an aspect to which insufficient attention appears to have been paid before the 1960s. Building the ends of floor joists into external walls provided a degree of restraint to those walls, just as long as other defects of a more serious nature did not cause the joists to pull away, since there was usually no fixing other than the occasional nailing to a wall plate (**Figure 10**). Those walls parallel to the way floor joists spanned, however, were usually left without any restraint whatever, with a consequential risk of leaning or bowing outwards. Even more serious consequences arose from situations where staircases were placed against a substantial area of an external wall where there was no balancing thrust from an adjoining dwelling (**Figure 11** and **Figure 12**; **210**). The end flank walls of terraced dwellings are particularly prone to bulging, particularly on a down-hill slope, either when staircases are positioned against them or when floor joists span from front to rear. The defect is also not uncommon with semi-detached dwellings, although not usually on such a dramatic scale.

Gabled flank walls can also be subject to leaning outwards, often necessitating rebuilding due to a lack of restraint. Floor joists below may be spanning from front to rear, as will the ceiling joists acting as ties to the feet of the rafters, leaving the gable wall virtually free-standing. Of course, if the ceiling joists are not acting as proper ties through a lack of nailing, there could be roof thrust at the top of the front and rear walls as well, causing them to lean out in addition – an unsatisfactory situation to say the least, likely to involve much work of correction.

In comparatively rare cases, instead of the floor load acting vertically downwards as it should in normal circumstances, light construction and overloading can cause severe deflection in a floor and exert an eccentric sideways thrust on a wall, causing a bow to develop. A similar problem could sometimes affect interwar semi-detached houses due to the presence of a tall chimney stack above the flank wall (**Figure 13**).

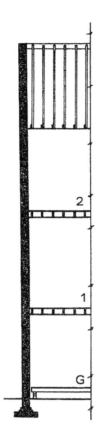

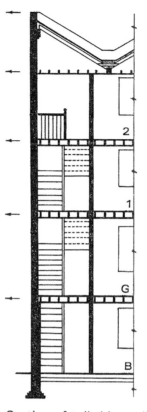

**Section of tall, thin wall
with no restraint**

'A'

Figure 10 (*above*) Expecting a one-brick-thick wall to do too much, as here, where no restraint is available from the roof or the floor joists which are spanning from front to rear with a load-bearing partition for intermediate support, will almost certainly induce a substantial lean. Such a fault is not uncommon in dwellings of the mid to late nineteenth and early twentieth centuries.

Figure 11 (*top right and right*) The effect of positioning the staircase against the flank wall of the end house of a terrace and supporting floor joists on a central partition leaves the flank wall with virtually no restraint, as here shown on section 'A' and the key plan. The butterfly roof provides no restraint and probably aggravates the situation. The consequences are often a substantial lean developing in the flank wall and long vertical cracks at the junction with the front and rear walls ('C' and 'D' on the drawing) widening at the top. This is not an uncommon defect in houses built during the eighteenth and nineteenth centuries, not only at the ends of terraces but also later when semi-detached dwellings became popular.

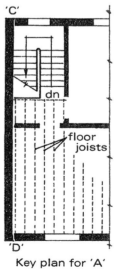

**Key plan for 'A'
Top floor**

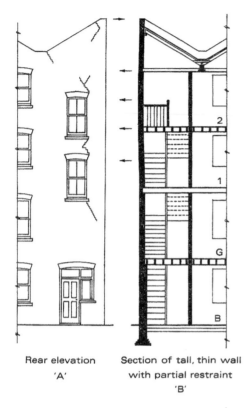

Rear elevation
'A'

Section of tall, thin wall
with partial restraint
'B'

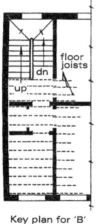

Key plan for 'B'
First floor

Figure 12 Where the staircase is positioned against the flank wall of the end house of a multi-storey terrace but floor joists are alternately running from front to back or from side to side as here on section 'B' and the key plan, movement in the flank wall can be complicated. If the bond of flank wall to front and rear walls is sound then cracks in the rear wall may be induced at the weakest points, namely the window openings, as panels of brickwork pull away, as shown on the rear elevation 'A'. Displacement of the flank wall in this instance may take the form of a horizontal bow as shown on the key plan.

Most of the taller terraced dwellings of the eighteenth and nineteenth centuries were provided with heavy ornamental cornices near the top of the front wall, with only a comparatively lightweight parapet to hold down the back edge. If there is no restraint from the roof on the front wall, there is a tendency for the cornice and the brickwork of the top storey and parapet to lean outwards (**Figure 14**). This could be dangerous if steps are not taken either to rebuild or tie back the feature, and construction detailing could even make matters worse (**Figure 15**). Such defects will be found in dwellings where there is little or no bond between the front and the party or separating walls. The inspector should look for vertical cracks at the junction of these walls, particularly where there is evidence of earlier making good which has subsequently reopened, indicating the likelihood of further movement.

The problems of inadequate wall thickness, lack of restraint and poor construction in the speculative building of terrace housing in the eighteenth, nineteenth and the early part of the twentieth centuries was often dealt with either by rebuilding or the provision of metal tie rods. These would be secured to disc plates on the face of the wall and fixed to a structurally sound feature internally or through to a wall, via the floors, on the other side of the dwelling that was moving in the opposite direction (**212**). Nevertheless, there are many tall terraced dwellings existing where recourse to either of these procedures has not been considered necessary, yet walls are clearly leaning outwards or bulged.

Figure 13 *(right)* The spread of the private sector semi-detached two-storey house in the period 1920 to 1940 with its typically one-brick-thick walls, bay window and hipped roof slope also saw the introduction of the small solid fuel boiler to produce hot water positioned in the kitchen at the rear. Flue linings were not considered necessary and the result was often a tall leaning boiler stack due to sulphate attack. If this was combined with a lack of restraint to the flank wall, often the case, an unhappy effect could be produced as the latter leant out and the stack leaned inwards. Many stacks have had to be rebuilt with a lining although the advent of balanced flue boilers has meant that many have since been taken down to roof level. Should this be the case, the inspector needs to check that ventilation to the flue has been maintained.

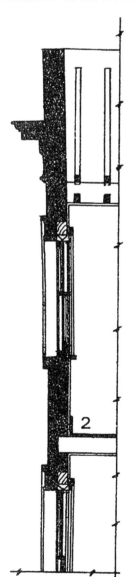

Figure 14 *(left)* Heavy ornamental cornices of stone or made up from oversailing bricks, tiles and cement were a feature of dwellings of the eighteenth and early nineteenth centuries, particularly after 1707 when wooden cornices were banned by statute because of danger from the spread of fire. If the top of the front wall is not weighed down by the roof, as here, there is a tendency for the upper part of the wall to lean outwards, often necessitating rebuilding and tying back.

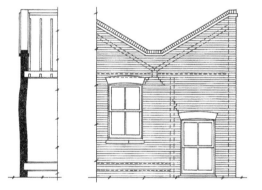

Figure 15 *(above)* Point loads on walls were not always adequately spread over the structure below by stone, or nowadays by concrete padstones. A case in point is at the ends of the beam carrying the valley gutter and the feet of the rafters forming the butterfly roof over many of the three- to five-storey terraced dwellings of the nineteenth century. One-quarter of the total loading from the roof is applied as a point load in the centre of both front and rear walls. At the front, this can often aggravate the situation shown on **Figure 14** by adding to the pressure pushing the wall outwards. At the rear, usually because of the presence of a staircase, the end of the beam may be exerting a point load quite near to the head of a window, causing fractures. If the bond to party walls either side is reasonably sound then a bulge could be induced in the pier between windows, as above on the left.

208 It is not often that the inspector will see such a clear indication of the front wall of a terraced dwelling leaning outwards. The front wall of the house in the centre, dating from around 1820, has been completely rebuilt from parapet down to ground level, leaving the front walls of both adjoining buildings projecting outwards. That on the left is more pronounced and clearly visible from the vertical shadow cast by the lean. The amount of out of plumb varies but it is at least 100mm at its worst. A surveyor on a visual inspection of the houses on either side, and possibly of any other house in the terrace where at least the top of the front wall has not been rebuilt, would need to advise the necessity of further investigation to establish whether it is safe to leave such a lean uncorrected. Regular monitoring to detect whether there is any further movement may be necessary.

Such dwellings have, in effect, reached a new position of repose; given continued domestic use in a quiet location, free of undue vibration or any undermining and avoiding major alteration, either externally or internally, these buildings will to all intents and purposes remain in repose for many years to come. An inspector can be more assured of this if decorations have not been renewed for some years and the signs of previous making good to cracks remains undisturbed other than perhaps from minor signs of shrinkage in the form of hairline cracks. This is so whether bulged or leaning walls have been provided with tie bars and disc plates or not and even whether adjoining walls of another terraced dwelling have been rebuilt vertically.

It is not thought that many of the restraining features such as disc plates, flat bar crosses or S-shaped braces on many older dwellings are really doing anything very much at all. In contrast with long channel sections that have been suitably packed out, they are often inadequate for the purpose, being too small and too far apart in position; and if movement had been continuing, the plates or bars would have been left behind in the brickwork as the remainder of the wall continued to move outwards around them. The explanation for the apparent success of many of the restraining features is that they were an unnecessary provision in the first place, the wall in question already having reached a new position of repose and not requiring restraint.

Faced with the situation outlined above, there is no reason why, following a visual inspection, the surveyor should not come to the conclusion – based on the age of dwelling, quality of brickwork, degree of lean or bulge – if this can be ascertained with safety by looking out of windows – oldish decorations and absence of cracks – that the dwelling has reached a new position of repose and that the movement is, therefore, of long standing and structurally acceptable and that further movement of a progressive nature is considered unlikely. The evidence and conclusion can then be reported to the client, but there should also be a warning which stresses the need for the absolute avoidance of major disturbance to the present condition unless proper advice is taken. All too often, if such a warning is not included, a buyer will assume that all is well and that builders and interior decorator can be brought in to start pulling the dwelling apart without professional advice. When things go wrong, the aggrieved buyer, looking around for someone to pin the extra costs on, will complain that there should have been warnings in the report.

Dwellings are sometimes redecorated before being put on the market for resale, not necessarily to deceive unwitting buyers. If the inspector encounters new decorations in a dwelling where there is visible evidence of leans and bulges, irrespective of whether any means of restraint have been provided, a visual inspection cannot provide the same level of assurance as can old decorations and sound making good. Reasonable evidence that movement is not continuing is lacking and, in the circumstances, it is probably necessary to advise the need for further investigation, possibly to include monitoring for a period of time. Advice of this nature will be considered unfortunate by both seller and buyer. The former might conceivably be able to produce documents in the form of earlier reports and details of work done. Whether these will be sufficient for the surveyor to conclude that they amount to adequate evidence to show that movement is not progressive will depend on the circumstances of each case.

Of course, in the case of older decorations, where the making good of cracks provides the surveyor with the evidence that the movement is

209 It is possible to appreciate an owner's reluctance to rebuild a front wall, especially as, in the case of this early eighteenth-century terrace house, much character including the wood cornice and door casing could be lost in the process. However, restraint of movement in this ugly and heavy-handed way does cast doubt on structural stability and must affect value to an extent even if, from appearance, it is doing its job.

progressive and of a continuing nature, the report can be more succinct. Considerable expense is likely to be incurred and any idea of the amount involved would need to await further investigation.

The advice to be given in the case of leans and bulges, referring to overall defects in brick walls which can be identified as originating above ground level, also applies to more localised defects. It may also be applicable to defects presenting similar characteristics but originating below ground level. The latter, however, are more often likely to be of a progressive nature involving diagonal slanting cracks, probably extending to ground level (as considered in Chapter 8).

210 The top floors of an end terrace property with basement and four upper floors where the entire flank wall has been rebuilt and is vertical, but where all still does not seem well.

211 The need for restraint is not limited to speculative housing. This stone-built Tudor building has been provided with metal ties to restrain further outward movement. Random rubble walls such as this may be liable to separation of the inner and outer faces if not properly bonded. A check on whether the ties are continuing to function or failing to hold the wall in position needs to be made by examining the adjacent external and internal areas for any indications of continuing movement.

212 Flank walls of the end houses of terraces are notorious for movement because they lack the support of an adjoining property. The flank wall of this late nineteenth-century terraced dwelling is bulged at top-floor level. Other dwellings in the terrace have butterfly roofs (of the type shown at **Figures 11**, **12** and **15**), which are prone to causing a lean or bulge in the flank walls of the end houses. Here, the original roof has been replaced with a new structure to provide additional accommodation, stacks have been built higher and the flank wall has been tied in with steel channels and disc plates – a belt and braces approach. Internal inspection should reveal whether progressive movement has been restrained now that the original cause has been removed.

Beams and lintels

Towards the end of the nineteenth and into the early part of the twentieth centuries, the smaller terraced dwellings tended to have splayed bay windows. These were often limited to the ground floor only, unlike the bow windows of the early nineteenth century, which like the later splayed bays on larger dwellings built either side of the 1850s, were carried through at least one more floor and sometimes to the top of the dwelling. Wherever a bay or bow window finishes, there has to be a beam to carry the walling above, and this is often found to be a source of weakness. Either the beam, invariably of timber, is inadequate for its purpose or neglect of the condition of the flashing, sometimes only a cement fillet, at the junction of the roof and the front wall allows dampness to penetrate and causes the beam to weaken through either wet or dry rot. The consequences would be a settling in the wall above, accompanied by distortion to window openings at a higher level and to sills (**Figure 16** and **217**). Depending on which way the joists are running, there could be movement in the floor above and possibly dry rot.

Of course, such faults and defects are by no means limited to beams providing support over the openings for bay windows. Every opening needs a beam to carry the walling above. While inadequacy is often compensated for, to an extent, by strength in the window frame itself, in the case of timber

Figure 16 Inadequacy in the timber beam supporting the wall above a bay window or neglect of the flashing at the joint of the roof to the bay window, whether pitched or flat, allowing dampness to penetrate and rot to develop is a fault and defect frequently found in the cheaper two-storey terraced dwellings of the late 1800s and early 1900s. The walling settles, windows are distorted, sills slope downwards and a crack appears in the ceiling of the room below. If there is no evidence of effective repairs having been carried out, the inspector needs to advise that the beam be exposed to check for rot.

214 A lead flashing has been inserted above the natural stone lintel of this Tudor window to prevent damp penetration to the interior. Even where damp penetration has been successfully stopped, the inspector will still want reassurance that any adjacent timberwork has not been adversely affected or that any remedial work was properly completed. If vulnerable areas cannot be seen and good evidence of appropriate work is absent, an inspector would need to advise further investigation.

213 Timber built into the external walls of dwellings of the eighteenth and nineteenth centuries as lintels and for bonding purposes causes much disruption to the brickwork when it becomes defective through damp penetration. The problems here are compounded by the soft and crumbling brickwork, unsympathetically renewed in places. A surveyor's facile advice to rebuild would be unwelcome since a buyer would no doubt be seeking charm and the cachet of living in a dwelling nearly 300 years old. Permission to do so would, in any event, be difficult to obtain unless potential danger could be demonstrated. A buyer needs a well nigh bottomless purse before embarking on the ownership of a dwelling of an age and type such as this, a point on which they sometimes need reminding if they are not themselves knowledgeable. Careful restoration seeking out suitable matching material but retaining as much of the original as possible would be the requirement of the conservationists – work which is always expensive.

lintels, these are often positioned within a half-brick's thickness of the outer face of the wall and in consequence, over time, are damaged by penetrating damp. This can induce wet or dry rot and ultimately the beam ceases to be effective in supporting the walling above, reflected to begin with by disturbance and movement in any brick arch.

Concrete lintels, much more likely to have been used in dwellings from the 1930s onwards, are not prone to such defects but do sometimes expand slightly along their length, causing cracks to appear in plasterwork at either end. This is a 'once and for all' defect, not to reappear if adequate making good is carried out. A more significant failing of some concrete lintels is when there is inadequate cover to the reinforcing bars which they contain. If the reinforcement is too near the external surface of the lintel, it will be vulnerable to dampness. The rusting that inevitably follows

215 Looking upwards and along brick walls will detect bulged, leaning, bowing or uneven areas. At times, viewing the property from a distance, perhaps with binoculars, can help to highlight such areas, especially when there is a direct comparison with the sills, lintels and brick courses of adjacent properties, as here.

216 Examining the sides of windows where the frames are set back and in recesses will often indicate that brickwork leans or that there is a bulge. In this example, the brickwork is nearly 200 years old and significant bowing away from the window has taken place. The resulting gap has been filled with mortar but there are signs of slight further cracking. The surveyor will need to decide whether this is simply due to shrinkage of mortar added many years ago, and therefore less of a concern, or to continuing movement which requires immediate investigation and remedial work. Even if the surveyor judges the wall to be stable, a recommendation for annual monitoring would be prudent in view of the amount of movement that has taken place.

217 The single-storey bay window was a popular feature of the smaller late nineteenth-century and early twentieth-century terraced dwelling. Lack of a proper flashing or poor maintenance can allow damp to penetrate at the junction of the front wall and bay roof. Flat roofs behind a parapet, such as this ornate example, are particularly vulnerable if not properly maintained. If any defects are not attended to promptly then damp penetration can cause rot in the beam holding up the front wall over the bay, causing a crack across the ceiling internally. Careful examination of the roof and flashings, the internal ceiling decoration and the sound or resonance of the floor above, together with any change in level, will be of help in deciding whether the matter has been satisfactorily dealt with in the past. If there is any doubt, the surveyor must recommend exposure works to allow a full examination.

218 A surveyor instructed to inspect either of these three-storey semi-detached houses of the mid nineteenth century will have food for thought. The front wall of the top storey of the house on the left has clearly been rebuilt, leaving the one on the right with its front wall at the same level still leaning out by at least 50mm in front of the rebuilt wall for a height of about 800mm from the top. By the line of rebuilding visible on the left-hand property, the maximum amount of lean must have been at the centre between the two windows, not on the line of the party wall as might have been expected. The question arises as to why the rebuilding was necessary. Parapet gutters, if neglected or even if badly laid out, are frequently a source of damp penetration affecting the feet of roof timbers, the ends of joists and sometimes the lintels as well. In the case of the house on the right, an inspector would need to consider whether, in the light of internal evidence, it is necessary to advise opening up to assess the true condition of hidden timbers and whether or not the front wall should also be rebuilt if movement continues.

will cause disruption and cracking to the outer face of the lintel, which will in turn expose the metal reinforcing to even greater moisture; a vicious circle is established that cannot be cured and will only end when the lintel is replaced.

Metal lintels in interwar or immediate post-war housing – fortunately less so with more modern buildings with higher specifications – can also suffer deterioration and rust expansion, leading to disruption of the surrounding brickwork in some cases; and once again, the only remedy is complete replacement. Brickwork above door and window openings which shows signs of rebuilding or slight colour mismatch with the surrounding walls may betray that this replacement work has already been carried out at some earlier time, and enquiries should be made to determine exactly what has been done.

12

Main walls – natural stone, clay and chalk construction

Types of building stone

Faced with a dwelling where it has been established that natural stone is visible, how is the surveyor to ascertain which type of stone has been used? Furthermore, is it likely to provide a new owner with trouble-free, and hopefully handsome, walling given proper maintenance over the years to come? The surveyor's local knowledge is vital towards providing an answer to these questions, but it is worthwhile taking a moment to consider the different types of rock that may be encountered during an inspection. Understanding their origins will assist in appreciating why building stones respond in different ways when exposed to the elements and the stresses of being incorporated in man-made structures.

There are three basic types of rock, defined by the way each came into being. Igneous rocks were formed when molten rock solidified. Some igneous rocks, such as granite, may contain large crystals but they have no 'bedding plane' or intrinsic natural alignment within the rocks. The second type, sedimentary rocks, come in two main groups for construction purposes. The first, sandstones, derive from small particles of eroded rock such as sand and grit which have accumulated and been compressed and cemented together. This process often, but not always, produces a distinct internal alignment – the bedding plane – which dictates how the stone should be used. Clay can be regarded as a precursor to sandstone in that it contains very small sandy particles which have been compressed but which have not yet been cemented together to create a solid rock. The second group of sedimentary rocks derives from the remains of marine organisms, whose shells or fragments of shells and corals have created limestones which often contains fossils. Limestones may also have bedding planes. Chalk is a soft form of limestone. The third rock type was formed when sedimentary rocks were exposed to extreme heat or pressure and transformed, or metamorphosed, into metamorphic rocks. In this way, limestones have been metamorphosed into marble and fine sandstones into slate.

This is a very simplistic overview and there are many variations within each of these rock types, each with its own distinct characteristics. Not all stone has similar properties, and while stone in general is a highly durable material, it is a natural material, so there are variations not only between different sources of a single rock but even within a single quarry. Factors such as poor stone selection, poor design of details, inadequate maintenance or poorly executed repairs can accelerate the rate of deterioration to stonework. Not appreciating the characteristics of types of stone can lead to problems whereas understanding these characteristics can assist analysis and interpretation; for example, determining that all limestones are vulnerable to acidic attack or that the different chemical makeup of sandstones and limestones means that they should not be used together in the same building.

If a dwelling's stonework is in reasonably good condition and the dwelling is of some age, say more than 40 to 50 years old, it might be thought that there is not much point in saying more, leaving it to an assumption that such circumstances will continue to be the case. However, the surveyor ought to be able to do more than this even if comparatively new to an area and lacking that local knowledge which surveyors ought to cultivate. Many do have local knowledge, of course, and this enables them to recognise on sight a type of stone particular to the local area. Area-based geological, archaeological and historical societies can be extremely useful contacts for the surveyor, and the interest of local historians is often expressed in papers and pamphlets obtainable from local libraries. In this respect, it may be that some stones are known locally by different names, sometimes reflecting their characteristics.

One of the attributes desirable in surveyors (mentioned in Chapter 2) is an interest in the older maps of the local area. These can provide information about what was on the site before construction of the dwelling being inspected, which could have a bearing on possible foundation problems. It may well be that these maps also show a quarry or, more likely, the site of a quarry. The inspector may know from personal experience over the years that such a site is a local hazard, because of steep cliffs and flooded pits, or that a site has become a local beauty spot as vegetation has grown back. Ordnance Survey 1:50,000 maps note quarries while the 1:25,000 maps can differentiate between a gravel pit, a sand pit, another pit or quarry, and a refuse or slag heap.

Construction with stone

Knowing or assessing the age of the dwelling can be a great help towards eliminating certain forms of construction when contemplating elevations faced with real stone. A dwelling built since the 1960s, if presenting the appearance of natural stone blocks laid in regular courses, whatever the finish given to each block, will hardly ever be of other than cavity construction. Real stone rubble walling is just not feasible in an outer leaf of 100mm to 120mm, and

for normal domestic purposes, it would only be in very exceptional circumstances that an outer leaf of greater thickness would be entertained because of
the expense. Matching a new extension to an existing much-loved and valued
dwelling in an area of severe or very severe exposure might be one such case.
Natural stone sawn and laid in courses in the outer leaf of a cavity wall will,
excepting intrinsic defects relating to the type of stone used, be subject to the
same structural conditions and potential problems as bricks or concrete blocks
used in a similar position. These have been discussed in relation to cavity
construction where the outer leaf is made up of bricks and need not be repeated
here (see Chapter 10). A dwelling of real stone constructed prior to the 1960s
will, on the other hand, be in one or other of the forms of solid construction
of long tradition, determined to some extent by the type of stone available
but much more so by the amount of money available for its working.

Using stone from as near the surface as possible and from the local quarry
produced the type of walling generally used over the centuries for the smaller
dwelling in rural, village and small-town situations, particularly in areas where
there was little timber or brickearth available. Rubble walling is the term which
is applied to many different types of masonry using the stones more or less as
quarried. Although this may look as though it has been put together somewhat
haphazardly, it should nevertheless, for stability, be built to follow certain basic
principles.

Using stones in a wall more or less as they come from the quarry,
removing only perhaps a few of the more inconvenient corners, produces the
appearance of walls shown in **Figure 17**. 'Uncoursed random rubble' (shown
on the left) needs to be laid, however, in order to maintain a bond both across
the thickness and along the length of the wall; otherwise, the stones will just
fall apart (**219**). Across the wall, the bond is maintained by the use of 'bonders'
– longer stones extending into at least two-thirds of the thickness of the wall
with not less than one for each square metre of wall on face. Viewing an
existing wall, it is not possible to tell which stones are bonders, but if a wall
is unduly bulged then an inadequate use of bonders is a possible cause. If so,
rebuilding would probably be required. Should bonders be allowed to project
right across the full thickness of the wall, it could be the cause of isolated damp
patches appearing internally.

More selection of the stones by the mason can produce a wall with
sections in more apparent order – 'brought to courses' (as shown in the centre
of **Figure 17** and in **220**). Walls built to courses are always stronger; here,
the stones are levelled up to courses at intervals varying from 600mm to 900mm
in height according to the type of stone and, often, to the custom of the locality,
the heights usually corresponding to those of squared stones used at quoins
and jambs. The same constructional principles for stability apply as for uncoursed random rubble, as they do also in the case of walls where the stones
are arranged according to height to form continuous courses – 'coursed random
rubble' (as shown at **221** and on the right of **Figure 17**). Here, again, the
height of the courses will probably be arranged to correspond with the stones

Figure 17 Types of rubble walling in elevation – the stones used are as taken from the quarry with a minimum of working: 'random rubble' on the left, 'random rubble brought to courses' in the centre and 'coursed random rubble' on the right, reflecting the amount of work by the mason on arranging the stones.

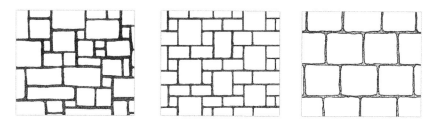

Figure 18 Types of squared rubble in elevation, the stones roughly squared before the mason sets them out in the wall: 'uncoursed squared rubble' on the left, 'snecked squared rubble' in the centre and 'coursed squared rubble' on the right.

available for use at quoins and jambs, resulting in courses alternately deep and shallow. More work by the mason on the stones themselves to make them roughly square can produce walls along the lines of those shown in **Figure 18**, progressing from the left according to how much arranging the mason carries out on laying. The least amount produces 'uncoursed squared rubble' (**222**). The stones are laid as 'risers' and 'stretchers' with, in general, the risers not being more than 250mm in height and adjacent stretchers not exceeding two-thirds of that height in length. Without those limitations on the size and proportion of the stones but introducing of a definite ratio of comparatively small stones not less than 50mm square – called 'snecks' – to prevent the occurrence of long, continuous joints likely to have a detrimental affect on stability produces the type of wall shown in the centre of **Figure 18** – known, not surprisingly perhaps, as 'snecked squared rubble' (**223**).

'Coursed squared rubble' as shown on the right of **Figure 18**, involves laying the stones rather like stretcher bond in brickwork in long, continuous courses using stones roughly squared to the same height and breaking joint from one course to the next (**224**). The courses can be of various depths ranging from 100mm to 300mm but usually averaging out at around 250mm; and the stones can be dressed, if required, to provide different types of finish – split or rock faced, for example.

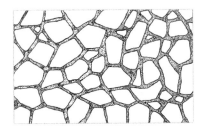

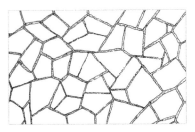

Figure 19 For stones that are hard and come out of the quarry in irregular blocks, polygonal shapes can be adopted for rubble walling. A minimal amount of work on the stones and wide joints produces 'rough-picked polygonal walling', as shown here in elevation.

Figure 20 Stones that, although irregular, are more positively shaped and can be arranged to fit together with comparatively narrow joints produce 'close-picked polygonal walling' as shown here in elevation.

Harder stones are more suitable for rubble walling and, in some cases, can be roughly worked into irregular polygonal shapes. They can then be bedded to show the face joints running in all directions as another form of rubble walling. In its basic form, where the stones are only roughly shaped and fitted with comparatively wide joints, it is known as 'rough-picked polygonal' walling (as shown in **Figure 19** and **225**). The stones can be more precisely shaped and the face edges more carefully formed so that the stones fit close together and then it is called 'close-picked polygonal' walling (as at **Figure 20**). Kentish Rag is a type of limestone that is particularly hard and difficult to work, frequently coming out of the quarry in irregular blocks, and is often used for this type of walling; dry stone walls in this form are common in many parts of the country (**226**).

Another form of rubble walling is produced by the use of flints or cobbles. The flints and cobbles are rounded nodules of silica, either dug out from gravel beds associated with layers of chalk and limestone or taken from the beach (**227**). Cobbles from the beach are sometimes set into panels in walls of other material and appear in dwellings near the coast, but more often the flints used are dug from the ground. While extremely hard and virtually indestructible, flints are brittle and can be snapped roughly in half to expose the shiny inner face as against the 'rind' of chalk that usually encloses the nodule.

Flint walling was extensively and continuously used in East Anglia and in southern and south-east England from Roman times; but because of their comparatively small size, unwieldy shape and impervious nature, bonding of flint walling can prove difficult. Traditionally, flint walls would be built in two parallel leaves between shuttering with a hearting of flint rubble and mortar and the longest flints used as bonders projecting into the hearting. As sufficiently long flints were scarce, more often the flints – either as dug or snapped – would be set in panels and backed and laced with brick or stonework, which would also form the quoins and the jambs around window and door openings. Whole flints, uncoursed and used in this way, are shown in **Figure 21** and **229** while snapped and squared flints, termed 'knapped', laid in courses and

Figure 21 Section and elevation of a brick wall with a panel of uncoursed flints, as dug, and with lacings and dressings of brick.

Figure 22 Section and elevation of a knapped, squared and coursed flint wall with a band course and dressings of cut and sawn stone.

in conjunction with stone dressings are shown at **Figure 22** and in **230**. Even when used in panels, flints can become detached from their backing and, although there may be no signs of bulging, it is always essential to carry out a sounding by gentle tapping over the whole of the surface to detect any hollow areas which could be a potential hazard by way of collapse and possible injury.

The finer-grained, more compact varieties of both limestone and sandstone, the two most commonly used building stones in the United Kingdom, are both capable of being sawn square and finely dressed to form blocks which can be laid in regular courses with the narrowest of joints; this is known as 'ashlar' (**231**). In the case of most limestones, the softer and more easily worked are also capable of being carved and chiselled into comparatively complicated shapes for cornices, architraves, columns, string courses and the like. With sandstones, this is not always possible or even desirable considering the exposure conditions in the areas where most of them come from or the incidence of pollution in the case of the more intractable of sandstones. For dwellings of good quality, ashlar would be the expected finish for owners with the wherewithal to afford it when building for themselves; this is also the case for developers building for the upper end of the market, the cities of Bath and Edinburgh having prime examples of such developments.

The surveyor will not be able to tell from the exterior whether the ashlar is part of a wall wholly of stone, probably with a hearting of rubble and mortar, or whether it is a facing skin blocked on to a backing of brick. From the interior, there might be a sight of the unplastered inner face in the roof space or a cellar, which could confirm a viewpoint either way. Wholly stone construction is perhaps more likely in dwellings built before the coming of the railways in areas where the availability of brick was limited. In the absence of any pronounced defects, it is perhaps a somewhat academic point to seek to differentiate. Overall in purely structural terms, it is probably better for an ashlar-faced dwelling to be built of brick with a facing of stone. Brick walls do not have the disadvantage of walls built wholly of stone where the mortar of the hearting crumbles over the years and drops down as dust with the rubble to the bottom of the area behind the stonework, often affecting stability because the outer leaves are deprived of their bond.

Apart from incorrect bedding of the more pronounced sedimentary limestones and sandstones, which is a frequent source of spalling on the face of blocks of stone, another problem with ashlar was that with the more easily worked stones, there was always a temptation to reduce the width of the joints to such an extent that there was hardly any mortar to hold them together, thus possibly depriving the outer facing skin of a proper bond. To overcome this problem, 'hollow' bedding was sometimes resorted to. However, this left such a narrow strip of stone near the face to form the thin joint to the stone below that when the mortar in the hollow was compressed as the structure settled, the narrow strip could suffer a compression fracture.

219 Uncoursed random rubble using gritstone.

220 Random rubble brought to courses, using sandstone with the bedding planes of individual stones clearly visible.

221 Coursed random rubble in limestone.

222 Uncoursed squared rubble in a modern sandstone wall.

223 Snecked squared rubble.

224 Coursed squared rubble.

225 Rough-picked polygonal walling – mainly limestone blocks but with some individual bricks and lumps of flint.

226 Dry stone walling using limestone blocks.

227 Cobbles from the nearby beach roughly arranged in regular courses.

228 These cobbles in the wall of a modern development are intended to replicate local vernacular construction. They might deceive the casual observer but it doesn't take much examination for the surveyor to see the regular horizontal and vertical joints of the rectangular concrete blocks in which they were cast to provide a decorative finish.

229 Uncoursed knapped flint walling in a property dating from around 1780.

230 Knapped and squared flints, some squared to a better standard than others, laid to irregular courses in a chequerboard pattern with limestone blocks.

231 Ashlar consists of cutting and sawing relatively soft and easily worked stone into precise rectangular and accurately dimensioned blocks and laying them in courses with minimum-width joints to produce, as here, a flat, uniform surface. Carried to extremes, this often meant that stability could be impaired. To overcome the problem, hollow bedding would be used whereby mortar would be spread into a hollow in the base of each stone. As this compressed over time, the stone on the outer edge of the joint would fracture.

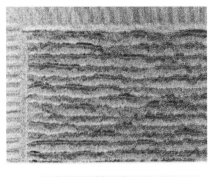

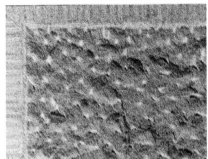

232 Apart from the usual sawn and rubbed finishes for ashlar, the stone is often dressed to a more elaborate finish to highlight certain features of a façade. Those shown here are 'broached' at the top and 'punched' below, both blocks having 'drafted' margins.

233 The dressed stone here has a 'droved' finish at the top while a 'herringbone' furrowed pattern tooling is shown at the bottom.

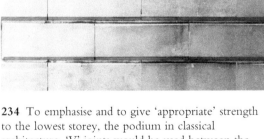

234 To emphasise and to give 'appropriate' strength to the lowest storey, the podium in classical architecture, 'V' joints would be used between the blocks of stone for ashlar to produce a pronounced pattern.

235 The importation of Caen stone from Normandy by sea to south-east England was comparatively easy compared with transporting the home-grown product by land. It is perhaps surprising that there are not more signs of erosion on the Caen stone of this two-storey dwelling, built over 400 years ago, though the location is clear of pollution, which has been the undoing of much of this type of stone elsewhere.

236 The incised stonework above an imposing doorway is only just over 100 years old, despite its classical style. The subtle difference between the dressing to the voussoirs here and those around the door opening on **327** (Chapter 16) can, with interest, be noted. The one here is 'vermiculated' – dressing with sinkings to form a winding, snake-like (verminous), continuous ridge – as against the 'reticulated' ridges on **327** – linked to form a network of irregularly shaped sinkings (reticules).

Some common problems

Where details such as copings to balustrades and cornices were provided with the thinnest of joints so as to appear in one continuous length, metal cramps were often used to hold the sections together. If made, as they usually were, of wrought iron then they rust over the years and in doing so expand considerably, invariably fracturing the stonework in the process. This is more likely to be found in connection with dwellings built of limestone. Sandstones, being generally coarser and less easy to work, would be provided with thicker joints with more mortar; but even here, for balustrades where there was a danger of them being pushed over, metal cramps would be used.

In general, limestones do not stand up to the weather – with its penetrating wind-driven rain and possible subsequent frost effect – to quite the same extent as most sandstones; and when exposure to the elements is aggravated by the presence of sulphurous compounds through pollution then sandstones win hands down. In the past, such pollution came about through the burning of coal for both manufacturing purposes and domestic use. Manufacturing was more extensive in the north where coal was more readily available, and it is perhaps fortunate that most sandstones are found in that part of the country while most limestones occur in the south.

The absence of smoke in the atmosphere should not, however, fool anyone into thinking that there is now less pollution to affect stonework. There is, in fact, more than ever since the burning of gas and oil for industrial processes and the production of electricity, coupled with the exhaust fumes from the internal combustion engines of motor vehicles, release vast quantities of sulphur dioxide into the atmosphere. This falls as dilute sulphuric acid in rain, and while the element of soot to blacken surfaces may now be much less, the acid

237 When damp from the ground brings up salts to crystallise behind the face of a stone that is incorrectly bedded, it is not surprising that it starts to spall.

238 At the time of cutting and sawing to prescribed sizes and regular shapes, it is not always very apparent where the stones' natural beds lie. This does, however, become apparent when, years later, damp penetrates over a period with its damaging acidic content – as it does to quite a considerable degree with rather porous stone as used here – and then freezes, blowing off the face of the stone.

239 Facing, as it does, into the teeth of the south-west prevailing winds, even more care than usual should have been taken to bed this stone correctly. Continual splashing from below added to wind-driven rain has fairly quickly revealed that a mistake was made.

240 Fractured and spalling stonework to a balustrade. It may be that the type of stone used here was too coarse-grained to be successful as sharp arrises and turned mouldings were required.

241 Metal cramps are often used to hold sections of stone together. If of wrought iron and they rust, the expansion of the metal is usually sufficient to fracture the stone. Here, the copper cramp is so much stronger than the stone itself that when structural movement has forced apart the two sections, the cramp has been left behind with the fractured sections of coping attached.

242 Weather-worn limestone after 90-odd years of exposure, revealing the fossil elements which are more resistant to the weather than the main body of the stone.

244 Erosion and spalling to the elaborate four-flue chimney stack on a Gothic revival dwelling.

243 The surveyor would be aware from local knowledge that this five-storey attached house is in an area subject to subsidence and it might be initially assumed that this is the cause of the defects seen here, but closer investigation will reveal that the disturbance to the ashlar is much more likely to be due to defects in the lintels.

dissolves the calcareous matrix – a carbonate of lime – holding the grains together. This is more prevalent in limestones than sandstones, the most durable of which are formed with a siliceous matrix that is comprised mainly of silica which is virtually indestructible. Even so, dwellings in the south and west are likely to be less affected by pollution since, in general, the prevailing winds blow any pollution from these areas to the north and east.

The differing composition and characteristics of limestones and sandstones mean that problems occur when the two types are mixed in the same building. Rainwater run-off of calcium sulphate from limestone to sandstone can cause rapid decay in the latter. It is rare to find original construction with the two types of stone mixed but not unusual to encounter localised repairs carried out with the wrong type of stone and leading to worse problems as a result.

Clay and chalk walls

BRE has pointed out that more than half the world's population live in dwellings built from the subsoil which is mixed with water and other unburnt local materials. We may feel somewhat superior in that the vast majority of people in the UK live in brick, stone, concrete or timber-framed structures, but a significant, though relatively small, proportion do also, in fact, reside in the former type of dwelling. One of BRE's first tasks in 1922 was to study building in cob and *pisé de terre*, and many such dwellings as well as others built similarly have outlasted dwellings built in far more sophisticated ways.

The approximate locations and wall thickness together with the names and the types of construction of dwellings built of unbaked earth are set out in the section 'Render faced' in Chapter 9. As stated there, most are likely to have been built in the eighteenth and nineteenth centuries, but all types and ages have certain needs in common and can suffer degradation and possible collapse if these are not met.

The most important need is adequate protection from damp and this was normally achieved in three ways. First, at their base, walls needed to be built up on a base of stone or brick to reduce the amount of rising damp, which, if in excess, could reduce unbaked earth to mud. While too much dampness is obviously a bad thing, it is just as important to ensure that there is sufficient moisture to avoid the wall of unbaked earth drying out and turning to dust. The necessary natural balance, established over decades or even centuries of a building's life, can quickly be upset by the misguided insertion of a damp-proof course. In contrast to buildings of masonry construction where an effective damp-proof course is essential, for a building of unbaked earth construction, the application of modern damp-proofing techniques can herald disaster. When inspecting a property of this construction, a degree of dampness can be expected to register on a surveyor's moisture meter but this should not stimulate a knee-jerk recommendation for damp-treatment works. As with any historic structure – and unbaked earth dwellings should certainly be treated

246 The freshly exposed interior of a cob wall in Dorset, clearly showing the lumps of chalk incorporated in the mix where the rendering has been removed from the exterior of a two-storey cottage in the process of refurbishment.

245 The exposed interior of an unbaked earth wall of clay and chopped straw, mixed in this case with a relatively small proportion of stones.

247 The inevitable consequences of neglecting a suitable protective coating on the face of a cob wall – here, a boundary wall and so exposed to the elements on both sides. The flint base and tiled top seem to have done their jobs properly but the residual piece of rendering clinging to the corner looks like non-permeable cement-based render; this will have accelerated deterioration, causing the partial collapse by preventing the wall from breathing. Prompt repair and application of a lime-based render might yet save the day, but this cannot be delayed.

248 An example of unbaked earth blocks used to build the walls of an East Anglian barn. The individual blocks and courses can be discerned, as can stones within the blocks. Known as clay lump construction, the blocks are prepared and dried before being used in the same way as bricks.

as such – it is essential that they should be assessed with the expertise and ideally the experience appropriate to the form of construction and not merely be appraised and found wanting against modern standards. When works are deemed necessary, they should be sympathetic and respect the original structure; simply imposing modern solutions has repeatedly been shown to be flawed.

The second means of damp protection is on the outer face of the wall construction where a sound form of rendering is needed. This must be permeable and based on lime, not Portland cement, to allow for a certain amount of absorption and subsequent evaporation (**247**). This is so that the interior of the wall can breathe and does not dry out to become dust or get too wet and turn into mud. Any decoration to the rendering must take this need into account. One aggrieved new owner was reported as having settled for a six-figure sum of compensation when his surveyor failed to notice that the cob walls of the purchased cottage had been painted with impervious plastic paint. Part of the cottage had collapsed during repair.

The third important need is for protection against dampness at the top of the walls; there should be sufficient overhang at the eaves to provide as much protection from damp penetration as possible: a minimum of 450mm is necessary and 600mm is preferable. This degree of protection was usually provided in the past by the thatched roofs with which such construction was normally associated. These can be fine but as they do not generally come with gutters, a check needs to be made that the discharge from the thatch is not liable to splash up against or near to the rendering at the base of the unbaked earth wall. However, many thatched roofs were replaced during the last century by other forms of roofing with a lesser overhang. If a new thatched roof cannot be provided, it might be possible to add sprockets to the rafters and extend the covering outwards, taking the rainwater away by means of longer swan-neck rainwater pipes. Needless to say, it is most essential to maintain roofs in good repair.

Fractured, damaged or neglected unbaked earth walls can be repaired using the traditional techniques or with stabilised blocks made of the same materials. Increasing interest in this form of construction along with a greater willingness to understand and retain vernacular building forms means that expertise is much more readily available to provide guidance than was the case even a few years ago.

13

Main walls – concrete construction

In situ mass concrete

In the second half of the eighteenth century, a technique called *pisé* that involves building walls by ramming a mixture of clay soil, straw and water between temporary shuttering was developed in France; this found favour in Scotland for use in dwellings on the east coast, south of Aberdeen and on the south side of the Moray Firth, east of Inverness. So the technique was by no means new when Sir Robert McAlpine used it again to build dwellings of dense *in situ* concrete in Scotland in 1876, and there are other recorded instances of dwellings built of this material before the 1914–1918 war (**249**). In the 1920s after the war, the first single-leaf no-fines concrete systems using removable shuttering were developed by Corolite, Unit Construction, Wilson Lovatt, Airey, and Laing, whose Easiform system evolved from 1928 into two-leaf cavity construction. The Boswell system of the late 1920s was similar, and this came with precast concrete corner posts to facilitate the setting out of the shuttering. Other types from the 1920s with external walls of cavity construction are Boot and Forrester Marsh (**250**).

Other systems utilising no-fines concrete, but cast with permanent shuttering, were also developed in the period 1920–1940. For example, 2,000 Fidler houses are recorded as having been built in Manchester and London, which had walls of *in situ* concrete poured between two leaves of clinker concrete slabs with a thickness of 50mm. Plastered inside and rendered externally, the overall thickness was roughly the same as an internally plastered one-brick-thick solid wall. Another 1,000 plus Universal houses are known to have been built in this period with solid external walls of about 180mm thickness of *in situ* no-fines concrete that was cast between permanent shuttering of asbestos cement sheeting, clearly posing a problem these days if repairs are required.

The period 1940–1960, with its post-war shortages of bricks and skilled labour and the urgent need for new homes to replace those lost to bombing and slum clearance, provided a boost to the construction of dwellings using no-fines concrete. Putting up, removing and repositioning temporary shuttering, mixing and pouring concrete made good use of what was available; and many local authorities – particularly the Scottish Special Housing Association in conjunction with the construction company Wimpey – soon had large programmes in hand, usually based on negotiated rather than competitively tendered contracts. Wimpey alone built around 300,000 dwellings of this type up to about 1980, claiming that their no-fines technique – although there was nothing proprietary about it – was used to build more homes for local authorities than any other system. In the 1950s, the company was building at the rate of 18,000 a year for local authorities and, additionally, a limited number for the private sector. Including a small contribution from other types such as the Incast and the BRS Type 4 with their flat reinforced concrete roofs, in excess of 1 per cent of the UK's total dwelling stock is provided with external walls of no-fines concrete.

Besides being an admirable use of resources at the time, the no-fines technique provided homes of a comparable standard to those being built of cavity brick construction. With external walls of cast *in situ* concrete, based on a mix containing no sand or aggregate from 10mm to 20mm in size, a honeycomb structure was produced. This meant that when plastered internally and with external rendering – often sprayed on – a good level of thermal insulation, for the time, was provided. An added advantage was that, in the main, if the rendering on general walling should degrade from exposure and become cracked, any water that penetrated to the concrete would trickle down through this comparatively open texture and run out at the base without finding its way to the inside face. Even when damp penetrates in any cracked areas above windows, it tends to be diverted out through the bellmouth invariably formed in the rendering immediately above the windows. Accordingly, with no-fines concrete dwellings, cracks in the rendering tend to have little significance other than for appearance; but if the bellmouth becomes cracked then it may need renewal or the cracks repairing with mastic.

Even though steel reinforcement was incorporated in the structure to support roofs and floors and over windows – later to be substituted by precast units of denser concrete set into the shuttering – the rendering provided greater protection to the reinforcement. As a result, the corrosion problems of other post-1940 dwellings, mainly those of prefabricated reinforced concrete construction, were not experienced to any great extent; and even if they were, they could easily be repaired. Thus, because of their satisfactory construction and the evidence of structural soundness, it was not found necessary to designate no-fines concrete dwellings under the Housing Act 1985.

It ought to be mentioned here that since the no-fines technique is very adaptable and is the same, structurally, as a traditional house in that loads are taken on external and internal load-bearing walls, the latter providing the necessary lateral support, it was not only two-storey semi-detached and terrace houses, based on two lifts of shuttering that were built. Bungalows and low-rise blocks of flats and maisonettes from one to six storeys in height were also built; although for the latter, a reinforced concrete frame would usually be provided, the no-fines concrete being restricted to the infill panels. Roofs and floors were of traditional construction in timber, the former in hipped, gable, monopitch and duopitch types, and dwellings were sometimes provided with porches and bay windows.

The normal rendered appearance of no-fines concrete dwellings might be varied with sections of tile hanging or weatherboarding and, of course, a no-fines concrete wall can form the inner leaf to a masonry-constructed outer

249 This late nineteenth-century mid-terrace property, above retail premises in the centre of a south-coast town, looks unremarkable but is notable for being a surprisingly early example of mass concrete construction.

250 More than 50,000 of these Boot houses are thought to have been built during the interwar period using *in situ* cavity construction, but the presence of precast concrete columns meant they became one of the types of construction designated defective under the Housing Act 1985. They are, therefore, unmortgageable unless improved to appropriate standards, like the one on the right of this pair.

251 These two-storey 'Airey Duo-Slab' no-fines *in situ* concrete terraced houses were built in the 1920s and refurbished in the 1960s.

252 This terrace of Wimpey no-fines concrete two-storey houses was probably built in the 1950s. The single-storey brick porch and store additions were not part of the original specification and have been added at a later date.

253 These Wimpey no-fines concrete two-storey houses from the 1960s, in a stepped terrace form, have panels of weatherboarding between the windows. The joint between the general wall surfaces and the weatherboarding and the area around needs examining to ensure that there is no damp penetration as such joints are often a weak spot. Although the rendering appears to be in fair condition, it is stained and appearance would be improved by the application of a couple of coats of masonry paint. Flat roofs, popular at the time, have proved a weakness and some local authorities have replaced them with pitched roofs.

254 This photo and **255** illustrate what can happen to 'BRS Type 4' cast *in situ* no-fines concrete local authority semi-detached houses built in the early 1950s. This example shows a pair of semi-detached houses as originally built, still with flat roof – although the one on the right has clearly been upgraded.

255 This shows the same property type as **254** but transformed with new pitched and tile-covered roofs and with walls encased in brick laid to stretcher bond. The surveyor would ideally need to obtain information from the renovation specification if asked to report on condition.

256 These *in situ* concrete properties with cavity walls, of 'Easiform' construction, have proved popular with occupiers and have been relatively free of the sort of defects that have plagued some other forms of construction used during the same period.

leaf. This was another way of introducing variety to the appearance, particularly for gable walls and where the considerations of wall tie failure would need to be taken into account in any report. The problems of detailing to prevent damp penetration where different materials meet need also to be borne in mind when tile hanging or weatherboarding have been introduced. Eventually, as in all cases where rendering is used, it will need renewal, the timing dependent on initial quality, exposure conditions and maintenance.

Precast concrete block

Building a dwelling with concrete blocks made to give the impression of stonework is not a new phenomenon. These blocks are fairly easy to identify in that they are about 100 years in age, cast using a mix of cement and stone aggregate, and provided with a textured finish to one face. There was no call at that time for blocks in a wide range of sizes so that a fair representation of a particular bond could be achieved; and, it seems, little conscious effort was made to obtain a particular texture or a particular shade of colour (**257**). Ways to differentiate the somewhat more realistic modern concrete blocks manufactured to resemble natural stonework from the real thing were considered in Chapter 9 in the section dealing with 'Stone faced' construction.

The need to match or at least to blend in with the surroundings provided a boost to the concrete block industry – or 'reconstructed stone' as the makers prefer to call their product – from the late 1970s onwards. For example, one manufacturer supplies blocks in eight different sizes with a consistent bed width of 103mm to provide the outer leaf for cavity walling, but which can be built in a variety of coursed and random designs following the traditional styles. The blocks can be obtained in three different colour shades – rather uniform greys and buffs – and all can be in three different textured finishes – tumbled, split and pitched. While it is said that water absorption is low, the blocks are not impervious to driving rain, the incidence of which is more likely to be experienced in areas where stone building predominates. For this reason, deep recessed joints are cautioned against and all joints are required to be completely filled, avoiding any 'tipping and tailing'. The latter aspects of workmanship are not, unfortunately, things which can be checked on site unless so grossly neglected as to be visible in places.

Where reconstructed stone blocks have been used, it is important that the units match the characteristics of those used for the inner leaf; otherwise, differential movement can cause delamination on the face and possibly cracks. Being a concrete product, movement joints need to be given consideration where the length of walling exceeds 6m and where the length of panels exceeds twice their height. Sometimes, differential moisture movement in the face of the blocks will cause crazing and occasionally spalling, though naturally the coarser textures are less prone to this type of defect. However, severe exposure will undoubtedly take its toll over the years.

257 The decorative blocks used at this 1930s house might appear, at first glance, to be of natural stone, but it doesn't take much examination to realise that each of the alternate quoin stones is identical and that there are a limited number of finishes to the other blocks because they are all of cast concrete. Nevertheless, they are weathering well and they are more convincing in appearance than some more recent efforts.

258 Concrete blocks in walling at least 100 years old. The blocks here, simulating coursed squared rubble, are in at least three different sizes, providing a slight variety; although in some places, not much effort has been made to break the long, near straight vertical joints. There is a fair amount of erosion to the top of the texturing on each of the blocks, but no doubt they will last for many more years.

259 Concrete blocks were used throughout the construction of this mid-terrace property dating from the early years of the twentieth century. The blocks were made to a much poorer standard than those intended for external decorative purposes and the elevations were fully rendered, giving the surveyor no clue as to the form of construction until entering the roof space.

Precast concrete panel walls

Many low-rise systems for houses and small blocks of flats were developed to boost the provision of new housing after the conclusion of the 1939–1945 war. Some incorporated storey-height precast concrete panels as the enclosing walls along with reinforced concrete columns and ring beams to support the floors and roof structure. One such was the Reema Hollow Panel system of the 1950s, which had wide, full-height panels (clearly seen in **260)** with an exposed aggregate finish on the external surface. The full-height panels enabled these dwellings to be put up quite quickly with the use of a small crane.

The Reema Hollow Panel system did not utilise metal angles to bolt the sections together but other systems did; the combination of large precast panels with bolted connections was developed and became the norm for the high-rise developments of the 1960s, when the demand for more new housing became even greater. Most of the systems for low-rise housing, including the Reema, eventually suffered deterioration from carbonation, the use of chloride accelerators and inadequate cover to the reinforcement, which led to problems of mortgageability. Eventually the houses, but not the flats, were designated as 'defective by reason of their design or construction' under the Housing Act 1985. The designated systems all comprised a frame in one form or another so they are dealt with at the end of Chapter 14 which deals with framed structures, as distinct from load-bearing walls; the names of the different systems are set out in **Table 3**, also near the end of Chapter 14.

For high-rise developments, as well as the low- and medium-rise blocks of flats which continued to be built – though now on the large panel and bolted connection principle – panels would, in the main, be cast off-site, complete with an external finish and with window and door frames incorporated. Tower cranes would hoist them into position, there to be fitted with joint seals and baffles and to be bolted together. One completed box of walls and floors would bear the weight of the ones above and transfer the load downwards to the ones below. The characteristic results are apparent from **261** to **264**, although **262** is unusual in that the panels were manufactured on site. The Reema system – not with hollow panels as in the 1960s version, but with panels a mere 160mm thick – was but one of around 30 different systems, among them Bison, Wates, Taylor Woodrow, Anglian, Jesperson, Carnus, Crudens, Skarne and Tracoda. In their original state, it was not easy to differentiate the systems visually; but this is not important now as long as they are recognised as being of 'large panel' construction – though often now this may only be deduced by enquiry in view of many being over-clad. It is estimated that over 1 per cent of the total UK housing stock is enclosed by walls of prefabricated reinforced concrete, though the statistics do not allow for separation of the dwellings of large panel construction from those with smaller panels fixed to a frame.

The panel systems suffered a setback in 1968 with the partial collapse of a block – Ronan Point – following a gas explosion in one of the flats, resulting in some loss of life. The blocks had not been designed for a gas installation in the first place. An extensive programme of modification and strengthening had to be put in hand for existing blocks, and changed requirements for new blocks were also put in place. BRE report that no major failures have occurred since but recommend that all blocks intended to serve longer than 25 years from the date of construction should be the subject of a full appraisal for structural safety and durability and that the initial appraisal should be followed by a visual inspection after 1, 2 and 5 years and, subsequently, at no more than 5-year intervals (**261**).

That a surveyor inspecting and reporting on a flat in a block of large panel construction clearly has a duty to have regard to the reports recommended by BRE and to consider what effect the content might have on his or her conclusions, was determined in the Court of Appeal case *Izzard* v *Field Palmer*

1999. In 1988 the surveyor was instructed to carry out a mortgage valuation of a two-floor maisonette in a four-storey block constructed on the Jesperson system, which the surveyor either identified or ascertained was built in the 1960s. The block was one of a number on an estate of 807 dwellings built by the Ministry of Defence but sold in 1985 to a developer who did some renovations and then sold the flats on 999 year leases from 1986. The flats were said to be attractive to first-time buyers even though it was known that they were expensive to maintain. It was held that the surveyor, who had valued the flat at the asking price of £42,000 and said it was readily saleable, was negligent in that: he left blank a box on the report form which asked for details of 'matters which might affect value'; he gave no consideration to the literature which was available on the type of construction; and, accordingly, he did not advise that further investigation should be made into the possibility of high service charges being incurred. Damages of £28,000 were awarded, this being the difference between the purchase price and a valuation produced by the plaintiff, not on the basis of missed defects but because no warnings were given and no account taken of the effect on value. All three Appeal Court judges agreed with the finding of negligence, but one disagreed on the amount of damages. In a strongly argued dissenting judgement, he considered that the plaintiff's valuation should be discounted, thinking perhaps that it was artificial and based on hindsight. Instead, he preferred to base his judgement on another valuation given in evidence of £40,000. This was on the lines of other comparable valuations on the estate at around the same time of between £38,995 and £41,000. The £40,000 valuation took into account the possible difficulty of resale because of the type of construction, advised that past accounts should be looked at but omitted taking sufficiently into account the risk of a higher future maintenance liability. For this omission, the judge took off £2,000 and considered that the proper valuation of the Izzard's maisonette at the time of purchase should have been £38,000, which would have reduced the court's award of damages from £28,000 to £4,000.

Interestingly, the dissenting Appeal Court judge in this case considered that, in the court below, the trial judge had been rather too impressed by the evidence of a distinguished expert surveyor, brought in from elsewhere; this expert had said in evidence that all the surveyors who had valued similar flats on the estate at around the same time had woefully underestimated the risk involved and, accordingly, valued too highly. While this may well have been true, it is, nevertheless, what people will pay that sets the tone of values and clearly he considered there was a market for the accommodation on offer, irrespective of type. The rub would come should the maintenance liability rocket, but there was no certainty that that would happen since, as the judge pointed out, having investigated Jesperson sites in both the north-west and south-east of England, BRE had given them a fairly good bill of health.

When reporting on flats of precast concrete panel wall construction, all the aspects brought out in the court case need consideration – as they do, for that matter, in respect of inspection and reports of flats of any type of construction. In addition, other findings by BRE on these specific types also need to be given consideration and most certainly brought to the attention of a client, whether prospective buyer, lender or having any other interest. The blocks are now all over 50 years old, the high-rise versions severely exposed to the weather; moreover, construction at the time was not always noted for its care and attention to detail. BRE found that much of the concrete used was of poor quality, affected now by carbonation and the use of chloride accelerators and providing insufficient cover to the steelwork. These faults have resulted in corrosion to the reinforcement, cracked and spalling concrete, failing surface finishes and deterioration to the seals and baffles to the joints, leading to a lack of weathertightness. There is also poor insulation. Even with over-cladding – which would overcome the problems of water penetration and lack of insulation, but which BRE recommends only after very careful consideration of suitability and all the implications involved – the view is that, in most cases, the concrete used should be looked upon as material with a limited life. It is to address this durability aspect as well as stability that the 5-yearly reports by structural engineers are so important.

Not for the first time, the surveyor inspecting such a property may feel there is a very difficult balance to strike between the expectations of the courts, guidance from RICS and the commercial realities of completing a timely and economically viable report for an intending purchaser. The court's expectations can obviously be met by undertaking an appropriate full investigation; and the surveyor fortunate enough to be instructed to complete a Level Three inspection can, to some extent at least, reflect the necessary research in the fee quoted and the terms and conditions. To report with full information, having regard to the BRE findings, the surveyor must see the initial appraisal and all subsequent 5-yearly reports before reporting. The lease should be seen to determine the demise and the extent of the joint liabilities and, not leaving it to others, adequate information on expenditure for maintenance purposes must be seen. This will enable a proper assessment, from both the reports and the surveyor's own observations, to be made regarding what expenditure is likely to be required in the future. By doing this, the surveyor is not – as some have done in the past – condemning all flats in such structures as unmortgageable but treating each case on its merits and ensuring that the limitations of the flat being reported are truly reflected in the purchase price and valuation, if such forms part of the instructions. If the reports are lacking or those available are considered to be inadequate for the purpose, the surveyor should advise the commissioning of a report and assist in the formation of the instructions so that the necessary information is obtained.

Such thorough and time-consuming investigation clearly goes far beyond the scope of the RICS descriptions of the Survey Levels One and Two, yet the Guidance Note *Surveys of Residential Property* states, with the implication that this applies to all survey levels, that the surveyor should 'give advice on obvious and relevant matters that may affect the client's responsibility for effecting repairs, and liability to pay towards their cost (whether potential or

260 A pair of 1950s two-storey semi-detached dwellings of 'Reema Hollow Panel' construction. These are enclosed by storey-height precast concrete panels which, with *in situ* reinforced concrete columns and ring beams to support the floor and roof structure and an exposed aggregate finish to the panels, enabled rapid construction. Often the joints between the panels are less obvious than in this photograph, and when houses have been painted externally, unwary surveyors have been known to assume they are conventionally built with rendered elevations – a serious mistake as this construction is designated defective and, therefore, unsuitable for mortgage lending.

261 The systems with full-storey-height precast concrete panels for low-rise housing of the 1950s became, with bolted connections, the types mainly used for the high-rise developments of the 1960s and early 1970s, as here for this 19-storey block of flats.

262 The concrete panels used to build this distinctive late 1960s block were manufactured using factory methods in a facility on the site.

263 The storey-height precast concrete panels on some system-built dwellings developed to speed housing provision will be seen on all types of development from the 1960s and 1970s – low-, high- and medium-rise – as on the three-storey block of flats shown here.

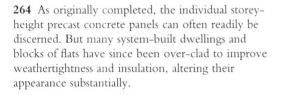

264 As originally completed, the individual storey-height precast concrete panels can often readily be discerned. But many system-built dwellings and blocks of flats have since been over-clad to improve weathertightness and insulation, altering their appearance substantially.

in respect of existing wants of repair)' (third edition, p. 12). Fortunately, this is helpfully preceded by a requirement for the surveyor to set out the limitations of any advice given in relation to the legal aspects of leasehold properties and it is followed by the statement that the surveyor should 'recommend that the client obtains independent legal advice on the interpretation of the lease before commitment to purchase' (third edition, p. 12). Nevertheless the Guidance also later states in relation to legal matters that 'it is important for the surveyor to identify apparent and specific items and features that have legal implications', adding: 'The range of identified matters will be the same for each level of service; what will vary is the explanation. A survey level two and three report will list these matters. At level three, the report will identify the matters and provide an explanation of the feature/issue' (third edition, p. 20).

For most conventionally constructed properties, even modest blocks of flats, this should normally be relatively straightforward. For blocks of precast concrete panel wall construction, however, the surveyor must be very wary of submitting a report that is either not fully informed or which fails to sound appropriate warnings about the investigations which should be undertaken to obtain all the necessary information before the client has a full picture of the dwelling in question. Squaring the time/fee/liability circle in such cases is likely to prove a significant challenge.

The surveyor certainly needs to hold the hand of the purchaser in these circumstances, particularly where Right to Buy is involved and the client may be dazzled by what seems to be a bargain, overlooking the fact that the discount is personal to the applicant, has nothing to do with the value of the flat and should be left entirely out of the reckoning. The lease to be granted will no doubt be on the local authority's standard form, and there is invariably a provision for termination should the lessor, for good reason, decide on demolition and redevelopment or even on a sale for substantial refurbishment by others. Whether the price quoted to the tenant–purchaser, exclusive of Right to Buy discount, takes all the aspects of age, type of construction, likely durability and liabilities fully into account needs to be given some thought. So, of course, does the asking price quoted by any private vendor, on this occasion no doubt supplied by the seller's agent in line with recorded sales of similar flats and not necessarily, as in the *Izzard* case, taking full account of all the circumstances. This is one area where the surveyor's advice can be very valuable, though it is not always welcomed by buyers in a hurry or desperate for accommodation. However, if such advice had been sought and taken at the time of many Right to Buy sales, the degree of shocked expression at some of the repair and refurbishment bills now being presented might be somewhat less.

14

Main walls – framed construction

Introduction

For a surveyor inspecting and reporting on dwellings with framed as distinct from load-bearing walls, there is the difficulty that the framing units – whether of timber, steel or reinforced concrete – are, with one notable exception, seldom seen. That exception is, of course, the oak framing of houses built roughly from the fourteenth century to the seventeenth century; but even with many of those, comfort and, at times, fashion dictated a cover-up. Showing off the old oak frame only became fashionable from about 1850 onwards.

What the surveyor does see is the form of envelope applied to the frame and, preferences being what they are, many framed buildings appear from the outside as traditionally brick built. With the arguable exception of the three decades following the Second World War, it has been the aim of most of the companies putting up dwellings with framed components during the last 100 years, especially since the mid 1970s, to make their products look as though they have been built with conventional load-bearing walls, particularly when supplying the private sector.

The procedure for ascertaining the construction of the external walls, wherever this is possible for dwellings, was dealt with in Chapter 9; but this may only provide the information that the structure is framed, not how or with which material. However, that in itself is significant as, coupled with a knowledge or an assessment of the age of the dwelling, it provides a pointer to further investigation. Times of shortage, particularly in the periods immediately following the wars of 1914–1918 and 1939–1945, provided the stimulus for much of the development of the systems used; and it is to the information about those systems and their characteristics that the surveyor will often have to turn. Unfortunately the large-scale adoption of systemised forms of construction after the Second World War utilised some technologies with questionable suitability and, necessarily, unproven durability for domestic properties. This resulted in a series of problems developing in the short term that were difficult and expensive to remedy, culminating in the designation under the Housing Act 1985 of certain prefabricated reinforced concrete (PRC) forms of construction as 'defective' and unsuitable for mortgage lending.

Arguably more significant, however, has been the still-lingering adverse perception amongst those working in residential valuation, surveying and lending that this debacle has had regarding anything which might be remotely considered a novel or innovative means of construction. No other construction sector has been similarly constrained and attempts have been made to promote nontraditional forms of construction for residential properties. For example, the government set targets for the use of 'Modern Methods of Construction' (MMC) in the social housing sector during the early years of this century. Despite this, builders in the private sector have proved resistant to change while mainstream lenders and the professional valuers who advise them still remember, and in some cases still bear the scars of, the post-war fiasco.

Finally, times and attitudes are changing. Motivated by the sustainability and climate change agendas, twinned with the demands of the national housing crisis, lenders and their valuers are finally recognising the need to be more open to nontraditional forms of residential construction. These pose great challenges to the residential surveyor and valuer and some current innovative forms of construction are reviewed separately in Chapter 15. Meanwhile, there is still a significant legacy of properties across the country dating mainly from the 30 years prior to 1975, which the surveyor may come across from time to time. Some have been described in Chapter 13 but the more problematic have tended to be those of framed construction, especially those with reinforced concrete frames. Many have since been successfully repaired under one of the licensed schemes which were created in the wake of the PRC difficulties so their dubious origins may no longer be apparent. Nevertheless, sufficient remain for it to be important that the surveyor is able to recognise them when they are encountered, if not by name then at least as being of non-conventional construction and thereby justifying great caution and thorough investigation. The later sections in this chapter consider the systems of steel and concrete frame construction, which had their origins in the years after the First World War and reached their questionable climax in the years following the Second. First, however, it is necessary to go further back in time and consider one of the materials used to form the frames and walls of dwellings since time immemorial: timber.

Medieval timber frame

It is always important to consider the structure of a dwelling in its entirety and not just as a diverse collection of parts. This applies irrespective of the materials used in its construction or the way it is put together. Walls depend on floors and roof for restraint and one wall upon another by reason of the bond at corners. Nevertheless, framed walls, as distinct from mass load-bearing walls, can be looked upon rather differently because if the casing is removed, whether it is brick or some other material, the frame still remains. This is true for modern steel- and timber-framed dwellings but, perhaps, even more so for the medieval timber-framed dwelling.

Medieval timber-framed walls are part of a series of interconnecting structural components, each highly dependent on the other. The framing of the walls is an extension of similar framing making up the floor and roof structure, and it is the continuation of those components which takes the loads of the floors and the roof downwards to the support provided at or near ground level. The cutting of, or damage to, one component can cause untold damage to the structural integrity of the dwelling as a whole, and the full effect can manifest itself quite some distance away. The installation of a brick chimney stack many years after the original construction is a case in point, leaving many a dwelling with a lopsided appearance. However, such is the strength, generally, of the form that there is seldom any danger of collapse, and careful strapping can do wonders to prevent further movement. A medieval timber-framed dwelling where most of the panels between the studs are ill-fitting or defective in some other way would be costly to deal with; but the cost would be nothing like as great as with a similar building where the frame was badly in disrepair through rot or beetle infestation. The hazards for the frame are, of course, from exposure to driving rain and the attentions of that most unwelcome visitor, the deathwatch beetle.

Repair of damage is costly, not only because the skills to do the work are scarce but also because most medieval timber-framed dwellings are either listed or in a conservation area, or both, and likely to draw the attention of the relevant authorities – namely Historic England, Historic Scotland, Cadw or the Northern Ireland Environment Agency – when repairs are required. Furthermore, such walls are often at their most vulnerable on account of their construction at ground level. Generally, they were provided with a brick or stone plinth on which a sturdy base plate was laid and to which the vertical studs were dowelled and pinned. From day one, there would be little or nothing in the way of a membrane to keep rising damp from the underside of the base plate. Sections attacked by damp from both below and from above, particularly on the elevations most exposed to driving rain, will eventually rot and need replacement, whether or not additionally ravaged by deathwatch beetle with its preference for damp timber. If replaced with timber, adequate lengthening joints have to be provided to connect new and old sections to maintain the structural integrity of the frame. If the lengthening joints are inadequate or, even worse, if brickwork is substituted for timber as is sometimes found to have been done in the past, the frame at ground level will become weaker and may not be able to resist sideways thrust from the studs. Essential checks need to be made in these circumstances. On other elevations not so much affected by driving rain, the top surface of the base plate may appear deceptively sound, necessitating the inclusion of suitable warnings in the report.

Depending on the required basis of inspection, the surveyor will examine all visible and accessible surfaces and there will, no doubt, be timbers which are visible but not accessible from the surveyor's ladder. Irrespective of whether there are signs elsewhere of rot or beetle infestation, the surveyor must advise that those inaccessible timbers be examined by whatever means necessary both for rot and beetle infestation and to check, as in other places, that there are no undue gaps between frame and panel which could provide easy access for water penetration. As to internal surfaces where the timber may or may not be visible, the surveyor must advise opening up if any indications whatever are seen to suggest that defects may exist in areas hidden from view. When there are no indications of any potential problems, the surveyor must include warnings of what might be going on in hidden areas and stress that the report

265 Jettied construction on both the first and second floors on a building dating from the first half of the fifteenth century. There is fairly close studding but it still requires some wind bracing. When such structures were built on either side of a narrow street, the occupants of the top floors could sometimes shake hands! The state of the street below was probably indescribable.

266 Close studding to the first floor of this dwelling, thought to date from around 1550, shows the culmination of this form of construction, which produces a very sturdy structure. Such profligate use of timber could not continue as supplies diminished so simpler forms started to appear.

267 This shows the complicated five-way joint between the external stud, wind brace, tie beam and rafter and the longitudinal wall plate, circled at the number '2' on the cross section C-C on the right-hand drawing in **Figure 23**.

cannot be regarded as providing insurance against the possibility of defects which may be present but hidden from view.

A line on a graph showing achievement in timber frame building in oak through medieval times would take the form of an arch. From fairly modest beginnings, a peak was reached in the decades around 1500, followed by a fall as the availability of the material declined either from shortage or appropriation for other purposes such as shipbuilding. At its peak, the achievement was impressive, finding solutions in the design of layouts and construction and in the design of joints to leave a legacy of sturdy and sometimes quite impressive

268 On either side of the peak period of achievement – from the mid fifteenth century to the mid sixteenth century – for timber framing in oak with its closely spaced studs and jettied floors, much simpler shapes prevailed. This well-preserved two-storey former dwelling is said to date from the early fifteenth century, before the peak, and is built on a much simpler box frame system. Some of the first-floor panels have outlines of brickwork just visible, but these do not look original and the more uneven finishes may hide earlier wattle and daub.

269 A sixteenth-century timber-framed property in the early stages of comprehensive restoration. The external rendering has been removed, exposing mainly wattle and daub infilling between the sections of frame. The unsightly metal-framed window will be replaced with something more appropriate as work progresses.

270 There are contrasting views on the inclusion of brick panels and their effect on the stability of medieval timber frame dwellings. BRE suggests that they provide added strength whereas others are of the view that they add far too much weight to a frame that can easily be weakened by rot and beetle infestation. The panels of brick, laid in herringbone pattern, do not appear to have had too much effect on the stability of this late sixteenth-century two-storey dwelling, built a little after the peak of achievement for timber frame construction. However, the surveyor would need to advise an inspection of timber out of reach from a surveyor's ladder even if an examination of timbers lower down revealed little or no evidence of rot or beetle infestation – which would seem unlikely from the appearance of the timbers here.

271 The flank elevation of this seventeenth-century dwelling, built well after the peak of timber frame construction, is of very basic square panels with an infilling of brick in a chevron pattern and a mere two wind braces. Sturdy timbers were employed which is more than can be said for many of the timber structures of the time. The arrangement of the timbers in the Queen Post-style roof can be clearly seen.

272 This four-storey property dates from around 1900 and it only takes a cursory examination by the experienced eye to question the arrangement of the mock timber framing, which could not perform any conceivable structural function. Nonetheless, even if they are only decorative, the condition of the external timbers must be considered and checked as far as practicable. Even though the cost of renewing a few planks would be relatively small, the need to provide access could add significant expense.

273 Modern timber-framed construction is usually encased with brickwork to look like other conventionally built brick dwellings. As fashions changed, medieval timber-framed dwellings often suffered a similar fate of being hidden behind claddings of either brickwork, tiles (including the mathematical variety that look like bricks), weatherboarding or lath and plaster. Even without the clue provided by the exposed timberwork on the property to its right, the cantilevered first-floor projection is a clear indication that, behind the subtly decorated plaster cladding, this building is a medieval timber-framed structure, probably dating from the fifteenth century.

dwellings, now 400 to 500 years old. Impressive not only in size but also for construction in an organic material that many would say carried the seeds of its own destruction.

The peak of achievement in the late fifteenth century and early sixteenth century is characterised by dwellings made up of substantial storey-height, closely spaced studs set on a massive base plate and connected together at their top ends by a longitudinal wall plate running both ways in the direction of the wall, with other timbers laid across as floor beams (**Figure 23**). The floor

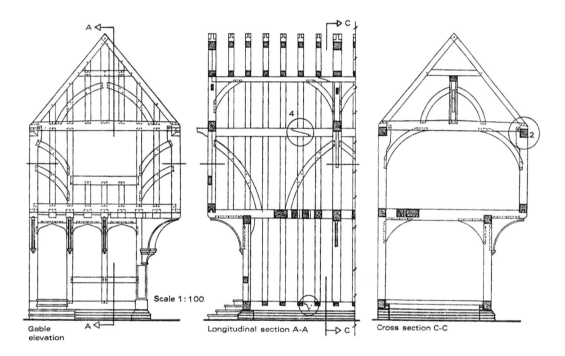

Gable elevation · Scale 1:100 · Longitudinal section A-A · Cross section C-C

Figure 23 Some idea of the strength and sturdiness of medieval timber framing at the peak of its achievement between about 1450 and 1550 can be gathered from these drawings of a gable elevation and two sections of a two-storey dwelling on a corner site with jettied construction to the first floor on the front and return elevations. It is easy to imagine what the dwelling would look like if all the panels of wattle and daub were removed, as here, leaving just the timber framing. It is not surprising that there is still a considerable number of such dwellings around, often showing hardly any signs of structural problems except, perhaps, where some ill-considered alterations have been carried out. On the longitudinal section A-A (shown in the central drawing), the number '4' encircles a scarf joint connecting two lengths of wall plate together. Almost directly below, there is another scarf joint of different design connecting two lengths of base plate. Joints such as these, which can often be readily seen, sometimes introduce dampness when the timber shrinks; this is more likely to be found on elevations with a southerly or south-westerly aspect. The number '2' on the cross section C-C (on the right-hand drawing) encircles the complicated five-way joint (shown in detail in **267**) between the external stud, wind brace, tie beam and rafter and the longitudinal wall plate. The wind brace is as it appears on the elevation of the gable (in the left-hand drawing), but along the side elevation, it would be replaced by an arched wind brace connecting stud and tie beam. When the joint was fitted together, there would have been very little indication of how it was arranged, apart from the position of the pegs or dowels holding the components together.

beams could be jettied outwards if necessary on all four sides of a dwelling in order to carry a further range of similar studs at first-floor level, the detail to be repeated at the top either as a basis for the roof framing or to carry yet another floor of jettied construction. Jettied construction all round produced a firmer structure and was a useful device in the narrow confines of medieval streets, enabling encroachment into the air space above the footpath so as to gain the maximum usable floor space from the site. Both the earlier and later periods of timber-framed construction depended on the box frame system. This was inherently weaker than the close studding of the middle period. Many of the dwellings of this form will exhibit the appearance of settlement, leans and bulges; yet even when looking as though they would fall down at any time, they often have a reserve of strength which should make the surveyor cautious of coming to hasty conclusions. A factual report on condition without speculation is required, although if unduly worried, a recommendation to consult a structural engineer specialising in the analysis of timber-framed structures should be made. In 1993, the RICS Building Conservation Group published the helpful guide *A Checklist for the Structural Survey of Period Timber-Framed Buildings*, which concisely sets out a logical method of inspecting such structures, illustrates many key construction features and concludes with a useful list of books from which to obtain more detailed information.

Post medieval timber frame

From about 1700, the oak forests in the UK became depleted and no longer available as a source for the substantial framing which is the characteristic feature of the timber dwellings constructed in the previous three centuries. In the main, the oak timbering was replaced by brick for the outer walls of dwellings in urban locations, more so because of the risk of fire than the shortage of the timber. Where timber had been used for roofs and floors, it was being replaced in any event by imported softwood from elsewhere in Northern Europe.

In rural areas not subject to the requirement to build in brick or stone, a tradition grew up for building relatively inexpensive dwellings almost entirely of timber. To roofs and floors in softwood would be added softwood framing for the walls, covered externally with weatherboarding (**159** in Chapter 9 and in **274)** and internally, first with panelling and later with lath and plaster. As an alternative to weatherboarding, lathes could be fixed to the exterior so the upper floors could be tile or slate hung and the ground floor rendered. Such construction would pass muster as reasonably resistant to the spread of fire for dwellings in the smaller towns where requirements might not be too strict.

Weatherboarded exteriors pose no problems in terms of recognising timber-framed construction, and the hollow sound when the inner face of an external wall is tapped will confirm this. The same hollow sound from the inner face of rendered or either tile or slate hung external walls will also confirm timber-framed construction. Such construction happily coexisted with brick and stone construction outside the major conurbations until about the 1920s. It received boosts at various times. One was the period from 1784 to 1850 when a hefty tax was placed on bricks. Another was in the late nineteenth century and early twentieth century with the influence of the philanthropists and the Garden Cities Movement who wanted to do something about the poor standard of housing in country areas. Various competitions were held to produce designs for 'cheap cottages', many of the entries being based on timber framing.

The timber-framed dwellings of the eighteenth, nineteenth and early twentieth centuries were rather basic. Foundations could be rudimentary and the frame lightweight and perhaps inadequately braced, leading to racking; weak floors could be overloaded and there would be an absence of damp-proof courses and certainly no insulation. They need the most attentive of visual inspections and may be reluctant to show on the surface any obvious signs of what could be going on well hidden from view. At the slightest sign of damp, rot or woodworm, the surveyor has little option but to advise opening up to expose vulnerable parts of the frame, particularly those near ground level.

Timber framing in softwood was given another boost in the aftermath of the First World War, when there was a shortage of building materials and labour costs were very high. This led to encouragement for the development and use of timber frame construction for the provision of local authority dwellings, including an attempt to import the parts to make up complete dwellings from Sweden which was scuppered by opposition from the trade unions. Even so, a number of authorities succeeded in constructing timber-framed dwellings using prefabricated parts and components from both Sweden and Norway. Other local authorities developed their own designs and at least two private companies became involved, supplying dwellings for the private sector as well.

The timber-framed dwellings produced in the 1920s again pose little problem for recognition, being mainly timber clad, but they are considerably more sturdy in construction. It is with these that the BRE comes to the aid of the surveyor with its studies of types and their characteristics as part of a wider study of most timber framing systems up to 1975, specifically built for local authorities since it is to these that BRE could gain access – although, of course, the information gained applies equally to those built privately for sale.

Three publications by BRE appeared in 1995 and are still available as a pack. The first, entitled *Timber Frame Housing Systems 1920 to 1975: Inspection and Assessment*, provides an overall view of the construction characteristics of 11 types built between 1920 and 1944, 9 types built between 1945 and 1965 when the system building programme began to get under way, and 34 types constructed between 1965 and 1975 when system building was being widely used. The vulnerable parts of timber structures are described in detail over seven pages, and a further seven pages are devoted to procedures for carrying out a site inspection to produce a 'structural survey' report where an initial inspection is followed by opening up for a closer examination of those parts

274 In the eighteenth and nineteenth centuries in quieter country towns and rural locations where there were no requirements to build in brick or stone to reduce the risk of fire, inexpensive dwellings were often built of timber framing with a cladding of weatherboarding (as on this eighteenth-century house) or tile hanging (as can be seen on the elevation to the left). The weatherboarding would either be painted or tarred.

275 This pair of two-storey semi-detached cottages from the 1860s have painted weatherboarding to the flank elevation and a rendered finish on laths to the front elevation, both on a timber frame.

276 The comparatively light framework of studs and diagonal braces that form the structure of the 1860s two-storey semi-detached cottages shown in **275** can be seen from this photograph where internal lath and plaster has been removed but the back of both the weatherboarding and the laths for the rendering can be seen. Careful consideration would need to be given on any refurbishment to the positioning of insulation, sheathing to strengthen the frame, and a vapour barrier.

277 An entry to a 'cheap cottages' competition held in 1905. The structure is timber framed with white-painted weatherboarding. Quite large for a cottage by the standards of the time, it could conceivably have been two originally; but refurbished and with a new roof covering, it is still giving good service after 100 years and, with proper maintenance, will continue to do so for many more.

278 A pair of two-storey semi-detached timber-framed and -clad dwellings built by a local authority in the north-east of England from prefabricated parts imported from Norway in the late 1920s.

279 A pair of two-storey semi-detached local authority dwellings of the late 1940s. These are prefabricated timber frame, clad with vertical timber boarding, and were imported from Sweden. They are reported as suffering very little deterioration over the years, and they are popular because of generous space standards and good insulation qualities.

considered vulnerable. There are illustrations and summary information on 55 named systems. The second publication is *Timber Framed Housing Systems Built in the UK between 1920 and 1965*, which analyses the details of seven of the systems built between 1920 and 1944 and nine of the systems developed between 1945 and 1965. The third publication, *Timber Frame Housing Systems built in the UK between 1966 and 1975*, completes the study by comprehensively investigating the details of a further 12 systems.

The BRE publications of 1995 contain invaluable information but not every known system is described in detail. For those that are, much of the information, as far as identification, details of construction and vulnerable parts are concerned, is derived from inspections involving invasive techniques that are not available initially on a purely visual inspection. For many residential surveyors, therefore, these volumes, while undoubtedly of interest, will probably have limited application unless instructions for work other than normal pre-purchase surveys on these early timber-framed systems are likely to be received.

Modern timber frame

Timber framing systems were the only types of system building which continued in favour beyond 1975, by which time about 100,000 had been supplied for the public sector alone. BRE estimates that between one-half and two-thirds of a million such dwellings – between 2 per cent and 3 per cent of the total dwelling stock – were built during the last century, the vast majority of them since the mid 1970s; a trend which continues. Of the systems used, 100 have been identified.

The inspection procedure suggested by BRE augmented the RICS/ TRADA guidance in their 1984 publication *Structural Surveys of Timber Frame Houses*, particularly where the latter leaves the condition of the dwelling in doubt. BRE pointed out that while many timber frame systems are essentially similar, each inspection and assessment must be tailored to the specific characteristics of the system being investigated. This seems to be a counsel of perfection in view of the number of systems which have been used and is, in effect, only valuable if you know what the system is before you start. However, a surveyor carrying out an inspection envisaged by this book will, generally, not know in advance whether the property to be inspected is even a timber-framed dwelling, let alone one constructed on which one of many different possible systems. The surveyor probably won't even know when entering the front door, since the vast majority of timber-framed dwellings are clad in brickwork. Most of those that are comparatively easy to recognise as timber framed because they are timber clad were built before 1965 for local authorities, and there were only about 30,000 of those.

As with all inspections, a surveyor has first of all to establish whether the dwelling is of traditional or nontraditional construction. This aspect has been dealt with in detail in Chapter 9, and the surveyor will hopefully have established by assessing its age, from the external material used, and by measuring and tapping that the dwelling is of a nontraditional type. Carrying forward the investigation at the site, there are certain features about timber-framed construction which, when encountered, will lead the surveyor towards the conclusion that this is indeed the form; although there are so many variations in use that there is now little expectation that a timber-framed system should be identified by name.

The indications of timber-framed construction are probably most noticeable in the roof space in the cases of terraced and semi-detached dwellings and those where the roofs are gabled as against hipped. The interior face of the gable wall of brick-faced, timber-framed dwellings may show the stretcher bond of the brickwork, which will need to be tied securely back to either a timber-framed spandrel panel or to a properly braced trussed rafter. Separating walls will generally be plasterboard faced on timber framing to produce the required level of fire resistance. The sight of brickwork or blockwork at this point is no indication either way in terraced or semi-detached dwellings where the roof is hipped because a separating wall in these materials can be used when the rest of the construction is either conventional or timber framed. While in the roof space, it is worthwhile to try and see the top of the external wall at the eaves. With the aid of a mirror attached to a long handle, it is sometimes possible to see the top member of the frame. In contrast to what can be seen in conventionally built dwellings, this is invariably planed as against being left sawn, as is the wall plate in other cases. In timber-framed construction, the cavity would be closed off usually by means of a timber batten or mineral fibre cavity barrier.

Viewed from the exterior, a dwelling with reveals to the windows and doors that seem deeper than usual is often found to be timber framed as both are usually fixed to the timber frame and not the brick skin. With other much thinner forms of cladding, this factor tends to make the windows seem set much further forward than usual while the overall thickness of external walls tends to be much less than would be the case if brickwork or blockwork had been used and dry lined. An unusual relationship between the level of the damp-proof course and the weep holes, draining the cavity of brick or blockwork external walls, will alert a surveyor to the probability of timber-framed construction. In conventional construction, they are invariably above the level of the damp-proof course, but with some timber-framed systems, they are set below.

Another external clue to timber-framed construction can be found in semi-detached or terraced houses where there are variations in roof height. If a cladding material, usually hung tiles, is present to the exposed upper part of a party wall above an adjoining roof slope, it can be a strong indication of timber framing, especially if the main elevations are brick faced. If the construction is masonry then the exposed upper part of the party wall would normally simply be left as brickwork or possibly rendered blockwork. A party wall of timber frame construction, however, requires an appropriate weather cladding to the exposed upper part (**282**).

If it is clear that the dwelling being inspected is of fairly recent construction, the sight of soft packing or space left below window sills is a sure indication of timber frame construction. This packing or space is usually around 6mm on the ground floor and 12mm on the first floor, and this is to allow for differential movement between the frame and cladding caused by the initial shrinkage of the timber frame. The shrinkage is usually completed within a short time of construction and, accordingly, if the dwelling is more than, say, a couple of years old then it may be that no gap is visible. The effect of an inadequate allowance for movement may be all too evident, however, in disrupted sills either broken, if tiled, or tilted upwards, if solid, and allowing water to flow backwards towards the window.

Low-rise timber-framed buildings of up to three storeys are routinely being constructed but six-storey construction is becoming increasingly common. There is the potential to reach as high as ten storeys and the limiting factor is, surprisingly, not the structure but achieving suitable fire protection; a seven-storey building, for example, requires 60 minutes. With taller buildings of timber frame construction, even more allowance for shrinkage must be made and this brings additional complications. An allowance of 21mm is needed at second-floor level and, to ensure weather protection, sealant will be needed; but the necessary gap of 21mm will not be achieved if the joint is filled with foam or mastic since these materials can never compress to zero. In addition to shrinkage due to moisture changes, taller buildings will suffer from compression due to loading. At ground-floor level, this could be even greater

280 These boxes from the 1970s – comprising large double-fronted two-storey local authority terraced dwellings – are timber framed, clad at ground-floor level with brickwork and at first-floor level with artificial slates, and have a flat roof of bituminous felt. The design, known as 'Silksworth', was developed by a consortium of local authorities.

281 A two-storey terraced development from the 1970s, developed by the same authorities as those in **280**, but with the provision of pitched roofs covered with tiles. The artificial slates are still present at first-floor level and the arrangement of windows persists, the openings much too near the party wall line.

282 A section of tiling on the line of a party wall in a stepped terrace is a good clue that the underlying construction is timber framed. If the construction was masonry, the exposed section of party wall would have been finished with brickwork supported by the lower part of the party wall; but a timber-framed party wall does not provide the necessary means of support for an upper section of brickwork.

283 This block of flats, helpfully dated 2010 above the front door, is of timber-framed construction but there are no immediate features to distinguish it from a conventional masonry property. Depending on which flat is being inspected, a surveyor may not have the opportunity to check within the roof space. Photograph **304** (in the next chapter) shows this block in the early stages of construction.

than moisture shrinkage. Roof flashing, service connections, pipes and duct-work all require consideration to ensure that they are not affected by shrinkage or settlement.

Internally, a hollow sound when the inner face of the external walls is tapped might immediately suggest timber frame construction; but the surveyor must not forget that brick or block walls can be dry lined on battens, which will also produce a hollow sound when tapped. The part of the wall above windows and door heads should always be included in tapping; paradoxically, dry lined walls will continue to sound hollow but timber-framed walls will sound less hollow because of the presence of timber lintels.

The presence of all, or some, of the above indications will lead the surveyor to conclude that the dwelling being inspected is timber framed. With modern timber-framed dwellings, it is doubtful whether any effort spent trying to identify a named system can be justified. There are many essentially similar systems and no particular system has been singled out as being structurally worse than any other. Irrespective of whether the name of the system is known and after an inspection of all visible accessible parts, the following will leave the surveyor with no alternative but to advise opening up for a closer examination of the structure adjacent to the areas of concern: any evidence of damp, whatever the cladding; any evidence of a shaky structure, such as popping of the nails of plasterboard or cracks, leans, bows or bulges in brick or blockwork cladding; any signs of rot, however slight, in cladding of timber. The same advice may need to be given, depending on circumstances, should there be any indications of alterations that may not have taken structural considerations fully into account. On the other hand, if there are no visible indications of defects, the surveyor has no grounds for giving advice to open up the structure.

Systems which have been developed since 1975 embody improved techniques, greater accuracy in assembly and a degree of quality control which was often lacking in the systems used before then, but production and construction techniques continue to be refined as system building processes become increasingly sophisticated. The expression 'modern timber-framed construction' has been appropriate shorthand for the specific techniques utilised in domestic construction during the three or four decades since 1975; but it seems probable that this will increasingly come to be regarded as an inadequate catch-all phrase for the full range of timber-based construction technologies likely to be utilised in the second decade of the twenty-first century and beyond. This theme is developed further in Chapter 15, which considers innovative and non-conventional forms of construction.

Steel frame

Like most of the systems used to boost the output of dwellings after the end of the Second World War in 1945, those of steel-framed construction owed their

284 An 11-storey block of flats built about 70 years ago on the coast in the all-white 'modern' style, fashionable at the time but now looking a bit grey and stained. Visual inspection will not reveal the material used for the frame; it could be either reinforced concrete or steel with perhaps a greater likelihood of the latter. In the absence of any signs of distress to the frame within the flat, in the common parts or on the exterior, there would be no need to advise obtaining a structural engineer's report on the stability of the frame or its durability. There are, however, ample signs that exposure to the elements has taken its toll on the exterior, and the internal inspection will no doubt reveal how the aged metal windows are coping. Maintenance costs and renewals are always high with this type of construction and, depending on the level of service being provided, the surveyor will either need to review the service charge accounts and the likely cost of future programmes or, at the very least, advise that this should be done before the client becomes committed to a purchase.

origins to developments in the 1920–1940 period. A publication by the Department of Scientific and Industrial Research in 1951, *The Corrosion of Steel Houses*, covered eight systems which were built on 36 sites between 1920 and 1927. The most interesting report for surveyors on steel-framed dwellings built prior to 1975 is one published in 1987 by the BRE, *Inspection and Assessment of Steel Framed and Steel Clad Houses*. It runs to 55 pages including a 30-page appendix describing 35 systems put up in both the public and private sectors and giving some of their locations. This report was specifically directed to building surveyors to enable them to carry out what was known then as a 'structural survey' – now described as a Building Survey or Survey Level Three – and to produce advice for prospective purchasers of steel-framed and steel-clad houses.

The 1987 BRE report indicated that about 140,000 steel-framed houses had been built – about 0.5 per cent of the total stock of dwellings – by the time of writing, the vast majority post 1940. However, BRE also noted that the locations of most were not known and that there were types other than those mentioned in the report. As to appearance, although a few are distinctive, it is probably these which are more likely to have had their appearance altered by recladding with other materials to extend their life, improve their performance and, where privately owned, to make them appear different from their neighbours. Many, however, were originally built with brick cladding, giving them the appearance of conventional dwellings, the disguise of some being nearly perfect in that it takes more than a superficial examination to detect that they are in fact steel framed. Fortunately, an inspection of the roof space reveals that most systems had steel roof trusses, the major exception being the Dennis Wild and Thorncliffe systems. These can be identified by other means, the former by their patented 'Wild Cradle' roof trusses based on 225mm by 75mm timbers and steel tie rods and the latter by their visible cast iron structural panels for the walls, bolted together. This leaves the remainder also identifiable as of nontraditional construction on account of their steel roof trusses and, accordingly, recognisable as steel-framed types.

With respect to advice on the procedure for carrying out site inspections, BRE suggested in its 1987 report that where the type of steel-framed house is not already known or cannot be ascertained by enquiry, it should be determined from Appendix 1 to the report. At the time, it was common for instructions to building surveyors carrying out a 'structural survey' to involve a certain amount of opening up of structures to seek further evidence on defects which might only be superficially apparent; the report includes the results of inspections where this procedure had been followed. This is useful from the

285 Some steel-framed houses were built in the 1920s. These dwellings, cased in brickwork with tile hanging to the first floor, are of the Dennis Wild type, much as originally built. They employed standard rolled steel sections with stanchions that were heavier than in many other systems. Their fairly wide spacing is apparent from internal inspection. Mostly, there was no protective coating to the steelwork though some were encased in concrete. Although looking like conventional brick-built cavity wall dwellings, they can be identified by their individual, patented, timber 'Wild Cradle' roof trusses, based on 225mm by 75mm principals with steel tie rods. Cavity widths vary but as the steel is close to or even touching the inside face of the outer leaf, driving rain can penetrate and cause corrosion in stanchions, necessitating repair and sometimes renewal of the outer leaf. Other versions of the same type were slate hung or pebble-dash rendered, both of which were on battens to the first floor, or wholly brick faced.

286 A pair of renovated steel-framed BISF houses, the type illustrated at **160** in Chapter 9. The large opening for the reception room window and the profile steel roof remain from the original building and provide good clues about the origin and underlying structure of these dwellings. A further example of the roof covering on a privately built bungalow can be seen at **64** in Chapter 4.

287 'Steane' steel-framed houses, the left-hand one as originally built and the right-hand one having been upgraded.

288 'Trusteel' Mk2 steel-framed houses of 1953 built for the private sector.

289 'Trusteel' 3M steel-framed houses of the late 1960s.

point of view of surveyors carrying out the visual inspections envisaged in this book, including those for Survey Level Three reports, where opening up does *not* nowadays form part of the normal agreed terms of engagement; now, it is the outward manifestations of those defects closely inspected, usually corrosion in the steel normally hidden from view, which are of initial concern. Each of the systems described in Appendix 1 of the BRE report has been classified, wherever possible, into one of the four major types of steel-framed dwelling. By referring back to the main body of the report which provides information on the major distinguishing features of each type that is derived from opening up the structures, the present-day surveyor can narrow down the identification of positions where serious corrosion *could* be present.

As might be expected, one of the most important factors contributing to the corrosion of the structural components in steel-framed houses is exposure to wind-driven rain. This is particularly so as, in many instances, there is only a thickness of material – whatever it might be – of around 100mm between the external wall face and the steel itself. This can allow saturated masonry or rendering to remain in contact with the steel for long periods, or it can lead to water being trapped in undrained or rubble-filled cavities. Places where this can happen need identifying wherever possible from the results of the BRE investigations. Should there then be opening up of a section, of course, a favourable outcome will only apply to the particular section examined; the surveyor needs to stress this aspect and indicate that only the balance of probability suggests that there is no corrosion or only superficial corrosion elsewhere. A purchaser needs to accept the risks involved in not exposing the entire frame. By 2010, most steel-framed dwellings were already around 60 years old and some even more, exceeding what is considered a normal lifespan of 60 years, and purchasers may need reminding of this fact. Corrosion is a progressive phenomenon and this combined with age has to be reflected in any valuation, perhaps more so when compared to some other types of system-built dwelling.

Against this rather pessimistic outlook, it is only fair to point out that BRE reported in its 1987 publication that about 60 per cent of all steel-framed dwellings inspected had revealed either no corrosion or only surface or minor corrosion. Many of those built in the 1920s were, after all, still in use. Those with rolled steel sections fared better on the inspections than those with the lighter twisted and bent sheet steel sections. The studies showed that the vast majority of steel-framed dwellings provided levels of performance not very different from many traditionally built dwellings of the same age. A small proportion of those in areas of severe exposure had suffered significant corrosion in parts of the structure. This was well defined and, once access had been gained, repair was possible. Provided the repairs are carried out, the envelope has been restored to a weathertight condition and is subsequently maintained, and assuming that there is no increased risk of condensation by inappropriate refurbishment – such as would be the case with the provision of insulation within the cavities – there is no reason why steel-framed or steel-clad dwellings should not give good performance into the foreseeable future on a par with the life conventionally assumed for rehabilitated dwellings of traditional construction. Bearing in mind that the BRE report itself is, by now, not far short of 30 years old and, consequently, the dwellings studied are that much older, it can be concluded that they need each to be treated upon their merits, taking into account the findings of that report and the more detailed reports on individual systems subsequently made available. The often encouraging levels of durability reported by BRE also bode well for the twenty-first-century incarnations of steel frame forms of construction considered in Chapter 15, enhanced as they have been with the benefit of an additional 50 years of research and development in materials science and building construction technologies.

Reinforced concrete frame

In situ *reinforced concrete frame*

In the domestic field, the surveyor is likely to encounter reinforced concrete frame walls in two forms. The first, and what could be considered now as more traditional, is where the frame has been erected *in situ* as the load-bearing structure for a small multi-storey block of flats. In fact, unless told, the surveyor would probably not know that the frame was of reinforced concrete. The alternative form of framing for this type of structure is of steel, but as this has to be encased in concrete for fire protection, protruding columns and beams look the same. Such a block in either material would be engineer designed, and unless there was any evidence of distress, it is unlikely that anybody would have been instructed to make an appraisal for strength and durability since the day it was built. With no sign of distress, the surveyor would have no reason to advise a client – whether a buyer, lender or investor – to have an engineers'

290 Recognisable as a typical reinforced concrete frame, brick-clad block of medium-rise flats of the 1950s. The remarks about the condition of the frame in **284** apply also to this local authority block. Service charges ought to be relatively modest as there is seldom a lift for a block of this height and each flat has probably, now, its own controllable central heating and hot water installation. Upgrades to windows and entry systems are always a possibility and the surveyor, apart from making his or her own estimate of needs, should press the freeholder for details of any future programmes.

report regarding a flat in the block. The surveyor's inspection and report can concentrate on examining what can be seen of the enclosure walls of the flat, the interior and the common parts, and the exterior of the block as a whole with a note on the structure being limited to a few comments. A significant amount of such construction took place from around 1920 to 1950, during which period *in situ* reinforced concrete construction in the domestic field was restricted generally to blocks of flats up to five or six storeys (**290**), leaving two-storey housing to be constructed by traditional means.

Prefabricated reinforced concrete (PRC) frame

The second form of reinforced concrete framed wall construction which the surveyor is liable to encounter is likely to be rather more problematic and follows the change which came about in the late 1940s when prefabricated units began to be used for framing low-rise housing. Eventually, the panels used for infilling were developed into the large panels with bolted connections for the high-rise developments of the late 1950s and 1960s. At the same time, prefabricated reinforced concrete (PRC) came to be seen as the panacea for overcoming the shortages that were compounding the housing problems of the time. The vast majority of system-built dwellings constructed between 1945 and 1975 were erected for local authorities, who also had responsibility for maintaining them as part of their wider tenanted housing stock. The introduction of Right to Buy legislation in 1980 meant that substantial numbers of local authority houses were rapidly transferred into private ownership as tenants took advantage of very significantly discounted prices. Some 200,000 were sold to tenants in 1982 alone; more than a million were sold by 1987. Naturally a representative proportion of these sales involved system-built properties for which the new owners took over the responsibility for maintenance and repair.

Before the introduction of Right to Buy, there had been virtually none of these system-built properties in the private sector. During the early 1980s, defects in design and construction were discovered in some of the system-built types constructed before 1960. Lenders, having financed tenant purchases, soon recognised their newly acquired exposure to the risks associated with these potentially defective systems and became concerned that similar defects might be present in other types. Lender concerns were compounded because residential surveyors and valuers advising lenders and private purchasers had little or no experience of inspecting or assessing these non-conventional forms of construction. Before long, the government of the time recognised the

Table 4 Types of prefabricated reinforced concrete (PRC) dwellings designed before 1960 and designated in the 1980s and 1990s by the government as 'defective by reason of their design and construction'

Dwelling type	
Airey	Shindler
Ayrshire County Council	Smith
Blackburn Orlit	Stent
Boot Beaucrete	Stonecrete
Boot Pier and Panel	Tarran
Boswell	Tee Beam
Cornish Type 1	Ulster Cottage
Cornish Type 2	Underdown
Dorran	Unitroy
Dyke	Unity Type 1
Gregory	Unity Type 2
MacGirling	Waller
Myton	Wates
Newland	Wessex
Orlit	Whitson-Fairhurst
Parkinson	Winget
Reema Hollow Panel	Woolaway

Note: Some other names of types appear on lists published earlier than the above BRE list of 2004. These are Butterley, Hawksley, Lindsay, Stour and Unit; they may be alternative names or variants of the above types. If encountered, dwellings with these names should be considered in the same way as those listed in this table.

embarrassment of promoting the sale of defective and potentially unmortgageable houses to former local authority tenants and of expecting the new owners to carry out major structural repairs. BRE research rapidly identified which systems were the main culprits; these were formally designated 'defective' in legislation, which instituted schemes for replacement or repair for their new owners or allowed the resale of 'defective' properties back to the original authority. The distinguishing characteristic of all the defective types was that their construction utilised prefabricated reinforced concrete (PRC), so it is not surprising that most of the systems affected were concrete framed. Since then, the 34 designated defective types have been, and still remain, unmortgageable while in their original form (see **Table 4**).

The discovery of the main problems with PRC dwellings followed a fire in an 'Airey'-type house. During reinstatement, the structural frame – normally hidden in this type – was exposed, revealing spalled concrete and rusting steel reinforcement in the columns. This was due, principally, to carbonation and the addition of chlorides to the concrete mix, intended to accelerate hardening in the rush to produce more prefabricated units at an even faster rate. Carbonation is a process whereby concrete absorbs carbon dioxide from the atmosphere and changes from being alkaline to being slightly acidic. The proportion of chlorides added to the mix tended to be widely variable but, in combination with the carbonation, produced an acidity far in excess of what would have been considered safe. Unlike the *in situ* concrete columns and beams used in multi-storey frames, the prefabricated units used for low-rise housing were relatively small. As, in many instances, there was little concrete cover to the steelwork in the first place, it was not too long before the expansion of the rusting reinforcement caused the concrete to crack and spall, compromising strength and allowing yet more oxygen and moisture to reach the steel and progress the rusting process even faster. There is no practical way of preventing this progressive deterioration in the PRC units made in the late 1940s, 1950s and 1960s, though its rate varies according to how much shielding of the units is provided by the design of the dwelling and the degree of maintenance.

For these reasons, the replacement of the original structural elements was thought to be the only sure way of ensuring that a dwelling would have a lifespan of at least 60 years, comparable to a dwelling of traditional construction, as required by lenders. For those being retained and not restructured, BRE recommended regular inspection for any signs of cracking, which, it says, should be seen as a warning, particularly at the corners where it considers the columns and the ring beam more likely to deteriorate earlier. For blocks of flats above two floors, appraisals of the structural stability and durability should be obtained from a structural engineer, which may involve the need for opening up, analysis of the concrete and examination of the steel.

In further research during the 1980s, BRE identified many other examples of system-built dwellings, publishing information in 1986 and 1987 on 30 which were steel framed, extending their studies to 54 timber-framed systems built between 1920 and 1975 and publishing this information in 1995. BRE acknowledged at the time that there were many more different types and, in 2004, published a book of 940 pages entitled *Non-Traditional Houses*, identifying the types built in the UK between 1918 and 1975. This book, a classic of its type and now available as a digital edition, details 450 house types categorised by form of construction along with notes for surveyors. Included are the comparatively rare and few in number as well as those which were built in quantity and are better known. Covering so many systems, generally by way of a two-page spread, this publication understandably contains less detail regarding some systems than was already available from BRE on the constructional aspects and the flaws which have become apparent over the years. Nonetheless, while reference might still need to be made to other material, the book provides extremely valuable information and can be regarded as the first port of call whenever a form of pre-1975 system building is encountered or suspected by a surveyor.

Any prospect of comparatively easy and instant recognition of types is complicated by the fact that so many were deliberately intended to look like other forms of construction. Difficulties are often further compounded by changes in appearance brought about by enveloping and over-cladding carried out by local authorities and by private owners seeking to make their homes look different from those of their neighbours. Another difficulty involves those properties originally built as one of the designated types and since repaired under a licensed scheme which, by definition, eliminates any contribution from the defective elements of the original building to the structure of the repaired building. Under some schemes, this could amount to complete demolition and reconstruction, but under others, residual parts of the original building could remain providing they did not act in a load-bearing capacity.

While the ability to recognise some of the more obvious types of system-built dwellings and their repaired replacements, where these are known to have been built in a particular area, is a useful attribute and would obviously save time, it is far too much to expect a surveyor to carry around a mental image of so many different systems of construction – even if it were possible to recognise their differences, which, according to BRE, in many cases it is not. Far more useful is the surveyor's routine of inspection, which should throw up the fact that a dwelling is not of traditional construction. This will trigger the search for further information that will be essential before the report can be compiled and may even involve a second visit to the dwelling.

All told, about 1 per cent of the total housing stock comprises dwellings that were system built using PRC components. Although 34 of the systems used for low-rise houses designed before 1960 have been designated as 'defective by reason of their design and construction' through the provisions of the Housing Act 1985, it is, in fact, only 19 of these types that account for 98 per cent of the total stock of PRC houses of this period. These 19 systems are: 'Airey', 'Ayrshire County Council', 'Boot Pier and Panel', 'Cornish' Types 1 and 2, 'Dorran', 'Myton', 'Newland', 'Orlit', 'Parkinson', 'Reema Hollow Panel', 'Stent', 'Tarran', 'Underdown', 'Unity' Types 1 and 2, 'Wates', 'Whitson-Fairhurst' and 'Woolaway'.

To an extent, the designation as 'defective' makes the surveyor's task somewhat easier since it means that unless the dwelling has had its structural components replaced by one of the approved licensed repair schemes – usually that certified by PRC Homes Ltd and provided with a NHBC Certificate – it is not mortgageable. This fact obviously puts off anyone who would require a mortgage, and a purchaser who could buy for cash would clearly need advice on value to take account of the need to do the work if wishing to make it mortgageable. The cost of the necessary work is substantial, as can be gauged by the maximum government grant then available of £24,000, which represented 90 per cent of the estimated cost of the work in the early 1990s to the worst-affected PRC houses such as Dyke, Hawksley, Orlit, Schindler, Waller and Wessex.

Periodically, unrepaired dwellings come on the market and some cash purchasers are prepared to take them on without having any intention of undertaking any of the structural replacement work, simply taking a chance on durability. The surveyor then needs to explain the risks and implications. Reference would need to be made to the findings of BRE set out in their report, following their 1980s inspections, on the structural condition of the particular type of PRC dwelling and due allowance made in any valuation which might be required.

It is not unlikely that from time to time a surveyor will turn up to inspect and report on a property not having been told in advance that it is an unreconstructed PRC dwelling and without the seller knowing the nature of the original construction. While it may be essential to fall back on the procedures for establishing whether a dwelling is of traditional or nontraditional construction and the necessary follow-up enquiry, it would obviously save time if the surveyor can readily identify some of the more common types.

The hazards of not making an appropriate identification are well established and £9,000 was awarded as damages in a Scottish court when a valuer failed to identify that the dwelling he was inspecting for a mortgage valuation in 1982, and whose report was disclosed to the purchaser, was of nontraditional PRC 'Dorran'-type construction (*Peach* v *Iain G. Chalmers & Co.* 1992). The house had been built privately in 1976 in a good location in the north of Scotland and had a swimming pool. The surveyor described the walls as 'concrete block built, harled' and valued it as a traditionally built house at a mere £1,000 below the asking price of £35,000. A pair of 'Dorran' system-built semi-detached houses are illustrated at **291**. This shows quite clearly the narrow, full-storey-height concrete panels, the projecting ring beam at first-floor level and the fact that the windows are set flush with the external wall face – all features typical of the system. In *Peach* v *Iain G. Chalmers & Co.* the surveyor's description of the walls as 'harled' could well have meant the joints of the concrete panels were obscured, but the other features were evident, particularly the windows set flush with the external face in walls very much thinner than walls of traditional construction. Although the BRE report on the structural condition of 'Dorran' system-built dwellings was not published until 1984, evidence was produced in this case to show: that problems with PRC dwellings had been well publicised in 1981; that in the locality where this one was situated, they were known for what they were; and that those sold in 1982, when the inspection took place, fetched between £19,000 and £25,000, according to location and condition, which was less than houses of traditional construction so as to allow for possible future problems.

The point to be drawn from this case is that even if the surveyor had not kept up to date with his reading and, on, perhaps, a fairly cursory inspection for a mortgage valuation, had not noted the features which distinguish a 'Dorran' type of PRC dwelling, a set routine of inspection ought to have shown up the house as of nontraditional construction and different from other harled houses in the area – if only because of thin walls – thus putting him on enquiry.

As it happens, the BRE report of 1984 gave 'Dorran'-type houses a comparatively clean bill of health. It did this along with others such as 'Myton', 'Newland' and 'Tarran', based on similar structural principles and differing only to the extent, where visible, on whether steel or timber was used for the roof structure. However, the report indicated that, like all PRC houses, they suffered from corrosion of the steel as well as cracking, spalling and rust staining of the concrete and, accordingly, they were all subsequently designated 'defective'. Nevertheless, at the time, BRE did not consider there was need for urgent action, just repair and proper maintenance. They were described as of a type where stability relied on the ability of the wall panel units to support the load, with the restraints provided at first floor and eaves levels by tie bars connecting the part of the ring beam in the front and rear walls and the party and gable walls. Early signs of any distress would become apparent through cracking in the ring beam units and bowing of the wall panels – defects always to be looked out for, according to BRE, by surveyors inspecting any of these types, perhaps for a purchaser where a mortgage is not required and where it is apparent that the original structural elements are still in place.

What of system-built dwellings that no longer look, at first glance, anything like they did when originally built? The illustration at **292** shows another pair of semi-detached houses built on the 'Dorran' system. The storey-height, distinctive concrete panels have disappeared and so has the projecting beam at first-floor level, while the windows and the whole appearance up to roof level have a characteristic 1980s look. Another look at the illustration, however, will show a further pair of semi-detached houses in the background, which could provide an obvious indicator for the inspector as to age and type. Most of the nontraditional dwellings are located on fairly large estates, so looking around nearby could provide the answer to a problem where the dwelling being inspected looks different from the others, as it does here.

On the other hand, what if all the houses on an estate are much the same? There are visible clues on the exterior of this pair of houses; both have hipped roof slopes and chimney stacks so a date of construction in the 1940 to 1960 period is much more likely than one in the 1980s. This should make the surveyor immediately aware that this is a dwelling which has undergone some refurbishment and was originally built some 20 to 30 years earlier. In some instances, the type of dwelling may not be immediately apparent, but there should be sufficient evidence from measurement and soundings and internal details, including features in the roof space, to determine that the house is not of traditional construction – that evidence to be verified from the BRE report when further enquiry reveals the type of system.

A distinctive type of PRC dwelling, shown at **293** as originally built, is the 'Cornish Type 1'. Another readily recognisable type is the 'Airey', illustrated at **298**. Reconstructed 'Airey' types are also shown at **298** as well as at **299**, and the approved schemes of reconstruction mean that they would be eligible for acceptance as security for a mortgage advance. 'Orlit' houses with flat roofs are shown at **300**, and **301** illustrates 'Orlit' houses with pitched roofs and new windows.

291 Original 'Dorran'-type PRC houses.

292 Refurbished 'Dorran' PRC houses.

293 In its original condition, probably the easiest to recognise of all the PRC system dwellings of the late 1940s and 1950s, subsequently designated as 'defective by reason of design and construction', is the 'Cornish Type 1' with its exposed frame and mansard roof; although some did have vertical hung tile first-floor walls and hipped roofs. The characteristic whitish appearance of the dense concrete is due to the use of washed quartz sand, a by-product of the China clay industry.

The structural units of the easily recognised original 'Cornish Type 1' and 'Type 2' dwellings along with those of the 'Dorran', 'Myton', 'Newland', 'Reema Hollow Panel', 'Stent', 'Tarran' and 'Wates' types are all visible and allow for the regular inspection advised by BRE for signs of cracking and spalling of the concrete. BRE said in 1984 that if no cracking or signs of movement were seen, it was reasonable to assume that the PRC components of those types were in sound condition at the time of inspection.

Those PRC dwellings where the components are hidden – 'Airey', 'Ayrshire County Council', 'Boot Pier and Panel', 'Orlit', 'Parkinson', 'Under-

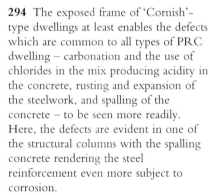

294 The exposed frame of 'Cornish'-type dwellings at least enables the defects which are common to all types of PRC dwelling – carbonation and the use of chlorides in the mix producing acidity in the concrete, rusting and expansion of the steelwork, and spalling of the concrete – to be seen more readily. Here, the defects are evident in one of the structural columns with the spalling concrete rendering the steel reinforcement even more subject to corrosion.

295 In many PRC dwellings, the cladding panels obscure sight of the structural frame. Here, it has been necessary to open up so that the serious cracking at the foot of one of the reinforced concrete columns can be revealed.

down', 'Unity', 'Whitson-Fairhurst', 'Winget' and 'Woolaway' – need opening up for examination of the structural units if any opinion is to be given; although the five of these which are rendered – 'Boot', 'Parkinson', 'Underdown', 'Winget' and 'Woolaway' – may show early signs of problems in the structural components by the presence of cracks in the rendering.

With all PRC system-built dwellings, if cracking or movement is found then the cause and significance need determining because deterioration is progressive and could be at a faster rate than with traditional construction.

A few of the individual characteristics of some of the PRC dwellings may help surveyors recognise immediately that the dwelling being inspected is of nontraditional construction. For example, 'Stent' dwellings – some of which were rendered and some left with the structural components visible – demonstrate a novel way of fixing the roof tiles with clips to a steel mesh which was laid on roof trusses of either steel or timber. Lightweight steel roof trusses distinguished Ayrshire County Council dwellings from the very similarly constructed Whitson-Fairhurst type, which had traditional timber cut and fitted roofs, though both, built in Scotland, had the typical timber ground floors and subfloor spaces even when, elsewhere, construction on a concrete base forming the ground floor was more common.

Another PRC dwelling type, although it might at a quick glance be confused with a traditionally built brick dwelling, could at least be distinguished from other PRC dwelling types. The 'Smith' system type was faced with brick

296 It was not only terraced and semi-detached dwellings of two storeys that were built using PRC systems. This designated 'Tarran' bungalow – still mainly as originally built – like its two-storey relations, has wall panels that are storey-height and 400mm wide; but here these are supported on a timber kerb. The timber tends to deteriorate over time as do the concrete elements, in the same way as with other PRC dwellings. The roof is of lightweight profiled asbestos sheeting as there is no concrete ring beam to support anything heavier.

297 The cost of removing the structural elements of a designated 'Tarran' bungalow and replacing them with a new structure may not be cost-effective. Here, the enveloping option has been taken; the wall panels have been provided with an external insulating render coat and the original asbestos roof replaced with lightweight profiled steel sheeting to give the impression, at first glance, of profiled tiles. The useful life of the bungalow could well be extended by 20 to 30 years given adequate maintenance.

298 Original 'Airey' PRC houses with their very individual appearance (on the right) and a reconstructed house (on the left).

299 Reconstructed 'Airey' PRC houses.

slips 30mm to 40mm thick, bonded to concrete slabs in a factory and forming a solid wall. The brick slips formed a pattern in stretcher bond with two-course soldier arches over window and door openings; but, as the panels were joined together on site, matching was not always carefully done and a mottling effect was sometimes apparent. Because walls were solid – 200mm thick at ground-floor level but a mere 50mm at first floor – damp penetration was often a problem.

300 Original 'Orlit' PRC houses with flat roofs.

301 'Orlit' PRC houses with new roofs and windows.

302 A pair of semi-detached two-storey houses built with the designated 'Unity 2' system of construction where the pattern of the concrete panels, resembling blockwork laid with continuous joints, can be discerned. This system was also used for flats. BRE considered that added strength was provided to the 'Unity' system by the fact that the flats had concrete floors, concrete stairs and landings, internal walls of blockwork and brick-clad gables. As a consequence, blocks of flats built with this system were not designated 'defective' and are, therefore, technically mortgageable – although, in practice, this must be in doubt.

303 A pair of two-storey semi-detached designated PRC houses, shown as originally built on the 'Woolaway' system where the structural components are usually totally hidden by a machine-applied roughcast rendering. In this photograph, the underlying structure is apparent on the house in the foreground, in the course of redecoration, but will be concealed again once redecoration has been completed. The design is based on storey-height aerated concrete columns at about 750mm centres with wall panels and ring beam of the same material – the only restraint, however, being provided to the front and rear walls by the roof. Wall panels are fixed to the 150mm square columns by bolts – corrosion of which can provide a distinctive pattern of staining – the columns from the window and door jambs, the wide column division between the two sections of the large windows to the main rooms at ground- and first-floor levels being a distinctive feature. In the 1980s, BRE reported that the columns showed the most advanced degree of carbonation, which was due to the use of aerated concrete. There was also some cracking to the columns and some further cracking in the panels. However, they considered that carbonation was more apparent internally since the rendering gave some protection to the surfaces facing towards the exterior. Note also the characteristic green colour of the copper-clad chimneys.

It will be noted that early Wates PRC dwellings, as originally built, had a single-flue chimney stack more or less in the centre of the rear slope. This was unusual for the time when it was common to have at least a two-flue stack or sometimes one of four flues on the party wall line. The single flue was to serve a central warm air unit supplying warmth for the whole house and which had a cast iron flue pipe. The units have probably long since gone but the stacks generally remain on those dwellings that have not had their structural units replaced, either still looking as built originally or merely enveloped to improve insulation standards.

15

Innovative forms of construction

Introduction

Innovation in construction is nothing new; even something as ubiquitous and traditional as brickwork didn't become common until the seventeenth century and concrete roofing tiles only came widely into use after the Second World War. The most concentrated period of innovation and experimentation in residential construction in the UK was probably during the years immediately following the Second World War. At this time, the nation's desperate need for housing combined with the necessity to convert manufacturing capacity from wartime production resulted in the wholesale adoption by local authorities of an almost indiscriminate variety of factory-based methods of mass construction for residential purposes. The consequences of this process, far-reaching and in many cases unfortunate, have been mentioned in several earlier chapters, most notably in the final section of Chapter 14, dealing with the construction system most seriously affected, that of PRC housing ('Prefabricated reinforced concrete (PRC) frame'). This experience of innovation in construction has coloured the views and perceptions of a whole generation of surveyors, valuers and lenders who have consequently regarded anything of nontraditional construction with considerable scepticism, if not outright hostility.

It is no exaggeration to state that as a direct consequence of the PRC debacle, the only significant innovation in whole-house methods of construction during the second half of the twentieth century has been the adoption by some builders of timber-framed construction. Even this suffered a serious setback for a period after adverse reaction to a *World in Action* television programme in 1983, which highlighted poor detailing during construction on some sites. The way the market responded to this programme only served to reinforce the concerns of residential valuers and surveyors about innovation in construction methods, a legacy that persists in this sector of the profession.

It would certainly be wrong to minimise the difficulties that PRC housing has caused and continues to cause: lenders, valuers and homeowners are all still, literally, paying the price for mistakes in identification. As recently as the early years of this century, the second edition author advised in a case involving this type of construction which ended with a six-figure settlement; so it is not surprising that the professional scars remain raw. Nevertheless, these issues are so deeply entrenched in the professional consciousness that it might arguably be more accurate to describe the reaction of many surveyors to nontraditional systems as a phobia than as prudent caution. Construction research and materials science have made significant advances in the half-century or so since the post-war era. It is taken for granted, for example, that cars and aeroplanes are not built now in the way that they were in the immediate post-war years; so it is unreasonable to simply assume that nontraditional houses can only be built in the same way and with the same intrinsic flaws as those which were constructed 60 or more years ago. Consequently, it is becoming increasingly necessary for residential surveyors and valuers to take a wider view and, where necessary, shake off some preconceptions.

In fairness, it should be stated that the surveying profession is not solely responsible for the lack of innovation in residential construction. In every construction sector, other than residential housing, off-site construction methods are now widely used; but the mainstream private sector housebuilders have, in the main, shown little interest in moving from their tried, tested and extremely efficiently managed business models using essentially traditional forms of construction. There has been increasing use of individual components manufactured off-site – for example, roof trusses and window units – but there has been a distinct reluctance, with some notable exceptions, to move to whole-house, off-site methods.

During the first decade of this century, government recognised that off-site methods of house construction offered a number of significant advantages over traditional methods. These include improved quality assurance, reduced time on site and less waste, all due to utilising manufacturing processes. For several years the social housing sector was targeted by government to increase their use of off-site construction, defining the desired technologies as Modern Methods of Construction (MMC). The term MMC has since become widely adopted as generic shorthand for non-conventional methods and its original distinction has been blurred. As a result, it is better to avoid its use altogether and, when not specifying a particular system, to refer more broadly to innovative, non-conventional or nontraditional forms of construction. The consequence of government pressure on social housing providers is that their readiness to utilise nontraditional forms is now significantly greater than is the case with builders in the private sector. Meanwhile, private sector builders were eventually challenged by the introduction of various codes that require sustainable construction, which they recognised could not ultimately be met without making significant changes to traditional methods of construction.

Finally, therefore, the housebuilding industry is beginning to come to terms with the new world of sustainable construction and higher levels of energy efficiency. The Green lobby have been promoting the desirability of these technologies for some years but they are now starting to be considered by mainstream builders. Providers of alternative forms of energy supply and innovative forms of construction have received formal recognition by government and the possibilities have started to be regarded, not merely as interesting or unusual curiosities, but as realistic alternatives worthy of serious consideration. Experience suggests that during the coming years there will be a 'shakeout' as some systems prove more cost-effective and durable than others and gradually the more favoured alternatives will become established, first as best practice and eventually as standard methods which will be widely adopted. Unfortunately, there is no way of predicting which these will be.

At the same time, there is an increasing recognition that the needs of the existing housing stock must be addressed. In many ways, this presents an even greater challenge than introducing innovation to new-build construction. With new build, everything can be done from scratch; but with existing housing, there are innumerable variations to deal with so standardisation is very much more difficult. With the existing housing stock, there is also the very practical problem of managing owners and occupiers and their lives while their properties are being upgraded.

So how does a professional surveyor or valuer inspect, make a judgement and report on these unfamiliar technologies while so many of them remain unproven? Regrettably there is no easy answer to this dilemma. The required measure of expertise will continue to be the one set down by the courts – that of the 'reasonable surveyor or valuer'; no more, but certainly no less. It is incumbent upon members of the profession to remain as up to date as possible by judicious reading of professional and technical publications and by attending seminars, conferences and presentations related to their relevant area of practice. There are no shortcuts, and those who try to take them will ultimately and expensively be found out.

Far from despairing, however, surveyors should be embracing the opportunities which these changes present. This is a field where the advice of an independent expert will be increasingly valued, and those who rise to the challenge of establishing this as one of their areas of expertise will be operating in an increasingly important and influential part of the profession. The political complexion of government may influence priorities or initiatives but the necessity of responding to the increasing costs of energy and the impact of climate change will ensure that dealing with developments in new build and implementing improvements to the existing housing stock (and, the cynical will add, remedying those efforts which prove unsuccessful) will provide meaningful and potentially lucrative employment for those who practice in this area.

Different systems

Most surveyors and valuers who routinely inspect residential properties will differentiate in their minds between the familiar 'traditional' or 'conventional' forms of construction and 'modern' methods or 'innovative' forms, about which they will confess to little in the way of expertise. If they reflect for a moment, however, those who have been practising for any length of time will recognise that over the years numerous innovations in construction have been incorporated in the new-build properties which they have been inspecting on a regular basis. Prefabricated timber trusses have been used for decades but were once revolutionary. More recently, plastic hot water pipes, OSB (oriented strand board) joists, and light gauge steel-framed internal partitions are only three of the building components which are now accepted without question when inspecting a residential property in the course of construction; yet these would never have been seen until the very last years of the twentieth century. There seems to be a, not unreasonable, assumption that if a novel form of construction has been utilised by a national builder then it must have passed some form of assessment and, therefore, comply with the relevant standards and requirements. When encountered by the surveyor for the first time, these innovations will raise a question or two; but when appropriate enquiries provide the necessary assurances, these will be accepted and added to the mental database of professional experience. In this way, there is already a creeping acceptance of innovation by surveyors and valuers, which is natural and understandable.

When confronted with a whole building constructed with an innovative system, however, the pressures being faced by the surveyor are very different. A single building component, if it proves to be fatally flawed in use, might conceivably be replaced – albeit with significant cost and disruption – but the building itself will remain standing and retain its essential value to the homeowner and mortgage lender. On the other hand, if an innovative building system proves fundamentally defective, the entire building and its whole value, the very asset held in mortgage or in ownership, may become worthless. In extreme cases, it might even acquire a negative value, becoming a net liability. Therefore, the issue facing the surveyor is fundamentally different when encountering an innovative form of a single building component for the first time compared with the professional challenge posed by an innovative form of whole-house construction.

Identification of nontraditional forms of construction can be problematic, to say the least, largely because finished homes are deliberately clad with materials which make them difficult to differentiate from standard brick and block construction. The main exception to this general practice is larger blocks of modern flats, many of which are deliberately designed to have a striking appearance. Smaller, infill flat developments, by contrast, are usually conventionally finished, such as **283** (Chapter 14).

Consideration of retrofit solutions to the existing housing stock is beyond the scope of this book, and the remainder of this chapter is only intended to give an initial overview of a sample of innovative systems. Those seeking more information on a wide range of systems and helpful guidance on identification are strongly recommended to obtain a copy of a small but excellent book by Keith Ross entitled *Modern Methods of House Construction: A Surveyor's Guide.*

Closed panel timber frame

Modern timber frame construction, as described in the previous chapter (in the section 'Modern timber frame') has been in use for several decades. It is not generally regarded as an innovative form of construction and is not differentiated from conventional construction by many surveyors and valuers (**304**). In fact, it is more accurate to describe this as open panel timber frame construction because what arrives on site are essentially sections of an open framework. The framed sections can be erected rapidly but they still require a significant degree of post-erection work on site: the placing of insulation, services and internal finishes, quite apart from the external elevation treatment. These processes are all vulnerable to the vagaries of the weather and subject to the same quality management issues as conventional brick and mortar construction. One way of reducing time on site, and exerting a greater degree

304 Open panel timber frame construction has been in use for several decades and is now, effectively, regarded as conventional construction. It reduces the build time compared to bricks and mortar but still requires a significant amount of finishing on site. This is the ground floor in the course of construction of the building shown finished in **283** (Chapter 14).

305 Closed panel timber frames are completed to a much further extent in the factory so site time is much less than with open panel timber frames and quality control is easier to manage.

306 (*left*) Significant elements of construction can be completed in the factory and brought on to the site in an advanced stage of construction. Here, a large part of a roof structure is lifted into position.

of control over quality, is to transfer many of these processes from the site to the factory and produce 'closed panel' units. These start life in the same way as open panel timber frames but then the inner and outer faces of the panel are fitted, enclosing all the necessary insulation and, in some cases, all the service runs as well. This means that whole walls arrive on site, to be erected by trained teams who ensure that junctions between panels are properly formed and airtight (**305**). In the same way, large floor units and roof cassettes can be fabricated in factories and lifted into position as single units (**306**).

Structural insulated panels (SIPs)

Use of structural insulated panels (usually referred to as SIPs) is form of construction requiring a similar level of teamwork during the site erection phase to that needed for closed panel timber frame construction. A section through an SIP is reminiscent of an ice cream wafer (**307**). The sheets of material on either side provide structural strength and the thicker central core gives each panel stability and insulation. The most common outer sheet material is OSB but cement-based board is also used. The solid foam core insulation is usually of expanded polystyrene, polyurethane or polyisocyanurate. Some commentators regard the greatest advantage of SIPs as being their ability to meet increasing insulation standards without significantly increasing wall thickness. The panels are fitted together on site and a variety of external finishes can then be added (**308**). The artificial slate shown in **12** (Chapter 4) was fitted to an SIPs' structure.

307 This small section cut from an SIP shows the two outer sides of OSB, which provide structural strength, with the insulation bonded between them, giving each panel stability. Storey-height panels are used during construction.

308 This multi-storey block of student accommodation in the course of construction is using SIPs to support the outer cladding – the rectangular sheets in various hues of grey which have already been applied to the lower parts of the building.

Light gauge steel construction

Although widely used in non-residential construction, the thought of a dwelling with a structural frame of light gauge steel is likely to bring a residential surveyor out in a cold sweat as the spectre of BISF and similar houses come to mind. Associations of rusting reinforcement, which was the death knell of PRC houses, will not be far away. It is understandable that there should be reservations about embracing new technologies, especially if they are in some way tainted with the problems of the past; but appropriate design and specifications can minimise the risk of damp penetration, condensation and rust so modern steel-framed construction need not pose the risks understandably associated with properties constructed during the post-war period.

Steel frame construction, like timber frame construction, can be fabricated as open or closed panel systems, factory finished to a greater or lesser degree, and likewise taking a correspondingly faster or slower time to complete on site (**309**). As with the other construction forms considered in this chapter, a range of external finishes can be utilised so, once the building is complete, there is unlikely to be any obvious indication of the main structure (**310**).

309 A steel flooring panel being lowered into position on a development of flats of light gauge steel frame construction. Across the centre of the photograph, in front of the figure, are sections of open panel steel framing which will form the internal walls of one flat. Although the material is obviously different, the structural appearance of the open panels is not dissimilar to the open panel timber framing in **304**. In the foreground and behind the figure are sections of closed panel steel framing, which will provide the external walls. Once the entire structural framework of the building is in place, the exterior of the block will be roofed and finished with whichever cladding has been specified for this particular development.

310 Houses of light gauge steel frame construction. The unconventional appearance may provide a hint that the construction is not traditional, but the brick, timber and metal claddings and the metal roof do not provide any clear clues about the nature of the underlying form of construction.

Volumetric construction

When the principles of light gauge steel construction are taken one step beyond the fabrication of individual closed panels, it is possible to construct a complete room with a base, four walls and a ceiling, fully serviced, fitted out and decorated to the necessary specification, all within the factory (**311**). This might be a hotel room, a student bed-sitting room or a kitchen/dining room, but these are just a few of a whole range of possibilities. These complete rooms can then be transported to site and placed side by side and stacked with similar rooms before being enclosed in a whole-building envelope (**312**). Because the rooms are fully fitted out, the services to each unit only need to be connected to a central service point or core. In a hotel, the services for each room connection would be run along the corridor connecting each of the rooms.

311 Volumetric modules being manufactured in a factory. After the walls and top have been secured to the base of the module, doors, windows and service runs will be installed and the interior will be completely fitted out, including decorations, to meet the requirements of the particular contract. The finished module will then be transported to site.

312 The finished volumetric module is lifted from the back of a transporter into its final position in a row of similar units, on top of others and with another row to follow above. When the modules are in place, external finishes and roofing will clad the structure and the service connections for the individual modules will be linked to the central supply runs.

313 This multi-storey block of student accommodation has been built using volumetric construction. The external detailing and cladding with a variety of materials successfully disguises the rows and columns of identically sized units which were used to build it.

Using this method, it is possible to erect substantial multi-storey structures very quickly. An 11-storey block of 289 modules, providing two- to six-bedroom student flats and studios, for example, was erected from ground-floor slab to completion in only 35 days. This form of construction is typically used for hotels or student accommodation blocks, but there is no reason in principle why it might not be utilised for residential flats (**313**). Individual houses have been built in this way, but the real advantage of this method is its use of repetition which is more challenging with houses, especially those at the upper end of the market.

Thin bed mortar joint construction

A construction method widely used in Europe, especially in Germany, utilises large extruded honeycombed clay blocks or large blocks of aerated concrete which are bonded with very thin layers of mortar (**314**). The mortar is only a few millimetres thick and acts more like glue than conventional mortar, binding the blocks to each other. As with most of the systems considered in this section, the need for accuracy in setting out and construction is paramount because there is no opportunity to 'level up' any unevenness by modifying the thickness of the mortar courses, as is possible with conventional brick and mortar construction. This need for precision is another cause of scepticism amongst residential surveyors and valuers whose expectations of the building industry are coloured by experience on residential building sites. In other construction fields, however, there is no automatic assumption that accuracy

314 The size of the clay blocks and the minute thickness of the mortar bed used in thin bed mortar joint construction are illustrated in this image.

cannot be achieved. Extremely complex civil engineering projects are completed with millimetre exactness; and with laser levels, there is no practical reason why similar precision cannot be obtained on the most modest residential construction site if the will is there.

Insulating concrete formwork (ICF)

Insulating concrete formwork or insulated concrete formwork (ICF) is, in some ways, the opposite of SIPs in that the structural component is in the centre of the finished unit and the insulation is on the outside. The method of assembly is also very different. Whereas SIPs are usually storey-height panels, ICF structures are made by slotting together a series of relatively small, hollow blocks of expanded polystyrene into a wall and then pouring concrete into the central gap, with or without reinforcement depending on the requirements of the specification. This is a popular form of construction with self-builders because the individual expanded polystyrene components are light and easily carried about and positioned on site prior to placement of the structural concrete. Once the structure has been completed, the desired interior and external finishes can be applied (**315**).

315 The main structural walls of this house, built using insulated concrete formwork, have been completed and await the external cladding. A pile of unused expanded polystyrene formwork blocks, with their hollow centres, is visible at the front corner nearest the camera.

Cellulose-based technologies

The increasing interest in and awareness of embodied energy associated with construction materials has led to certain crop-derived materials, such as straw bales, being examined as potential low-impact building materials. A research paper on this subject, *Cellulose-based Building Materials: Use, Performance and Risk*, was published by the NHBC Foundation in 2013. As an aside, it is worth mentioning that the research paper was also partly in response to the renewed interest in the utilisation of thatch and cob, and it made some interesting and helpful observations about these historic and very traditional construction methods though they can hardly be described as innovative. Arguably, the most interesting and certainly innovative development in this area, however, has been in the use of dried hemp, grown in East Anglia, combined with lime in a mixture known as hemp lime or hempcrete. This material can be worked as a building material by being either sprayed, cast or moulded into blocks, and it is used for walling either as *in situ* construction (**316**) or as factory-produced panels, usually with a timber frame.

Hemp lime has good insulation qualities with some thermal mass but it is also a low-density material so it results in a relatively lightweight construction that imposes reduced foundation loads. A demonstration house using hemp lime was built some years ago to Level 4 of the Code for Sustainable Homes at the BRE in Watford. It has since been used both residentially, including in a social housing development (**317**), and commercially, with number of high-profile clients – notably Marks and Spencer for their flagship carbon-efficient store at Cheshire Oaks, Ellesmere Port.

316 This wall of hempcrete, a proprietary form of hemp lime construction, has been left unfinished for demonstration purposes and shows the face of the material after removal of the formwork. This would normally be concealed with an external rendered finish. The horizontal line shows the junction between two lifts as the wall increased in height.

317 A social housing development of houses and flats built wholly using hemp lime construction.

Accreditation and warranties

The training, background and experience of residential surveyors and valuers – not to mention their traditionally conservative mindset – mean that many are likely to find it a challenge to embrace these new forms of construction technology. Most, however, while not welcoming them will at least be prepared to come to terms with them. This process of adjustment will be greatly facilitated if the technologies are supported and underwritten by one or more of the various warranty schemes that lenders require to be in place before being prepared to lend on new-build homes. The market leader is NHBC but one of their prerequisites for external elevations is that there should be a ventilated cavity to reduce the risk of damp penetration. Some innovative forms, such as insulated concrete formwork, do not require a ventilated cavity for the integrity of the structure, but one can be added to meet NHBC requirements. With other technologies, for example hemp lime or cob, the addition of a ventilated cavity would add significant complications, which can actually work against the method being used.

Alternative warranty schemes have differing requirements but their starting point for assessment is usually based around conventional forms of construction, so it may be necessary for builders and developers with innovative systems to modify their basic product to meet a particular scheme's demands. One scheme, however, the Buildoffsite Property Assurance Scheme (BOPAS), has been specifically designed with innovative systems in mind, with the objective of determining whether any given construction method can be assessed and accredited to standards approved by the Council of Mortgage Lenders (CML) and the Building Societies Association (BSA). Under this scheme, the development of which was facilitated by RICS, innovative providers have their construction technology subjected to a rigorous assessment process by BLP (Building Life Plans), which is underwritten by Lloyds Register. If the technology meets the necessary standards, it is accredited as 'suitable in principle' for mortgage finance by mainstream lenders and will be listed as such on the BOPAS website (www.bopas.org).

Once a development utilising one of the BOPAS 'in principle' approved technologies starts on site, a further assessment regime checks that construction is following the specification which was accredited. The BOPAS website operates at a plot-specific level so information about individual properties, along with details of their construction and the technology being used, is readily accessible to surveyors and valuers. Those with the communications capability can access the BOPAS website while still at the building site with the property being inspected. This overcomes the perennial problem faced by valuers and surveyors when confronted on site with a new construction technology: the difficulty of not being able to obtain necessary information about what they have been instructed to inspect. The website will confirm the form of construction, while the mere presence of the plot on the website means that the construction technology has already been approved in principle for mortgage lending. The final decision on suitability of any individual property, however, will of course be subject to the valuer's completed valuation report following an inspection.

A further advantage of the database supporting the BOPAS website is that it is not simply intended for use at the first sale of a new dwelling but will continue to be maintained for the full life of the property. This means that it will not only be of value in assisting assessment of suitability for mortgage finance during every ensuing transaction but it can also be used by surveyors who are asked to advise on any aspect of the property – such as alterations or extensions – to gather information on the construction and any implications for structural modifications. Similarly, surveyors undertaking pre-purchase inspections will be able to determine whether any specific maintenance requirements have been complied with during the life of the property or, perhaps, whether extensions or alterations have been properly carried out, bearing in mind the underlying form of construction.

16

Wall details and applied surfaces

Wall details

Arches and lintels

Openings for windows and doors in mass load-bearing walls of both solid and cavity construction are a source of weakness. If they are numerous and too close together then thin, weak, narrow piers may be formed. If the openings are not adequately spanned across at the top by an arch or lintel of brick, stone, timber or steel, there may be settlement and undue thrust on the supports on either side. This is not quite the case where framed walls are involved since openings are normally provided in the panels at the construction stage. If additional openings were found to be necessary later, it would be foolish to cut and remove part of the frame, although it is not unknown for this to be done. An error such as this is more likely to have been made internally to form an opening in a partition where it was not realised that the partition was framed in such a way as to be virtually self-supporting. Sagging in overloaded floors can be the result.

Brick-built dwellings are usually provided with brick arches, which can take various forms, some more effective than others. The misnomer of a brick lintel being called a 'soldier arch' is unfortunate, and the difference between a brick arch and a brick lintel should be recognised by surveyors and the correct descriptive term used. Both have near flat horizontal soffits, but the true arch has wedge-shaped voussoirs while for a brick lintel, the 'voussoirs' are set truly vertical with the bricks generally laid unshaped on end but, occasionally, on edge. The span of a brick lintel should not normally exceed about 900mm unless provided additionally with a steel bar or angle. The flat brick arch is not greatly stronger than a brick lintel but will safely span an opening roughly half as much again; the length can also be increased by the use of a steel bar or angle.

Both flat arches, in gauged and rubbed brickwork, and brick lintels will be encountered frequently, as will segmental and semicircular arches. Less frequently encountered will be pointed arches in, among others, their Venetian and Lancet forms and the weaker, four-centred Tudor form. Semi-elliptical arches are considered the weakest for taking the load from above while those that take on more of an equilateral shape, where the rise approaches the span, are considered the strongest.

For dwellings built up to around 1920, one of the most frequently used forms of construction employed either brick arches or lintels in combination with timber lintels. The arch or brick lintel, for the sake of appearance, would carry the first outer half-brick thickness of walling with timber lintels to take the remaining load, the number of lintels depending on the total thickness of the wall. This brought the outer face of the timber to within 100mm to 120mm of the exterior. Exposure to driving rain would often mean that the lintel would be affected by damp penetration through the joints of the brickwork and, with little ventilation in such a confined space, rot would eventually set

318 This window opening in a multi-storey late Victorian dwelling is spanned partly by dual timber lintels, the outer face of which is positioned a mere 100mm to 120mm from the exterior. If the wall was in an exposed position, driving rain could easily penetrate to the timber and induce rot. In such a confined and unventilated space, it would almost certainly be dry rot. Most of the load from the one-and-a-half-brick-thick wall above the opening is carried by the lintels, but the outer half-brick thickness is supported by the top section of an artificial stone window surround. This is made up from partly hollow clay units placed in a mould and cast in sections with a stucco type of cement mix, part of which can be seen on the right of the photograph before replacement.

319 The flat brick arch above the lower window, partly revealed by the fallen rendering, has distorted badly, probably due to rotted timber lintels. This has led to the brick courses above becoming out of level and the sill of the window in the floor above to fracture. The movement seems to be old but telltale signs of long-term leakage down the line of the nearby replacement downpipe suggest a possible source of dampness.

320 The rebuilding of brickwork to the left of the lower window on this illustration suggests that major problems have been affecting stability. These have disrupted the setting of the arch leading to fractured brickwork above, the crack extending through all brick courses and fracturing the sill of the window on the upper floor.

321 The very severe movement, perhaps due to differential settlement, causing the long, vertical crack, which is wider at the top and which passes round the left-hand windows at both levels, has so disrupted one of the arches that a voussoir has dropped at least 75mm with consequential disturbance to the brickwork above. As a matter of serious concern there is a possibility that the dislodged shaped brick could fall out, to the considerable danger of passers by, if not dealt with urgently, even before due consideration is given to other repairs.

in. The timber lintel would originally be sharing the load from above with the brick arch or lintel; but when weakened by rot, the bulk of the load would be transferred to the brickwork, upsetting the stability. In the likely event of the arch settling, or being pushed outwards, brick courses above will settle and cracks may be induced.

Very popular in the period up to 1920 were both real and artificial stone mouldings used for decorative purposes around windows and doors. Should the supplier's advice not have been followed in that the mouldings were expected to span too great a distance, there will be a tendency to fracture and they will almost certainly do so if there is settlement or subsidence in the piers between openings or if there is an inadequate bearing at each end.

Band courses, cornices and window surrounds

A dwelling with little or no decoration at all in the form of projections of any sort can look somewhat dreary. Housing of the immediate post-war period is a case in point (**322**). However, on the other hand, it has to be admitted that such projections can sometimes be the cause of problems (**323**). Ideally they should all be covered with a flashing if damp is to be prevented from entering the main structure and appearing internally. Another factor is that they can be a danger, if neglected, to the extent of falling off and causing injury.

To a degree, inspecting the exterior of a dwelling in the pouring rain, while unpleasant, can help to identify the places where damp penetration could be occurring. If inspecting on a dry day and there are damp patches on walls, the cause of which is not readily apparent, then a suggestion to this effect can sometimes yield good results.

322 Having additional decorations and projections on the elevations of buildings adds expense at the time of construction and may have implications for durability and maintenance during their lives. On the other hand, the plainest buildings, such as those built in the years after the Second World War, have little to commend them architecturally, having an eminently forgettable appearance which adds little to the streetscape.

323 Ornate decoration can look attractive when new but if neglected it can become an unsightly and expensive liability, if not potentially dangerous to those underneath.

324 Ideally, the tops of all features projecting from the face of walls should be protected by a flashing, tucked into the main wall at or above the splashing level – normally considered to be about 150mm – dressed over the feature concerned and finished with a drip on the outer edge to throw water clear of the wall below. Such flashings are seldom found to the degree of those subsequently provided to this late nineteenth-century five-storey block of flats, perhaps suggesting that a lesson has been well learnt from past experience.

Cornices, in themselves, by their sheer size and weight can be a source of problems. In the past, all sorts of materials and methods were used in their provision, among them stone, corbelled brickwork and building up with tiles and cement, the latter two usually being stucco rendered. Often some iron would be used for tying back and additional support and to cramp sections together. This can eventually rust and expand, causing severe cracks. Very few were provided with proper flashings and the weighing down of the back edge by the parapet would be inadequate to prevent the outer edge sloping down and pulling the top section of the front wall and the parapet forward. Where maintenance of the dwelling has been neglected and access for inspection is limited, the surveyor must advise that a close inspection be made.

Terracotta

It is appropriate at this point to consider a material known as terracotta – from the Latin meaning cooked earth – found on the elevations of buildings from the mid eighteenth century until the twentieth century as a decorative detail, especially around openings and on gables and cornices. The fine clay from which it is made contains a higher proportion of silica than that used for brick making – 75 per cent as against 50 to 60 per cent – and less alumina – 10 per cent as against 20 to 30 per cent. These proportions permit a hard compact vitrified skin to be formed when fired at a high temperature and less shrinkage than is the case in the making of bricks. The material can be cast and moulded to produce crisp ornamental details for cornices, string courses and decorative panels, and because it is so hard, it is also ideal for casting sectional window and door surrounds and the components for balustrades – indeed anything which requires repeating and would otherwise have to be carved if made in stone (**326**).

Italian craftsmen were brought in during the sixteenth and seventeenth centuries to produce ornamental details to adorn some of the brick dwellings of the time. Cardinal Wolsey's early sixteenth-century palace at Hampton Court is one such example. In 1722 a factory was established in Lambeth to produce a range of terracotta designs to be bought off the shelf from a catalogue. Other factories were set up – as many as 400 at one time. But when the Lambeth works fell on hard times, it was taken over by a Mr and Mrs Coade who, by eliminating the iron oxide from the clay which gave terracotta its shades of red, adding a further material – nobody is quite sure what, but possibly the white mineral feldspar – and increasing the firing temperature, produced what was marketed as 'Coade stone'. This proved to be superior to other makers' products and from 1767 to about the middle of the nineteenth century was much favoured for architectural details, being virtually indestructible in whatever weather conditions, unlike much of the natural stone used previously for carving similar details. Because it is a cast product, mouldings for the production of the details could be prepared in any particular architectural style, from classical to Gothic revival (**327**).

If terracotta is found to be in good condition and on a dwelling built from the late eighteenth century towards the middle of the nineteenth century, it is probably from the Coade works. Unfortunately, the raw material used by others was not always well selected or well prepared and under-burning was fairly common. Consequently, some terracotta had a tendency to disintegrate fairly rapidly – particularly if the near impervious skin was damaged by knocks, cutting to make the pieces fit or becoming chipped in the process of fixing – thus allowing damp to penetrate to the softer interior.

As with brick making, the proportion of other materials, particularly iron oxide, determines the final colour of the terracotta, from red through pinkish shades to the later more common buff. In the mid nineteenth century, terracotta was also cast in large sectional hollow blocks, usually filled with a light concrete on fixing. These were intended to resemble, and were used as a cheaper alternative to, the finer work in stone employed on the more expensive build-ings as ashlar. In this form, it was popularised in the mid nineteenth century by A. W. Pugin and Sir Charles Barry, both heavily involved in the rebuilding of the Palace of Westminster following its destruction by fire in 1834, together with the influential writer and traveller John Ruskin. Their advocacy was taken up by many architects such as Alfred Waterhouse who used terracotta for the Natural History Museum in South Kensington and the RICS Headquarters in Parliament Square.

At this time, terracotta was sometimes also provided with an enamel glaze and given a second firing at low temperature to produce large tiles and smaller details in different and often vivid shiny colours; this is known as faience. By the late nineteenth century, such vividness had gone out of fashion, but the more subdued shades of terracotta continued to be much used on com-mercial buildings well into the twentieth century. For domestic purposes at this time, terracotta, if used at all, was confined to the more expensive houses and blocks of mansion flats, generally left unglazed and employed in combination with brick.

While terracotta from this period, manufactured by the larger reputable firms, in the main seems to have worn well, it can bring about the same weather-resisting problems as brick – the dense impervious surface producing the 'raincoat' effect as distinct from the 'overcoat' effect and permitting water to run down the surface and enter through the joints. Accordingly, there can be a build-up under the impervious surface, causing it to spall. This can be severe if the water is carrying salts from the atmosphere and can cause blistering and powdering to the surface. Of course, such salts can be derived from other material forming the backing to details – from brickwork, for example, and from deposits of the products of combustion in chimney flues. Such defects will first become apparent around the edges of blocks where joints occur. If fixing of the details is by metal cramps and these rust due to water penetration at the joints, major cracks can be induced by the expansion of the metal.

While the impervious surface of sound and well-produced terracotta resists most aggressive atmospheres, the less sound and much lighter in colour

325 Light buff terracotta used as blocks for dressings and sills around windows with brickwork as general walling on an expensive house, looking well and long-lasting after around 100 years of exposure.

326 Terracotta could be used to good effect when contrasting plain and highly ornate sections. The detailing here looks as crisp now as it did when erected over 100 years ago.

327 Typical moulded and cast clay Coade stone details around an attractive fanlight above the main doorway of a 1770s house. Such details could be bought from a catalogue much more cheaply for a whole terrace of houses than if carved in stone, with the added advantage that they were practically indestructible.

328 For repetitive work, as here in this balustrade, and as coping stones to the one-and-a-half-brick-thick walls, terracotta was often used as a cheaper alternative to stone in the latter part of the nineteenth century and the early part of the twentieth century for the more expensive dwellings.

under-burnt surfaces will become soft and pitted because the surface will not be properly formed. Lamination can also occur if the material is not pressed adequately into the mould, and warping and cracking can become apparent if the material is not sufficiently dried before firing.

While it is no longer possible to buy terracotta details and panels from a catalogue, there are still firms who will produce replacements to match old and damaged sections; though inevitably, of course, the price is high and probably only economic if details are required in quantity.

Damp-proof courses

Somewhere near the base of the external walls is where the surveyor will hope to find a sign of a dwelling's damp-proof course (dpc). The incorporation of a dpc has been a requirement for all new dwellings since the passing of the Public Health Act 1875. It is no earthly use, however, for the surveyor to assume that because the dwelling being inspected is either known or assessed to have been built since that date, it necessarily has a dpc. What the law required was not always carried out and the surveyor needs to follow a routine, which may involve a bit of dashing in and out, to establish whether there is a dpc and, if so, whether it is effective or not, irrespective of the age of the property.

Old dwellings can have a dpc installed. Installed dpcs are not always effective; some can wear out and others can be rendered ineffective by subsequent action. Since it is the external walls that are being dealt with in this section, it is what can or cannot be seen in the way of a dpc from the exterior which will be considered initially. Effectiveness is a matter of observation and demonstration with the moisture meter, but it is salutary to be reminded that the presence of dampness can render a dwelling 'unfit for human habitation' in law. Effectiveness is, strictly speaking, best dealt with from the internal viewpoint, but as presence and effectiveness are inexorably linked, it is appropriate to deal with the latter aspect at the same time.

Damp-proof courses were by no means a novelty in 1875; some dwellings would have already been provided with one from the time of their original construction before that date. They would probably have been mainly those dwellings not provided with basements. With principal rooms on the ground floor, the marks of rising damp would have been all too evident. For dwellings with basements, the provision of a dpc would be considered not all that important. The servants worked there and some might even sleep there, but they would be expected to put up with any dampness while its effects would be less likely to reach the ground floor.

Some of the earliest types of dpc are usually the easiest to spot from the exterior. Two courses of either blue or red engineering bricks laid in strong cement mortar are easy to spot. Even the red have a different texture and are probably a different shade of colour than other red bricks used for general walling. Two courses of slates laid to break joint in cement mortar are equally easy to spot. Neither is perfect for the purpose of providing an adequate barrier against the rise of dampness from the ground. The bricks and slates themselves are fine and impervious to moisture although some poorer-quality slates will deteriorate with age, just as they do on roofs. The mortar in which they are laid, however, can transmit damp – more so if the perpends between the bricks are filled with mortar. Both brick and slate dpcs can be rendered ineffective should there be any subsidence causing them to fracture.

Another early type of dpc which is easy to spot is of asphalt laid on a level bed of mortar. This will produce a joint thicker than the remaining brick joints and, even if some of the asphalt has not been squeezed out to be highly visible as often happens, the thick joint in that position is a good indication of the presence of an asphalt dpc. However, the provision of an asphalt dpc near the base of all the walls of a dwelling does provide a plane of possible slippage, and movement of the structure horizontally above the level of the dpc has been known to occur, particularly on sloping sites. This can show up as a slight overhang on one side of the house while the opposite effect will be apparent on the other side.

A strip of lead makes an excellent dpc and has, of course, been available for the purpose over a very long period indeed. It is thin enough, however, to be embedded in a brick joint without anyone being aware that it is there at all. It is only when it is arranged so that it projects slightly beyond the wall face that it can be seen. However, builders seem reluctant to allow this and prefer, for the sake of appearance, to leave it so that the joint can be pointed in the same way and at the same time as all the other brick joints. Lead used for this purpose in new or near new dwellings needs to be coated in bitumen or tar to protect it against the action of the lime in the mortar and, of course, should be of a thickness appropriate for its purpose.

The thin nature of many of the other materials used for dpcs, both past and present – such as those of bitumen with a hessian, fibre, felt or asbestos base, whether or not incorporating a core of lead sheet – can make it difficult to detect whether they have been used or not. There is a further drawback to not leaving any indication of their presence. This is because pointing the outer edge of the joint can undo the protection that a dpc installation provides because of the bridging effect, which will allow rising damp to bypass the dpc and to appear higher up the wall, albeit perhaps to a limited extent. The most popular modern material is plastic and this is usually allowed to project slightly beyond the wall face, overcoming the bridging hazard without any adverse aesthetic impact.

At about the height in the external walls where the surveyor would hope to see signs of a dpc, there may instead be a line of holes, either well or badly made good (**329**). These are indications that internal dampness has been noted at some time and result from attempts to prevent dampness continuing by drilling holes and injecting a chemical dpc. Experience suggests that a well-installed chemical dpc probably has the greatest chance of the various alternative methods of being successful as long as all necessary additional works, such as replacing affected internal plasterwork, have also been properly undertaken. Sadly, experience also suggests that many installations have not been well carried out. Frequently, plasterwork may not have been replaced or unsuitable plaster may have been used. On occasion, unscrupulous operators have been found to have merely drilled a series of external holes and not injected any chemicals at all! The ultimate test comes down to whether any dampness is still present internally. Guarantees for treatment should ideally be available for inspection, and the most reassuring will be those from companies which are members of a reputable trade association such as the Property Care Association (PCA).

Other methods of damp prevention have been advocated from time to time. Electro-osmotic treatment involves a copper band secured to the walls and extending right round the house. Another method that is claimed to lessen the effect of rising damp consists of installing absorbent hollow tubes at low level that ventilate to the outside air. These tend to be regarded with great scepticism by experienced surveyors, few of whom seem to have encountered a successful installation.

When the surveyor is reasonably certain that the location of a possible dpc has been found in the wall, even if the type cannot be identified with any degree of certainty, that position in the structure should be traced around the house and its relationship at all points to the ground or paving levels should be determined. While doing this, any features that could be a source of bridging – such as steps, driveways, patios or the frequently encountered earth banked up by a keen gardener – should be identified and noted for checking internally; and the risk they pose should be flagged in the report with a recommendation for removal if appropriate. Strictly speaking, evidence of damp caused by bridging of the dpc is in the category of penetrating damp, not rising damp, and the surveyor should distinguish between the two. It is usually a great deal easier and cheaper to eliminate the causes of penetrating damp than rising damp due to a lack of dpc or a defective one.

BRE are of the view that the incidence of true rising damp is less than is generally thought and that most damp in the lower part of the walls of dwellings at ground-floor or basement level is due to other causes. Bridging has already been mentioned but another cause is the presence of hygroscopic salts in the plaster. These could originally have been deposited in the wall by rising damp that has subsequently been eradicated, but without renewal of the affected plaster. On damp days, the salts take up moisture from the atmosphere and the plaster looks, feels and registers as damp on a moisture meter. Renewal of the plaster is usually necessary for a minimum of a metre above floor level and is a vital part of the process of installing a dpc in an existing wall. The situation internally will not necessarily be improved until this has been done.

Irrespective of whether the surveyor finds evidence of the presence of a dpc in the external walls, it is a standard requirement to test with a moisture meter for any evidence of damp within the dwelling. If there is a dpc, this use of the meter will usually demonstrate whether it is effective or not. In many dwellings, the lack of a dpc will reveal itself fairly readily by high readings on the meter but, in others, it may not be so clear-cut. Steps to hide the rising damp or lessen its incidence may have been taken. Lining the walls with metallic foil-backed paper is one method that nullifies the use of the moisture meter by short-circuiting the probes. Securing bitumen-impregnated lathing to the walls, plastering and decorating is another, although tapping will reveal a sound less solid than when the inside of the external walls are tapped elsewhere.

The surveyor needs to be aware that the standard surveyor's electronic moisture meter does not automatically indicate the presence of dampness but simply registers the presence and level of a small electric current between the

329 Evidence of the probable insertion of a chemical dpc in a house built some 20 years or so after legislation requiring the provision of a dpc came into force but where there is no evidence that one was ever provided. Even now, the effectiveness of injected dpcs can be a bit hit and miss at times.

330 This unusual pattern of dampness could not be due to rising dampness because it is at first-floor level. It was caused by a persistent leak around a shower tray.

two probes. If the probes are touching any conducting material, whether foil-backed paper or materials such as carbon dust within wall plaster, they will register a reading even if the wall is completely dry. Equally, even if dampness is present, the readings themselves will not differentiate between rising dampness, penetrating dampness or condensation, and the surveyor must use other evidence to draw the correct conclusion. Rising dampness, for example, tends to have a distinct level on a wall at which readings change suddenly from very high to nil, whereas condensation tends to have readings which vary more gradually. If false readings are being caused by conducting materials in a wall, they are likely to be present on the whole area of a wall, which would be unlikely for either rising damp or condensation.

There are methods of installing chemical dpcs which, when carefully carried out and all areas disturbed are made good, are virtually undetectable. Clearly, there is a need to question the seller if there is no sign of damp and no evidence of a dpc after exhaustive use of the damp meter. Of course, it is highly probable that the seller, having seen all this going on, will be only too happy not to wait to be asked and will be trumpeting the success of whatever damp-proofing system the dwelling possesses. Details of the system and particulars of any guarantee which accompanied the installation will need to be obtained so that the surveyor can express an opinion on its likely long-term effectiveness.

The surveyor cannot be expected to test for dampness every few millimetres near the base of all walls in a dwelling, but experience will indicate, having regard to the age, type and general condition of the dwelling coupled with the inspection, where the most likely places for rising damp could be. A careful note of all the positions tested for damp should be made on a diagram in the surveyor's site notes. If the terms of engagement require testing at certain intervals then that requirement must be fulfilled. Unlikely though it may seem, there are dwellings with no dpc and no evidence of damp at or near the base of the walls; but they are relatively rare. What could be lengthy procedures are more often short-circuited by the finding of damp. Dampness, or firm evidence that it is merely being hidden, is a category of defect requiring urgent attention and leaves the inspector with no option but to advise further investigation into the most appropriate form of eradication.

Depending on the age and construction of the property in question, if the surveyor finds no dpc in a dwelling yet, after appropriate investigation, can find no evidence of dampness, it should be remembered that it is the presence of dampness and not the absence of a dpc that can render a dwelling unfit for human habitation. Nevertheless, it would be as well to ensure that the report reflects the extent of what has been investigated and discovered during the inspection and not simply to become a hostage to fortune by making some sweeping statement about freedom from dampness which cannot be backed up by the facts of the inspection and supported by the site notes.

Plinths and airbricks

The relationship of four of the items at or near the base of the external walls of most dwellings – dpc, ground level, plinth and airbricks – is important and requires some consideration by the surveyor. Any plinth should be below the level of the dpc to avoid bridging. In turn, the dpc should be at least 150mm above the level of any paving, the generally recognised limit of splashing from rainfall. It should also be below the level of any timber built into or attached to external walls as well as below any wall plates on top of sleeper walls used to support a hollow timber joisted ground floor.

This puts the ideal level for the top surface of a hollow timber joisted ground floor at about 350mm above external paving level, roughly two steps up to the front door. On older houses, this provided an ideal opportunity for a long, narrow airbrick to introduce a through current of air below the entrance hall, always provided, of course, that there is a corresponding airbrick at the rear. The introduction in 1999 of requirements for disabled access at new-build dwellings means that gently sloping access paths and level entrance thresholds eliminate the former convenient location by the entrance steps, so alternative provision is needed if a suspended timber floor is present on newer housing. To ventilate the general area of hollow timber flooring satisfactorily, other airbricks should be positioned immediately above the dpc in sufficient numbers for an adequate amount of through ventilation. Testing whether there is a through current of air is necessary and can be done with the aid of long matches; when lit, on all except the very stillest of days, these should show some indication of air movement. Departure from the ideal relationship at the base of the external walls can introduce the risk of problems and the surveyor needs to weigh up their significance and assess their effect.

331 The relationship of items to each other at the base of the external walls, such as dpc to ground level, to the level of a hollow timber ground floor and its essential ventilating airbricks and to any plinth is vital to the satisfactory performance of walls and floors at this point. The surveyor needs to give them full consideration and departures from the ideal should provide a spur to further investigation into any observed defects. For example, there is no sign of a dpc in this square bay window and the plinth, extending above the level of airbricks, appears damp ridden. It would be very surprising if a clean bill of health could be given.

332 Too many homeowners regard airbricks as a cause of unnecessary draughts rather than a source of essential ventilation to preserve the health of the subfloor structure. Blocking them like this is a recipe for potential disaster.

Of course, all the foregoing presupposes that there is indeed a hollow suspended timber ground floor; but there are quite a number of dwellings, particularly some of those built in the late 1940s and 1950s, which were provided with solid floors. Here, the ideal arrangement on a level site is a damp-proof membrane in the floor lining up and lapped with the dpc in the external walls, bringing the top of the floor to a little more than 150mm above outside paving level. On a sloping site, matters become more complicated and as the dpc changes level so there should be a vertical dpc to link floor membrane and wall dpc. There is more opportunity here for miscalculation, errors and subsequent changes in ground level by the owner, all of which can lead to bridging and be a possible cause of damp patches appearing at low level on external walls. While the dpc can usually be seen externally, the integrity of the internal junction with the floor membrane cannot be visually checked without exposure works; though it can reasonably be assumed if internal dampness is absent. More recently, suspended solid floors have gained in popularity and these should have provision for ventilation of the subfloor void as well as a dpc in the walls, properly linked to a damp-proof membrane in the solid floor.

Applied surfaces

The inspection of the external elevations is discussed in some detail in Chapter 9, which considers how to determine the construction of the property being surveyed. Usually, the external finish will simply be a consequence of the main structure of the building but there will be times when it has been applied as a superficial cladding, in which case it may warrant further attention. In positions of severe exposure to driving rain, masonry construction is likely to be vulnerable to penetration by dampness and additional materials have often been utilised as an external weatherproofing layer. Approved Document C of the Building Regulations describes claddings as being either impervious (including metal, plastic, glass and bituminous products), weather resisting (including natural stone or slate, cement-based products, fired clay and wood), moisture resisting (including bituminous and plastic products lapped at the joints, if used as a sheet material, and permeable to water vapour unless there is a ventilated space directly behind the material) or materials which are jointless or have sealed joints (which need to allow for structural and thermal movement). However, the provision of additional weather protection is not the only reason for a building being clad in some non-structural finish. Sometimes social pressures have encouraged owners to apply claddings to alter the original appearance of buildings for no other reason than the dictates of fashion. During the post-Second World War years of experimentation, a whole range of external materials were utilised as claddings, often because they imposed relatively lightweight loadings to contemporary forms of construction or simply to provide variations in external elevation finishes. Once again, an awareness of the age of the original building will often assist the surveyor in identifying when a cladding material has been added for important protection or for some other reason, whether it is an original or later element of construction, and how it can be expected to perform.

Tile, slate and shingle hanging

In exposed positions, and frequently elsewhere, tile, slate and shingle hanging will be encountered frequently as the external cladding of dwellings, particularly above ground-floor level. It will come as no surprise to learn that the durability of individual components will be very similar to when they are used on roofs so the relevant comments made in Chapter 4 also apply here. Provided they are kept in good condition and the base upon which they have been fixed is sound, they should give satisfactory service for many years. One of the main problems with the vertical hanging of tiles, slates and shingles is that they can easily be the subject of impact or vandalism at ground-floor level. For this reason, the form of construction is usually accompanied by a ground floor of masonry.

333 Tile hanging to the first floor of a two-storey terrace house of the late nineteenth century, with brickwork to the ground floor in stretcher bond. This suggests timber frame construction which might be confirmed by a sight of the inside face of the gable wall. The damaged area of tile hanging on the corner needs repair with angle tiles to match the original.

334 The opportunity for attractive patterning is not so great with slates as it is with tiles, and special slates for turning corners are obviously not possible. Damage here needs urgent repair to restore protection against damp penetration.

335 An attractive slate cladding in a very exposed coastal location. The surveyor cannot examine the underlying support and can only enquire about the standard and specification of the works in the hope that satisfactory answers will be forthcoming. Some of the questions to be asked are: Were the battens treated? What is the underlay? What type of nails were used? Several small sections of pointing have started to become loose and fall away near the corner on both elevations, centre bottom, and repairs should not be delayed if more serious and progressive deterioration is to be avoided.

336 Shingles hung vertically to the gable end of a timber-framed semi-detached house, apparently periodically treated so they have not weathered to their characteristic silver grey shade. When backed by a 'moisture-resisting layer', shingles, vertically hung, are a form of wall construction considered capable of coping with exposure conditions categorised as 'Very Severe', quite possibly a requirement in this Scottish location.

337 Red cedar shingles hung vertically on the first floor of a detached dwelling built in the 1930s in an 'olde world style' but somewhat let down by the rather mean overhang to the thatched roof and the characterless windows. The remainder of the elevations are covered with overlapping horizontally hung waney edge boarding; and as long as all the walls had been provided with a moisture-resisting layer beneath the cladding, these would, perhaps surprisingly, be classed as satisfactory construction to cope with exposure conditions in the 'Very Severe' category according to BS 5628: Part 3: 2001.

'Mathematical' or 'brick' tiles

It is appropriate to mention at this point what are known as 'brick' or 'mathematical' clay tiles, the existence of which is the reason why it is necessary to tap the outer surface of a brick wall to see if it might sound hollow. The tiles which, when properly fixed and pointed, look exactly like bricks in a wall, were developed for reasons of fashion in the eighteenth century and were still being produced in the early 1980s. Even though an owner might be comparatively well off, occupying a timber-framed house came to be regarded as socially undignified in the late eighteenth century. If the owner could not afford to rebuild, having a makeover by covering the dwelling with mathematical tiles could quickly and economically transform a timber-framed property into what appeared to be a brick dwelling in the latest fashion. It is gleefully recorded elsewhere that even the inspectors compiling the Statutory List of Buildings of Special Architectural and Historic Interest were deceived by the construction in a number of instances, believing it to be brickwork. To be fair, the exact moulding and narrow joints can deceive the most experienced surveyor.

For makeovers of timber-framed dwellings, it was a fairly simple matter of covering with softwood boarding, battening and then nailing and bedding the tiles to the battens and each other in lime plaster, and then pointing carefully the fine joints between the tiles with lime putty. However, it was also possible to bed – nailing them where possible – the tiles on to an external rendered or plastered surface applied to brickwork or stonework. It is recorded that at least two large country houses were covered over with mathematical tiles of

338 This 'brickwork' is in fact mathematical tiling of the late 1850s in stretcher bond – quite late for this form of finish. Although convincing enough to deceive some experienced professionals – especially as this is not in an area associated with this construction – there are clues that all is not as it seems: the detailing around the window is 'not quite right', there are inexplicable vertical sections of rendering on the corners and individual bricks are not quite flush (bottom right). These and other slight peculiarities suddenly make sense when the truth dawns. Tapping the tiles produces a hollow sound when the tiles are fixed to battens, as here – often the case when used as an eighteenth- or early nineteenth-century makeover to an older timber-framed dwelling. Not so, of course, if the tiles had been bedded on to solid brickwork as can be the case in other instances.

339 The junction at this corner shows alternative methods of facing a timber-framed dwelling dating from around 1600. The brick face to the right was added first, possibly around 1700. When the building was extended in about 1800, mathematical tiles were used to provide a uniform facing to the enlarged elevation to the left. The shape of the tiles, their bedding on lime mortar and how they fit together to create the illusion of brickwork can clearly be seen on this exposed corner, a detail frequently concealed by rendering, as in **338**. The underlying original timber frame means there is a distinct hollow sound when the face of the tiles is tapped.

a yellowy white shade in the 1780s to conceal the, by then, highly unfashionable red brick of the original construction.

It is sometimes said that mathematical tiles came to be used as a means of avoiding payment of the tax on bricks, first introduced in 1784 at the time of the Napoleonic Wars. However, the tiles were well in vogue by then for the purposes mentioned above, and their use was first recorded in the 1740s. Although tiles were not included when the tax was first introduced, nor when it was increased in 1794, they were when it was again raised in 1803. The tax on tiles was wholly repealed in 1833 but that on bricks, not until 1850. The use of tiles was widespread across the southern and eastern counties of England until it became fashionable again, around the mid 1850s, to show off old oak timber frames. The tiles were produced in all the colours normally available for clay bricks in the locality required, both as stretchers and headers. Some in the latter part of the eighteenth century were provided with a lead-based powder before glazing to produce shiny black tiles thought to be a protection in a salt-laden atmosphere. Generally these were laid in header bond since it was a time when bow windows were fashionable and headers fitted better together round the curve.

340 The upper part of the front elevation of a neglected 1790s four-storey dwelling faced with mathematical tiles. The settlement in the window head at first-floor level has allowed the mathematical tiles to slip forward, loosening their bond to the lime and sand rendering to which they should be attached. Although held in place at the moment, a loosened tile could pull away from the face of the underlying brickwork with the serious danger of then falling onto someone beneath.

Where the bedding of the tiles is in lime plaster rendering, they often continue to adhere even though the nails have rusted. Even so, unless there is evidence of recent refixing, they must be considered a danger to occupants and passers-by, and suitable warnings need to be included in the report. Of course, if they do fall off then they are almost certain to break and replacements may need to be specially made – no doubt at a very special price as well – unless it is possible to achieve a reasonable match with brick slips or, in the case of those which were originally glazed, with ceramic tiles.

Weatherboarding

A weatherboard, or clapboard, cladding to a softwood timber frame in white pine, hemlock or spruce has been a favoured method of weatherproofing since around the end of the 1600s. In the early 1900s, western red cedar became available for the purpose and has the advantage of being naturally resistant to deterioration. Examples of weatherboarding on timber frame dwellings have already been illustrated at **159** (in Chapter 9) and **274–276** (in Chapter 14), but other examples are included here. It has the advantage over the likes of tile, slate and shingle in similar exposure conditions of not being prone to impact or vandalism at ground-floor level. Painting or the application of preservatives, traditionally tar, to the bare wood is an essential measure to protect the timber from the ravages of the weather. Flaking paint and bare patches of timber are a source of damp penetration and will eventually set up wet rot in the timber.

Early forms of plastic boarding, popular from the mid 1960s, failed to live up to claims for durability, fading and becoming brittle; but improved materials have led to this being reinstated as a favoured material. Indeed, some plastics can prove to be difficult to distinguish by sight from timber, and it may require the judicious application of the surveyor's sharp probe to confirm whether the cladding is natural or manufactured.

341 Not usually suffering from the disadvantage of impact or vandalism at ground-floor level, as is the case with tile, slate and shingles hung vertically, weatherboarding over a 'moisture-resisting layer' is capable of coping with exposure conditions in the 'Very Severe' category when properly specified and fitted. This two-storey dwelling from the late eighteenth century was re-clad with cedar boarding in the early 1970s, but the specification adopted at the time is unlikely to be known.

342 Weatherboarding covering all the elevations of a bungalow, probably dating from the 1930s. This is unlikely to have an underlay of a type which would bring it up to the standard of a dwelling able to cope with 'Very Severe' exposure conditions but is, nevertheless, still able to provide weatherproofing superior to that of inferior brickwork.

343 Shiplap boarding, stained black, on a detached two-storey house of the 1980s. Contrasting panels of boarding were a fashionable design feature from the 1960s onwards, often providing problems at the junctions with other materials unless carefully detailed. Here, the boarding covers the bulk of the elevations above a partial brick ground floor – rather more than a plinth and, therefore, probably precluding the dwelling's construction as being suitable for an area where exposure conditions are 'Very Severe'.

Ceramic tiling

The use of ceramic tiles to clad buildings was popular with designers during the 1960s and 1970s. Unfortunately, in many cases, such claddings more or less routinely fell off after a very short time, often leaving expanses of cement and sand substrate with the pattern of where the tiles were once pressed into place before grouting as a continuing reminder of the failure. A surveyor with the task of inspecting a dwelling encased substantially in ceramic tiles needs to be very wary indeed. Water streams down the impervious surface, soon to find a way in through the joints to the back of the tiles where it freezes, expands and forces the tile off. Once one tile has fallen off, the process becomes unstoppable. Often, it is started by the bedding down of the newly completed structure. The slightest movement is usually sufficient to cause fracturing in

344 Ceramic tiling to three of the six floors of a block of flats built in the 1990s. The tiled first, second and third floors are devoid of projections, even to the extent of the sills for the windows. This means that a continual stream of water will run down the face of the impervious surface during rainfall, seeking a path through the joints, which because of the size of the tiles selected, are even more numerous than would be the case if impervious glazed bricks had been used. The surveyor would need to give warnings not only of possible damp penetration but also of tiles potentially being loosened because of frost action and becoming a danger.

the tiles. The surveyor might be able to do some judicious tapping to see if sections of tiling have become loosened but, as with some of the other applied finishes, it may be a case of warning of what might be happening at inaccessible points and advising a closer inspection.

Mock stone

The idea of aspiring to be able to afford a stone-built dwelling, as distinct from one of brick or some other material, was prevalent in the latter half of the eighteenth century and the first half of the nineteenth century, reappearing again in the second part of the twentieth century. Whereas the earlier manifestation took the form of rendering the brick and then making it look like stone by incising a pattern of joints on the surface, the more recent incarnation took the form of applying thin panels of a material imitating stone on to brick or other surfaces. The panels are moulded to look like squared rubble stonework with individual 'stones' sometimes given a contrasting colour.

Some panels are more convincing than others in what they purport to represent, but it is unlikely that any would fool a surveyor into thinking the mock stone was real. That might just be possible by applying the panels to a totally detached house but, generally, they stand out like a sore thumb when applied to a dwelling that is either semi-detached or in a terrace.

While some mock stone panels have been of fibreglass, the sheets merely stuck on with adhesive, others are in the form of a mix of sand, cement and pulverised fuel ash with a colouring agent pressed on to a cement and sand coating. Whenever materials for walling that have a comparatively smooth face are used, there is the same likelihood as with shiny bricks or ceramic tiles that water will run down the face, finding its way through the joints to affect the interior or to freeze, expand and blow off the applied material. This has happened at a number of locations, leaving the surveyor with no option but to advise inspection of those areas of applied mock stone which are out of reach. As with rendered coatings, applied mock stonework should be tapped at regular intervals to establish whether it remains soundly bonded or has become loose – a failure characterised by a hollow resonance.

345 Mock stone applied to a brick-built two-storey semi-detached house of around 1890. The routine for relating details at, or near, the base of the external wall should reveal whether in applying the veneer, airbricks have been covered over or the damp-proof course bridged – faults reported from other locations. It is curious to see that a representation of an arch with stone voussoirs the size of bricks is provided over the windows – highly unlikely to be encountered with genuine stone construction. The rendering which has been applied to the other half of the pair, although less obtrusive, is no less uncharacteristic of the date of construction. Both houses were originally finished with brick elevations, like the dwelling on the right.

346 Perhaps the ultimate of inappropriateness is the addition of mock stone to the first and second floors of this four-storey brick-built end-of-terrace dwelling of the early nineteenth century; along with other alterations, this completely changes its character.

Rendering

The use of an applied coating – rendering – has a long history. It has two main purposes: first, as an aid to keeping the weather at bay and, second, as a way of improving appearance. It is probable that all medieval timber-framed dwellings started life with the panels filled but the frame left exposed. Where there was an ample supply of oak and the frame was of storey-height closely spaced studs, the basis for the filling would be on split oak laths that were sprung and wedged into grooves left in the sides of the studs. If the supply of oak was running out or in areas where it was sparse, the panels were more square in shape and oak staves would be sprung into holes left in the horizontal members to provide a framework for hazel boughs or laths to be woven through and interlaced to form a close basketwork pattern of wattle. Panels would then be daubed with a mixture of wet clay with added flax, straw or cow hair thrown on from both sides in layers and, when dried, completed with a coating of smooth lime and hair plaster. The panels did not always end up draughtproof and, as a result, over the years many were subsequently plastered all over, including the frame, in order to improve comfort. The coats of plaster were much thicker than is common today and lent themselves in some areas to quite fanciful decoration known as 'pargetting'.

Other than on cottages, rendering went completely out of fashion for the brick and stone terraced dwellings of the late seventeenth century and eighteenth century but came back into favour in a different form for about 50 years into the first half of the nineteenth century, exemplified by the Regency terraces and their successors. Initially smooth, it would be incised to imitate the joints of stone ashlaring to cover the front elevations and sometimes the whole of the terraced dwellings of the period, some of which were popularly known as 'plaster palaces' on account of their overall smooth stucco rendering. Some of these dwellings were painted early on in their lives and this practice became common. The upkeep of the rendering and the cost of the regular painting required can be a hefty burden on owners; but if neglected and damp gets behind the rendering then large, hefty chunks can become loose and fall to the danger of owners and passers-by alike (**351**). The surveyor needs to be liberal with warnings as, inevitably, there will be substantial areas which are inaccessible without long ladders, scaffolding or hoists.

Later in the nineteenth century, stucco rendering became popular, confined in the main to the lower part of the front elevation of the larger terraced dwellings at ground-floor and basement levels. At the same time, the Arts and Crafts movement of the late nineteenth century and early twentieth century favoured a return to cottage styles of building. For these properties, roughcast – still based on lime but with a coarser aggregate to produce a rougher texture – became fashionable again; although it had continued in use as a

347 In some parts of the country, the thick coating of lime plaster on the panels, or even covering the whole of many timber-framed dwellings, was often modelled and relief designs formed of foliage, heraldic figures, animals, and the like – a technique known as 'pargetting'. Here the pargetting and the adjacent shell door canopy is dated 1690, presumably to celebrate the arrival of William and Mary on the throne, but is worked on a dwelling of a much earlier date.

348 Pargetting on this two-storey timber-framed but wholly plastered dwelling was probably carried out in the eighteenth century when the exterior was remodelled to include double-hung sash replacement windows. The panels are mainly of geometrical patterns in basket weave and sunrise designs with a more elaborate panel of foliage between the windows.

protection of the structures in more exposed areas – for example, the use of harling in Scotland.

The inclusion of Portland cement in the mix for renderings in the period 1920 to 1940 allowed the application of much thinner coats than hitherto, resulting in the characteristic appearance of much interwar housing with pebble-dash and smooth-coated areas or elevations. At times, too much Portland cement was added, making it too strong for the substrate. Possessing a stronger shrinkage element, if it cracked and fell off then it would often bring with it parts of the brickwork to which it had been applied. Checks for hollow-sounding areas are always required where rendering is present. Localised failure may be acceptable, but larger areas should be replaced before they fall away naturally with obvious risk to areas and people beneath.

349 In the first half of the nineteenth century, there was a strong demand for smooth rendering incised to imitate the joints in stonework to cover over cheap brickwork or, as here, unfashionable timber framing.

350 The full effect of painted stuccowork is reflected in the long lines of four-, five- and six-storey terraced dwellings typical of the mid nineteenth century, of which this is a well-maintained five-storey example. The full array of cornice, band course, window surrounds, full width first-floor balcony, reticulated ground floor and columned portico is present. Compared to some, however, it is comparatively light on decoration – no ornamental balustrades, for example – but expensive to maintain even so.

351 When elaborate stuccowork is neglected and dampness penetrates, it can lead to sections being forced off, as here. This is clearly a danger to passers-by and, until assurance can be given that there is no danger, the surveyor should advise that protective scaffolding be put up.

352 A good period for the quality of rendering, both as to materials and workmanship, came in the early twentieth century before the general practice of incorporating Portland cement in the mix came about. The Arts and Crafts movement favoured cottage-style dwellings with a coating of a thick, rough-textured rendering, as here.

353 The considerable expansion of the suburbs in the period 1920 to 1940 with the typical semi-detached three-up two-down dwelling and the continued use by many housebuilders of one-brick solid walling for the external walls saw a considerable number finished at first-floor level with pebble-dash rendering. With a mix containing a high proportion of Portland cement, there was usually a considerable amount of shrinkage, which caused cracks and allowed damp to penetrate and assist any tendency to sulphate attack. Once this got under way, large areas could become defective. On this pair of semi-detached dwellings, not entirely typical of the period, the rendering has had to be totally renewed.

354 Rendering on corners and in narrow passages, as here, is particularly vulnerable to impact damage; but the original mix at this property seems to have been incorrect, making the damage more extensive than ought to have been the case. Unless repaired soon, this will only deteriorate further.

Rendering is once again being used on new-build construction, though fortunately there does appear to be a greater understanding of the types of rendering most suitable for the different categories of substrate.

Painting

Painting the outside of the external walls of a dwelling, as distinct from the woodwork, is fairly easy to do but rather difficult to reverse. Once done, the burden of repainting at regular intervals becomes a fact of life. It was almost automatic that the right sort of paint would be used when there was a long local tradition handed down of what to do and the same person who painted the exterior walls would probably have done it again. Limewash would be applied over limewash; oil paint, on to oil paint. Now many paint formulations are changing as the use of certain constituents, such as solvents, are being discouraged or, like lead, have been banned. Stuccowork, painted at regular intervals over the last 150 years or so, may not take kindly to being repainted with some water-based products, and the disastrous results of painting a cottage built of cob with an impervious coating of plastic paint have already been touched upon. Old, established professional decorating firms will know of these 'wrinkles', but much repainting is a DIY activity or, at best, carried out by small firms who quoted the lowest price and who may not be aware of the pitfalls.

Now that plastic is being used so widely to replace original window frames and gutters and to cover external joinery to the eaves, soffits and gables of so many houses, the burden of external redecoration has been significantly reduced for many owners. Nevertheless, exterior painting is still necessary in many cases, and surveyors should be able to put their finger on what has gone wrong when they see the effects of ill-considered decisions regarding what paint to use on both old and newly painted surfaces. They should also be able to identify another effect of automatically accepting the lowest estimate – that of failing to check the specification which hopefully accompanied it and, if satisfied, seeing that it was carried out. Generally, even on the smallest of jobs, the painter will provide a little more information than just the estimated price; and in most cases, this will cover, where necessary, the woodwork of windows and doors and the metal of pipes and gutters as well as masonry. The owner may not give this specification the attention it deserves but, at the very least, it should include the cleaning of surfaces, their preparation, types of paint and number of coats.

Defects in the form of flaking, bare patches or the separation of coats are easy to recognise and, if accompanied by a fading of colour or dulling of the sheen where gloss paint has been used, would indicate that the interval for renewal has been exceeded; as a result, additional costs for preparation will be involved when that renewal is carried out. BRE puts renewal for solvent-borne paints at around 4 to 6 years and for water-based paints at slightly longer – 5 to 8 years – including the water-based cement paints normally applied to masonry. If paintwork is in good condition but not looking as though it has been applied recently, the surveyor needs to make an assessment of when the next maintenance repainting should be carried out, differentiating between different areas if necessary.

Should flaking and separation of coats of paint be evident on surfaces which appear to have been redecorated comparatively recently, indicated by the colour being still fresh and an absence of dulling to the sheen, the surveyor should enquire from the owner when the work was carried out and ask for a sight of the specification and the estimate given at the time. It may then be possible to identify the cause of failure.

Where painting has been newly carried out and no flaws are apparent, the surveyor can do no more than report the fact. The basic problem in these circumstances is that, however ill-considered the decisions which may have been taken about what paint to apply and regardless of the standard of workmanship, the surfaces will probably not show any effect if they were wrong or inadequate for about a year or 18 months. It is essential, therefore, for the surveyor to ask for the specification and estimate covering the work to be produced if possible to enable a comment on the likely performance and durability. If the surveyor is dubious about the contents of the documentation, appropriate reservations can be included in the report. Even when satisfied on both specification and estimate, the surveyor needs to make clear in the report that he or she was not present to supervise the work as to preparation, number of coats, etc.

Vegetation

There is quite a bit to be said for the odd climbing plant, loosely wired but not entwined to a sturdy trellis attached to a batten that is fixed to the external wall of a dwelling. The colour of foliage and contrasting blooms, some with a heavy scent on a warm summer's evening, can be captivating. The advantage of a trellis screwed to a batten is that it can be taken down when required for the repainting of external walls and then put back. Some plants can run riot without any help by way of support; and although they may, arguably, look attractive, they do have a cardinal disadvantage – not just to a surveyor endeavouring to carry out an inspection. They all help to retain moisture against the wall, reducing the opportunity for evaporation, and they can clog gutters and rainwater pipes. Some are even destructive, especially if allowed to spread across a relatively fragile cladding, such as hung slates or tiles.

One of the prettier ones – the Virginia creeper – attaches itself to masonry by suckers, turns attractive shades in autumn and dies down in winter. However, even if the main growth of the creeper is removed, the small suckers remain attached to the surface and are difficult to dislodge, remaining clearly visible on many finishes. Virginia creeper and other plants such as wisteria which climb and can attach themselves by entwining around anything that is available – pipes, for example – need managing and it is better to control their

355 A gable end wall might be regarded by some as a blank canvas for decoration with a climbing plant but, quite apart from any concerns about the brickwork and pointing, when ivy reaches roof level there is a serious risk of damage leading to damp penetration.

activity than allow them to run riot. Ivy is a great deal tougher with roots that burrow into the joints of masonry, growing all the time and able to eventually dislodge whatever gets in its path (**355**). The surveyor's advice must be to get rid of it. If it is not already causing damage, it is only likely to do so in the future.

17

External joinery

Windows

Reporting on windows usually requires a brief summary of the overall condition of all windows at a property, and normally this is readily achieved. More difficulty may be encountered when the construction or types of windows varies or the condition of individual windows markedly differs throughout a property. The level of detail in the report, and indeed the proportion of windows which need to be individually checked, will be determined by the type of inspection which is being undertaken. While all windows may not be checked by physically opening, it is nonetheless important to consider each one from the exterior, physically at ground level or by binoculars if higher, and also internally. Windows should be inspected from the aspects of safety, security, condition and performance. While undertaking the internal inspection, every opportunity should also be taken to use the windows not only as a vantage point to examine external areas – such as the exterior elevations immediately to the sides, above and below the window – but also as a vantage point for features not visible from elsewhere such as the roofs of extensions or neighbouring land beyond the boundaries.

The word 'joinery' in the title of this chapter reflects the historic supremacy of wood as the primary construction material for most external features other than the walls and roof covering. In recent decades, however, plastic has become increasingly dominant both in new build and whenever components such as windows or eaves and soffits need to be replaced; it is not unusual for the surveyor to find very little external timberwork in evidence. Poor-quality softwood frames used during the 1960s and 1970s kick-started the replacement window industry, which then responded to the energy crisis by offering double glazing as an energy conservation measure. Initially, aluminium-framed replacements were quite popular, but the early versions quickly fell out of favour as their exterior surfaces deteriorated and their interiors suffered from condensation due to thermal bridging. Once uPVC frames were being manufactured with resistance to deterioration from ultraviolet light and with the additional promise of maintenance-free frames, the plastic window frame revolution achieved an irresistible momentum. The *UK Housing Energy Fact File* published in 2013 by the Department of Energy & Climate Change reported that, in 2011, 83 per cent of homes had at least 80 per cent of their windows double glazed. By comparison, the figure in 1983 was a mere 9 per cent. It is little wonder that the surveyor, glancing up and down the road for dating evidence of construction period, often encounters difficulties finding a house with the original windows still present – an admitted inconvenience to the professional but, more significantly, another loss to the urban landscape.

More recently, there has been a resurgence of timber frames with modern preservatives meeting the needs of durability while reducing the previous time and cost demands for maintenance with three coats of paint. There has been an increasing realisation that uPVC may not be as maintenance-free or as durable in the long term as sales teams led prospective purchasers to believe. The ecological credentials of uPVC have been challenged and its whole-life costs in comparison with timber may be less favourable than initially assumed. But it is likely to be some time yet before the current balance in its favour is upset.

Other materials, of course, have also been used for window frames. Wrought iron was used for a while in the seventeenth century (**356**) and again after the Industrial Revolution on a small scale, but it was not until the 1920s that windows made from rolled steel sections began to encroach on the dominance of wood. Steel was popular during the immediate post-war period, but its poor insulation capabilities and propensity to rust meant it fell out of favour. However, modern steel windows, meeting all relevant standards, are now available.

Manufacturers of double glazing units usually provide warranties of 5 or 15 years, so that the surveyor needs to ascertain from the unit, or units, the date of manufacture and ideally examine the warranty. These are very often conditional upon installation being carried out in accordance with the maker's specification and subsequent maintenance again carried out in accordance with the maker's requirements. Failure to follow what the manufacturer requires can vitiate the warranty. Fixing can be difficult in older windows because of the 16mm and 20mm gaps which are industry standard and which are often too wide for existing rebates. Furthermore there are preferred glazing methods. For example, double glazing units in wood windows often fail because of the absence of drainage and ventilation provision.

Approved Document L (Conservation of fuel and power) defines standards which must be met by glazing, whether newly installed or as replacement for original windows. The document has been amended several times since it was made mandatory in 2002, and the date of construction or installation will determine what is applicable at any given property. It is, therefore, important for surveyors to remain up to date with the standards and to have an awareness of what was relevant at different times. Since 2002, the replacement of windows has been notifiable work under the Building Regulations.

When replacement windows are encountered, the surveyor should check – or recommend that the client or client's legal adviser checks – that the works were either granted the appropriate approvals or, most commonly, that they were carried out by 'an appropriately qualified person' under one of the relevant Competent Person Schemes, such as FENSA (Fenestration Self-Assessment Scheme) or Certass. Ticking this box ought to ensure that the technical details are compliant with the relevant regulations; although experience suggests that this cannot simply be taken for granted and the surveyor should still keep a wary eye open for areas of non-compliance.

As the demands for energy conservation continue to rise, there is a gradually increasing interest in triple glazing. At present, this is only likely to be found in the highest specification properties but it seems likely that it will become more widespread in future years (**357**). If so, the surveyor will need to ensure that the form of glazing is correctly identified and reported.

Adequate windows, to provide both light and ventilation in most instances, are of course a requirement and have been since the Building Regulations were introduced in 1963, and before then under bylaws in most areas. There will be times, however, when the surveyor may consider it necessary to calculate whether the correct proportion of window to floor area or the right provision for ventilation has been provided – often an issue where extensions have been added. The report should reflect the findings and express a view as to whether both light and ventilation are adequate. BRE contrasts the efficiency of some older windows with those of modern times in this regard. The 1800s double-hung sash window with frame takes up about 15 per cent of the opening while the framing of many windows nowadays takes up around 35 per cent. Lowering the top sash a little provides adequate ventilation above head height in most circumstances as do the small opening top-hung fanlights in casement windows, whereas a horizontally pivoted window lacks such fine adjustment, providing a draught rather than room ventilation. The surveyor should comment in appropriate cases.

Since 2013, Approved Document K has defined the requirements for protection from injury from glazing in England; Approved Document M, which did that previously, now deals solely with Wales. The document sets out such matters as the areas where safety glass should be fitted, how to guard against people walking into full-height glazing, or how to safely clean opening windows where falling might be dangerous. This is an aspect of reporting which has often been overlooked but which a surveyor needs to bear in mind. While there is no requirement to bring older dwellings up to standard, there are many windows and doors – particularly from the period 1960 to 1980 – glazed down almost to floor level, which can pose a serious hazard to occupiers, especially children and the elderly. The surveyor must routinely search for the distinctive markings that indicate the presence of safety glass and point out the implications where these are not present at crucial points (**358**). If the report format has a specific section dealing with risks, warnings should be repeated in the part dealing with risks to people (see the relevant section in Chapter 29).

Another field which has received comparatively little attention in the past, but which has grown in recent years to assume much more importance, is that of security. Aspects of security are very much related to the occupiers, their possessions and the location of the dwelling. There is also the question of insurance, which no doubt takes the same factors into account. The surveyor will take note, even if only mentally, of the closing and locking devices as a part of an inspection of windows and external doors, as much because of safety as security. Every window should have a catch to prevent opportunist entry and every external door should have, in addition to a latch, a deadlock operable

356 Windows in medieval timber-framed dwellings would be provided in the panels between the oak frame, and themselves might comprise wrought iron frames containing leaded lights, sometimes, as here, in complicated patterns. Some would be arranged to open.

357 As BRE points out, the proportion of framing to the size of the opening is much greater for windows of the present time than it was in the nineteenth century – something like 35 per cent as against 15 per cent. The size of frame required for triple glazing is even greater than that for double glazing as this cutaway section of a timber-framed opening light shows. To have sufficient strength, the frame sections have to be substantial enough to incorporate the thickness and weight of the glazing unit and may also need to allow for means of drainage and ventilation.

358 Markings in the corner of each glazing unit mean that glass of a suitable standard has been used in vulnerable locations. If markings are missing where safety glass should have been fitted then the surveyor should recommend checks in the interests of safety.

by a key together with a bolt. A lack of any of these items should be mentioned. Some early uPVC replacement windows were found to have a fundamental flaw in that removal of the external glazing bead allowed the whole glazing section to be taken out – making access for intruders unusually straightforward! This defective design should no longer be encountered.

At the opposite extreme from unauthorised intrusion, the surveyor should also bear in mind the extent to which windows may facilitate or hinder an emergency exit from the property. Tragic loss of life has occurred when double-glazed windows fitted with safety glass and having inadequately sized opening lights have prevented occupiers from escaping from rooms when trapped by a fire. The availability or otherwise of adequate means of escape from first-floor rooms should always be noted during the inspection and a comment included in the report when appropriate, always bearing in mind that rooms above first-floor level, including loft conversions, are required to have a fire-protected internal means of escape.

Wood-framed windows

The durability of timber-framed windows, like all others, is very much related to their degree of exposure to the elements. This is mainly dependent on their positioning in the building and within the opening in which they are situated. In other words, windows at high level, facing the direction of the prevailing wind and set well forward in the opening in line with the face of the wall are more likely to need repair sooner than those at low level, facing the opposite direction and set well back.

The principal use of the double-hung sash was between 1700 and 1920. The vast majority of the surviving dwellings built during that time are terraced and consist of basement, ground and two or more upper floors. A fair proportion of the work carried out by surveyors practising in localities where these buildings predominate will inevitably consist of inspecting that type of window – many in quite exposed positions. They present quite a lot of surface, visible and accessible for inspection, and it can be time-consuming if each requires detailed inspection. For such windows in high exposed positions in dwellings built in the very early eighteenth century when frames were set flush with the wall face, the bottom of the sash boxes and the sills need judicious probing for rot along with the bottom rail of the lower sash. Probing may not be necessary if the appearance is such that rot can safely be considered present, not only because of flaking paint and bare patches but also because of the presence of cracks, opening joints, putty cracked and part missing from the glazing and discolouration of the wood.

Double-hung sash windows obtained more protection from the weather when, in 1709 in London, they were required to be set back 100mm from the face of the wall – the length of half a brick; and even more so in the 1770s when the sash box not only had to be set back that amount but also accommodated in a recess, leaving only a thin strip of wood visible from the outside.

This made the necessity for the junction of masonry and wood to be kept watertight even more essential because if damp got round the back of the sash box, it would remain in contact with the wood for very much longer than if surfaces were exposed, with a corresponding increase in the risk of dry rot. Legislative changes affecting windows imposed initially in London were soon followed elsewhere through the publication and availability of pattern books, although local custom also took precedence on occasion. For example, with horizontal sliding sashes in Yorkshire and the West Country and in Scotland, it was always the custom to set windows well back in the opening. In contrast, in softwood timber-framed dwellings, it was usually necessary to set double-hung sashes well forward to give them support from the frame.

The condition of windows will depend much on regular maintenance of the paint film to both wood and glazing putty. Undue neglect has been known to lead to dangerous conditions. Both glazing and whole sashes have been known to fall out when they have become jammed and force has been used to attempt to open them or make them close properly. Another aspect of safety with double-hung sash windows concerns the cords. If one breaks, all four should be renewed with the appropriate strength of cord as a matter of urgency. If this is not done, fingers could be trapped and injured when the next one snaps. Brittleness in the cords indicates that they are old and nearing the time for renewal. It might be better to replace with chains if the sashes are very heavy.

Wood casement windows, as with double-hung sashes, when well maintained and of reasonable quality will last a long time. Neglect will lead to warping, twisting, open joints, cracked and loose glazing putty, rot and rusty hinges and fastenings. There was a good period for joinery in the late nineteenth century and early twentieth century when casements began to oust the supremacy of the double-hung sash. The quality of both wood and workmanship at this time was high. The period between 1920 and 1940 saw a reduction in the size and a weakening of the strength of window joinery timbers, meaning they were not so able to withstand the unfortunate neglect of the 1940s when resources were otherwise engaged. Even so, those timber-framed windows which remain in place in interwar houses are often in relatively sound condition. If the original windows are still present in the typical semi-detached dwelling from the 1920s onwards then they must be tested for rot, especially if they have recently been repainted. The sills and bottom rails, particularly at the joints, should be checked; this can often be best done with a probe from the underside, frequently left unpainted but still visible and accessible though not quite so visible as to raise the ire of the owner.

The 1960s saw a paring down of sizes to an even greater extent in the rush to build more dwellings so that, for example, components such as sills were built up from small sections. Combined with poor-quality wood with too much sapwood present, the slightest fall-off in maintenance rapidly led to extensive deterioration and the need for replacement. Few original timber-framed windows from this period remain.

360 There is no need for the surveyor to use a probe to identify extensive wet rot in these boards above the entrance to a garage. The worst affected must be replaced, and even though localised repairs may be possible to the lower boards, they are likely to continue deteriorating so complete replacement is a better idea.

359 In the very early eighteenth century, softwood double-hung sash windows would be set flush with the face of the wall in a plain reveal below a wide rubbed brick flat arch, four courses high, as here, or a segmental arch. When positioned in upper floors facing the prevailing wind and rain, regular maintenance would be essential to keep the paint film intact and the junction between the brickwork and the wood moulding well pointed. Here, the sashes have been renewed but not in the style of the original. The glazing bars are too thin for the period and the horns are more appropriate to sashes with larger single or twin squares of glass, not likely to be generally available for at least another 150 years or so.

361 Casement windows of very high quality in a two-storey square bay window to a stone-fronted semi-detached dwelling of around 1890. The bow fronted extensions to the square bay have curved glass and ornamental stained glass leaded light fanlights. The stone front of the dwelling is of composite construction: the flank wall wholly of brick and the upper floor rendered. Brick courses line up with the stone – two brick courses to one of stone.

362 Wood casement windows set in square bays to a three-storey block of flats of the late 1980s. The windows are double glazed and most are fitted with a trickle ventilator. Timber-framed double-glazed units of this design have often proved prone to condensation developing between the panes when inadequate drainage at the base of the frame leads to failure of the glazing unit seal.

Much in the way of design and construction of both wood casement and double-hung sashes of the past seems to anticipate the need for maintenance and repair. Most can be maintained in a straightforward way, which is not just about slapping on a coat of paint but also involves making sure they are eased regularly and work efficiently. If repairs are needed, they can be taken apart, new sections inserted and then put back together again. There is ample evidence that this has been common practice with the result that many old windows are providing good service to this day. Admittedly, the work is labour-intensive, but much can be done by the homeowner exerting a modicum of skill, which is more than can be said for windows in some other materials.

Metal-framed windows

The rolled steel window with its bronze fittings and slim Z- and T-shaped cross sections – sometimes set in a wood frame but appearing even slimmer when set in a metal frame with sections of the same shape – was dominant for many years. It was the archetypal window of the latter half of the 1920 to 1940 period and, again, much used in the following 20-year period along with similar metal-framed doors. Projecting hinges, to facilitate cleaning from the inside, came to be fitted in the 1930s, enabling steel windows to be used in flats; continual developments have produced ranges on different modules, top- and bottom-hung windows, pivot-hung – both vertical and horizontal – folding and sliding windows. Aluminium on a smaller scale came to be used for similar types, but because aluminium is an even poorer insulator than steel, a plastic thermal break needs to be incorporated within the frames to prevent undue heat loss to the outside and condensation in the inner face.

Before the advent of hot-dip zinc galvanising, the main problem with steel windows was rust, which could develop rapidly if the paint film was damaged. The resulting rust expansion and build-up of corrosion could often lead to characteristic cracking to the glass (**364**). Galvanising largely dealt with this problem but neglect and rough treatment can, nevertheless, introduce the possibility of damage to the zinc coating and rusting to the steel below. Similarly, the experience with aluminium and anodising led to dissatisfaction with performance, and it is only with the introduction of factory-applied white polyester powder coatings for both steel and aluminium windows that the situation has now improved. Despite the availability of more durable units, however, there is currently little popular appetite for materials other than uPVC.

One of the reasons why metal windows tended to suffer unduly from corrosion lay in the fact that to derive full cost benefit from the use of metal frames and sills, windows were set well forward in the openings – a mere 50mm in from the face of the wall. While this was popular with occupiers since it provided a more generous internal sill, it subjected the putty used for glazing, which was not always the correct type suitable for use with metal windows, to the full force of the elements. This led to shrinkage, creating gaps which allowed rainwater to penetrate and lie for lengthy periods in the angle of the sections against unprotected metal.

There are degrees of corrosion, of course, and the surveyor should avoid blanket condemnation of metal at the slightest sign. The preservation of metal windows is perfectly valid in the context of dwellings from the periods when they were much used. This is particularly so with interwar dwellings of the Art Deco or so-called 'modern movement' variety with white walls and often with flat roofs or brightly coloured tiles (see **28** in Chapter 4). These properties are characterised very much by the slim lines of the frames and glazing bars of steel windows. Light rust can be rubbed down to firm smooth metal which can then be treated with a rust-inhibiting primer and repainted. Even where the metal is pitted, it can be repaired with an epoxy-based car body filler and repainted. It is only when there is hardly any metal left on some sections that it may be necessary to take out the window and send it to a specialist workshop for repair. The repairer will hopefully possess a number of the many metal windows tossed out from dwellings that have uPVC replacement windows

363 Steel windows are prone to rust when the paint film is neglected. These seem to be in a relatively good condition and may only require a thorough rubbing down, treatment with rust inhibiting primer and repainting to maintain their condition. But the poor insulation qualities of the frames make them prime candidates for replacement with double glazing.

364 When metal-framed windows rust, the resulting expansion can cause stress fracturing of the glass, which will gradually increase in size as the expansion continues. This is a clear indication that the frames are deteriorating and require treatment.

365 The period around the 1960s saw the adoption of enormous single-glazed windows without a thought given to heat loss. Successive hikes to oil prices brought home the folly. The centre horizontal pivot-hung windows provided to these five-storey terrace houses are inefficient as providers of ventilation as they are too large. There is either too much or too little. Some centre-pivoting designs have also proved a liability, being implicated in accidents when people overreached while attempting to clean the outside, losing their balance and being tipped outside, sometimes with fatal consequences; so surveyors should warn of the need for care.

366 Metal replacement windows for wood double-hung sashes in late nineteenth century two-storey terraced dwellings. They represent a reasonable compromise in that the overall proportions are not greatly different from those they replace. This is particularly so of those in the left-hand dwelling with thicker sections, the upper 'sashes' almost giving the impression of being set slightly forward. Metal-framed windows should include thermal breaks to prevent condensation on the inside of the frame – a common problem with many earlier fittings.

and from which can be cannibalised appropriate sections for welding in as replacements. Even bent or twisted sections can sometimes be straightened out in a workshop.

If corrosion beyond reasonable possibility of repair is present and it is not possible to renew sections, it is still possible to purchase rolled steel windows in the shapes and sizes used in the 1920s and 1930s. Now, they have just those slight variations to the sections to provide for double glazing and draught-proofing; and if polyester powder coating is not required, then stove enamelling can be provided for a glossy, everlasting, maintenance-free finish. Modern double glazing with thermal breaks should cope with the condensation which always used to be a problem with metal windows and where there was seldom any provision for draining it away, sometimes leading to the erroneous conclusion that the problem was a lack of weathertightness.

Plastic-framed windows

Plastics of various types have been the favoured choice of the homeowner since the 1970s for replacing whatever type of window previously graced the dwelling. Until 2002, there was no control, except in conservation areas or in respect of listed buildings, about what that replacement consisted of, and there was no regard as to whether it matched in style what was there before or whether it went some way towards complementing what already existed in the locality. This was curious as virtually any type of window shape is possible with plastic since the hollow extruded sections are cut to size and the corners heat welded to form any desired shape. It was government concern over energy consumption that prompted the introduction of control.

For the homeowner, good overall performance has been achieved with the added advantage that no maintenance is required. The quality of the material has been improved over the years with additives to reduce brittleness and the effect of ultraviolet sunlight. Nevertheless, problems can arise, and the welded corners can be a source of weakness if quality is poor and if insufficient allowance has been made for the expansion of uPVC in sunlight.

The introduction of woodgrain effects in the 1980s, which involved dyeing the white uPVC material, resulted in darker shades which heated up to a greater extent in the sun. This could lead to deformation if additional reinforcement was not provided in the hollow sections and the increased movement had not been allowed for in the design. Dyes, of course, should last the lifetime of the window, but on poor-quality material, they have been known to fade fairly rapidly. Failures from these sources mean that uPVC-framed windows are predominantly white, despite the risk of more intrusive appearance.

Some plastic frames are simply made of moulded plastic but others incorporate aluminium or steel strengthening within the frames. The bay windows in interwar-built housing usually provide a structural component to the building so it is important that any replacement frames are capable of providing equivalent support. Plain plastic frames are unlikely to have this capability so surveyors should be alert to any signs of failure which may develop.

367 This apparent four-section double-glazed window is actually a single unit with decorative dividers between the inner and outer panes of glass. The seal of the unit has failed, causing this severe internal condensation. The defective unit can be replaced without great difficulty, but if more than one or two at a property are found to be faulty then doubt is cast on the likely durability of all the others, which has more serious implications.

Another routine check which should be made to all double-glazed windows is for signs of condensation moisture within the sealed panes. This indicates that the seals have failed and the panes require replacement (**367**). This used to be a particular problem of early double-glazed windows, especially those with timber frames. Once one or two panes are found to be affected at a property, it becomes increasingly likely that other units of the same age at the property will also fail. Individual defective panes are usually not especially difficult or expensive to renew, but when multiplied by the need to deal with every unit at a dwelling, this can become a significant issue that may need to be reflected in any valuation. Less commonly encountered defects are faulty glazing beads or defective latches on opening lights; but the same principle applies. If only one window is found to be affected, this may not be significant, but once more than a couple of windows have these defects then it starts to indicate the likelihood of a more general failure and becomes a much more serious matter.

Perhaps the greatest disadvantage that uPVC suffers in comparison with other materials, especially wood, is that repairs are usually impracticable. If a section of a timber frame is affected by rot, it can often be cut out and replaced; but once part of a plastic frame has a fault, it usually means complete replacement. Until the use of poor-quality materials during the 1960s and 1970s, timber had proved its longevity over centuries of use. By comparison, modern plastics have only been in use for 20 or 30 years, so it will be interesting to see how durable they really are in the longer term and what the response will be if, as some suspect, they fail to live up to owners' expectations for maintenance-free and long-lived fittings.

Outside doors

Houses have a front door as a means of entrance and final exit, often designed to impress visitors, and usually a rear or side door. Many also have separate doors to the garden; up to the 1960s, these were often in the form of French windows and since then, large, glazed patio doors. Older terraced dwellings could also have doors in the form of long windows giving access to a balcony at first-floor level. Flats may have an entrance door leading to the exterior, if provided with balcony access, or to an internal staircase or lobby. There may also be a door to a private balcony.

A door opening inwards from outside is difficult to make weathertight to anything like the same degree as a window, although a great deal can be done in this respect. This is one reason why a porch or canopy with side cheeks extending to a minimum projection of 750mm is recommended, certainly for the front entrance door. The other reason is, of course, that it gives some protection from the pouring rain to visitors waiting for the door to be opened and to occupants fumbling for their keys.

The dominant material for the construction of external doors has always been wood, totally so up to the Industrial Revolution, but comparatively small numbers of metal-framed doors have been introduced from the 1920s, usually in combination with metal-framed windows and employing a considerable proportion of glass. As far as the aspects of corrosion and repair of older metal doors is concerned, the same factors as for metal windows need to be taken into account.

Although some wood-battened doors – usually of the framed, braced and ledged variety – will be found in cottage-style dwellings, by far the majority of external doors are framed and panelled. Some will be of hardwood in sturdy form but, unfortunately, as the years have progressed so the thickness of the framing of softwood doors has tended to lessen, leaving many installed since the 1950s insufficiently robust to withstand even ordinary use, let alone any assault on them by intruders. A kick will often break the door itself even if the fastenings are able to hold secure. With the use of too much sapwood at the time and a lack of preservative treatment, many external softwood doors succumbed to wet rot in as little as 10 to 15 years.

BRE recommend that both main and subsidiary doors are of solid wood to a minimum thickness of 44mm. This applies to the panels as well as the framing. Should a flush door be present, it should be of solid core construction as was common in the 1920s and 1930s, not the hollow type which tended to be used later. Tapping will distinguish between these types. Glazing should be in such sizes and positions that if broken then it should not be possible to reach latch-type locking devices, but deadlocks meeting appropriate security standards should always be fitted to exterior doors. Patio doors should also be fitted with locks of the same standard as other external doors and may need to incorporate anti-lift mechanisms.

Many external softwood doors provided for dwellings built up to about the 1940s will be found to be of adequate thickness so far as the framing is concerned even though some of the panels might be a bit short of the full thickness. Regular maintenance of the paint film is essential, as with wood windows. Sometimes warping will occur, making doors difficult to secure and allowing draughts to enter. Doors can be flattened in the workshop, but moving the door stops and draughtproofing can often be a ready solution. Aspect plays a part and panelled doors facing south-west or due west are particularly prone to shrinkage in hot summer sunshine and swelling in damper weather. This eventually causes the glue to fail and gaps to open along the shoulder of the joints, often within a year or so of repainting. The only real solution usually is to take down the door and re-glue and clamp the joints, but some regret that the simpler solution of providing a sunblind for such doors has gone out of fashion.

The arrangement at the threshold is important to the performance of a door, particularly those giving access to balconies above ground-floor level. Even when a canopy is fitted, driving rain can often reach doors in these positions and run down the face. A weatherboard near the foot of the door is the first essential to throw clear as much water as possible. Rebates, drips and water bars, formed and fitted correctly in the right places, are the last line of defence – though not always provided – to prevent rainwater getting under the door and penetrating to the interior by capillary action. It is not unknown for this to happen over a period of years, setting up wet rot and even dry rot in adjacent areas of flooring. Depending on the evidence from the door and adjacent areas, it may be necessary to advise opening up for closer inspection if, for example, fitted carpets are damp stained.

More recently, the double glazing industry has introduced plastic and metal doors. These have more effective weather seals and provide greater levels of insulation. Their latch mechanisms meet the higher standards for security

368 An attractive panelled entrance door with beautifully detailed stained glass leaded lights and similar treatment to the rectangular fanlight or transom window above, for this early 1900s house. Leaded lights are a security risk with a basic night latch so it is good to see a modern deadlock has also been fitted at low level.

369 Doors from the 1950s onwards were made insufficiently robust in many cases. This flush door is very flimsy, not helped by the large glazed panel. It needs attention but does at least have a weatherboard.

purposes. On older properties at least, fewer doors than windows have been replaced, and it is a pleasure to encounter an attractive original door still providing good service when arriving at the start of an inspection.

Other joinery and finishes

The surveyor's external inspection of the elevations will encompass not only the major components – such as the walls and windows, the roof and the rainwater disposal arrangements – but also a range of features which are important to the construction – such as joinery to gables, eaves and soffits – and others, the primary purpose of which is essentially decorative – such as finials or mock half-timbering. In this context, the expression 'joinery' can be interpreted quite widely, not least regarding materials, and may include a variety of metals and plastics as well as a wide range of claddings and boards (**370**). All need to be noted and their condition assessed and recorded, with the aid of binoculars for those at higher level, including their decorative condition where applicable.

370 Features such as this oriel bay have become increasingly popular in recent years. Materials other than timber are widely used but reporting on such elements seems to fall conveniently within the section of the report dealing with 'joinery'.

371 In the decades prior to about 1990, asbestos cement boards were widely used. In many houses similar to this one, built in the 1970s, they were used to line the garage ceiling and for the eaves soffit boards. While in a sound condition, they can be left in place; but surveyors should make appropriate comments in their reports to make sure their clients do not unwittingly expose themselves to risk by disturbing them.

Plastic-boarded eaves and soffits will be frequently found, but on older properties, the plastic often simply covers much older original timberwork and there is usually no ready way of determining the condition of the underlying timber. As long as the timber was sound at the time it was clad, the plastic covering will provide protection from the elements. But it is as well to point out to the client the limitations of the inspection, and if suspicions about the condition of the concealed timberwork are aroused then recommendations for more detailed checks should be made.

In some properties, especially those built in the decades prior to about 1990, soffits and similar claddings were often of asbestos cement boarding. When these are seen or suspected, appropriate warnings should be sounded (**371**). As long as asbestos boarding is sound and in good condition, there is no need to recommend removal; but the implications of managing or removing asbestos-containing materials should be pointed out, especially if any damaged sections are apparent or it is known that the client is proposing to undertake any work in the relevant areas. It should also be ensured that appropriate recommendations relating to one aspect of the building do not inadvertently lead the client into some undesirable action elsewhere. In one case, for example, a surveyor's recommendation to improve ventilation of a roof space resulted in a client's handyman blithely cutting a series of ventilation holes through the asbestos cement soffit boards with an electric hand drill, oblivious to any potential health implications.

At times, the external 'joinery' features will be quite substantial and may even comprise structures in their own right – external staircases, balconies or terraces, for example. The surveyor will need to decide how to report on such elements; they can in some cases be incorporated within a more general section of the report, but in others they will need to be treated separately. A balcony above living accommodation might be regarded as an ancillary roof, for example, while one projecting beyond the elevation and supported by columns or legs might be treated as a separate but attached structure (**372**). Whichever seems appropriate, the surveyor should not be tempted to dismiss such features as unimportant 'extras' which do not relate directly to the main building; the implications of failure or disrepair can be significant. Conservatories have often suffered this fate. They are often given the most cursory of inspections and are reported on in very superficial terms as 'outbuildings', but they have become increasingly important to homeowners and should be given proper consideration. They are dealt with separately in Chapter 27.

372 How should the imposing balcony to the top centre flat be reported, where the roof and the columns extending from ground level on either side seem to be integral parts of the structure? Is the flat owner solely responsible or is there a shared liability for this feature? The surveyor may not be able to discover the answer but the questions need to be posed so that there are no nasty surprises when the time for expenditure arrives.

PART 3

The interior

18

Roof space and pitched roof structures

The inspection

Since the major revision of the RICS *Residential Mortgage Valuation Specification* in 2011, there has been no expectation that valuers will routinely undertake any inspection of the roof space as part of a mortgage valuation inspection, unless having specific instructions to do so from their lender client. However, when conducting a survey – at whichever level – an examination of the roof space and structure is necessary. At Survey Level One, the Practice Note specifies only a 'head and shoulder' inspection from the roof access hatch, but for Levels Two and Three, the requirement is to enter accessible roof spaces and carry out a visual inspection. As acknowledged in the first chapter of this book, individual surveyors vary in their preference for the timing of the roof space inspection within the sequence of their normal inspection routine. Each routine has its merits and there is no single right approach; but the wisdom of starting the inspection with a brief walk around the whole property to note the accommodation and form a preliminary view about overall condition has already been proposed. This preview should include scanning the ceilings of the uppermost accommodation to locate the roof access hatch and to see whether there are any obvious signs of damp penetration or structural problems associated with the roof.

Access to the roof space is vital to the surveyor as it provides the opportunity to determine how the roof covering is supported. Firm support is necessary; otherwise, the covering can be dislodged from its predetermined position and this could interfere with its proper performance. While most of the coverings for pitched roofs will tolerate and accommodate a small amount of movement before leaking, undue sagging in the support or sideways displacement of timbers – causing unevenness of the covering in either direction – can lead to intractable problems of water penetration, particularly when rainfall is accompanied by high winds. In most properties built before the 1960s, the framing of the roof structure would, in turn, often require support not only from the external walls but from internal load-bearing partitions as well.

With access for inspection, all these aspects can be checked if there are any doubts about the effectiveness of either covering or structure.

If there is no access point or if any trapdoor provided cannot be opened, it amounts to a serious limitation on the inspection to which the client's attention needs to be drawn, together with any appropriate elaboration regarding the possible consequences of basic structural problems or disrepair which may lead to leaks. It will also compromise the surveyor's ability to report on matters like services within the roof space, such as electrical wiring and plumbing.

The Level One 'head and shoulders' inspection will obviously be much more limited in extent than that possible with full physical access. Even with the most powerful of torches, it is often difficult to see the character of construction and the detail of defects from the access trapdoor alone, and the view will often be further restricted by stored materials. With terrace or semi-detached dwellings presenting the appearance of brick, often the surveyor's first indication of modern timber-framed construction will be the nontraditional appearance of the party – or separating – walls in the roof space. Unless steel trusses have been used for the roof structure, it may not be possible to identify one of the 150,000 or so steel-framed houses which crop up all over the country, built in both the private and public sectors from the 1920s onwards, many of which, from the outside and the rooms inside, look exactly like a conventional house. That they are steel framed can only be ascertained from a close inspection in the roof space. The 'head and shoulders' inspection also poses the risk of a debate at a later stage about what could and could not be seen from the access hatch. This matter was settled in person by the judge in *Ezekiel* v *McDade* 1994 when hearing a claim in connection with a mortgage valuation. To resolve the matter, the judge made his own inspection and declared that he could see a defect even though the surveyor maintained he could not.

This significant limitation should be pointed out to the client during the preliminary discussion about the type of inspection to be undertaken, and it will often be a persuasive element of the discussion that convinces many clients to opt for the more detailed Level Two survey. Bearing in mind also that the Level One inspection is intended primarily for very straightforward and usually modern properties, this restriction ought not to pose too many difficulties for the surveyor in the majority of cases. In those unusual circumstances where there is real uncertainty or a suspicion of serious problems, the surveyor can always make a decision to go ahead and enter the roof space; but this should not be the default position. It is acting outside the mandatory Practice Note – though such actions can be taken if they can be justified by the surveyor – but more generally, the clear distinctions between the different survey levels start to become blurred.

Having gained access to the interior of the roof structure, the surveyor must assess its condition, its capacity to support the external roof covering and its long-term prospects for keeping the property weathertight. A prerequisite for this assessment, however, is to understand how the roof has been constructed and how the different timbers – or indeed other materials – relate to each other in order to know whether what can be seen is as it should be and to recognise when something has gone awry.

Dwellings built before the 1960s comprise two-thirds of the total housing stock, and the vast majority have some form of traditional roof structure. Most historical buildings built before about 1700, many of which are timber framed, have roofs of hardwood construction; but since then, pitched roofs have been based, in the main, on 100mm × 50mm common rafters, considered over many, many years to be the most economic size of softwood timber for the purpose. Until the 1960s, most domestic roof structures were formed on site by carpenters cutting and fitting the timbers together to designs prepared on the basis of the required span, the structural support available from below and the desired roof covering. Since the 1960s, the quest for efficiency and speed during construction has resulted in the successive development of a variety of engineered and prefabricated components being used within roof structures, exemplified by the ubiquitous trussed rafter. The following sections briefly describe the most common forms of pitched roof structure likely to be encountered by the surveyor during the inspection of a dwelling.

Roofs of early timber-framed dwellings

Most dwellings built since about 1700 have softwood roof structures; but oak, a hardwood, was the favoured material prior to that. In what are commonly identified as timber-framed dwellings, the framing extends to both walls and roof. Oak is usually found in these historic survivors but when that came to be in increasingly short supply during the seventeenth century, black poplar or elm was often used. Because the roof framing is an integral part of the whole structure – not just placed on top of a masonry wall – it is important that both wall and roof framing are considered together since the former is a continuation of the latter from one side of the dwelling to the other. Cutting or removal of one principal member can often have considerable effect at another point in the structure.

Three typical early timber-framed roofs are shown in **Figure 24**. The top example, a Crown Post roof, is illustrated in **373** while King Post and Queen Post roof trusses are illustrated in **374**. The simple collar roof at the bottom of **Figure 24** is not as simple as it looks and is really a truss. Besides the purlins housed into the principal rafters, which support the common rafters (shown dotted), there will usually be extensive wind bracing between the principal rafters. This type of roof is sometimes varied in that the purlins can be set into the top of the principal rafters and the collar provided with bracing from the top of the wall plates, producing an attractive arched ceiling from below. Many variations in the details of all these types of roof will be seen and, of course, they were all developed, over the years, from cruck-framed roofs of even earlier times, the shape of one of which can be discerned from the flank wall of the dwelling shown at **375**.

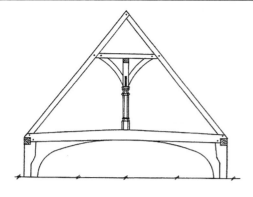

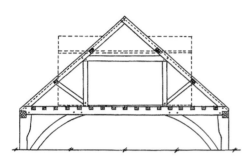

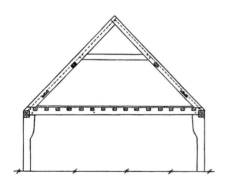

373 Early timber Crown Post roof truss with medial purlin and braces.

Figure 24 Some roofs of early timber-framed dwellings. Top: Crown Post roof with medial purlin and braces – the most frequently encountered type, although the details may vary. The carved Crown Post indicates that it was originally meant to be seen. Very often, however, a ceiling will have been formed subsequently at tie beam level, leaving it smoke blackened, unseen and marooned in a newly formed roof space. Centre: Queen Post roof truss, which enables rooms to be formed in the roof space. Bottom: Simple Collar Roof, a truss in all but name where rooms can also be formed in the roof space.

It is fairly unusual to find early timber-framed buildings in other than a deformed state although this is, of course, part of their attraction to many a purchaser, even if they do not quite realise what they are letting themselves in for (**376**). Such dwellings were normally constructed from newly felled green oak, which subsequently shrank. Although iron is found occasionally, most joints were oak pegged and the majority consisted of splayed housings that took up the strain when the wood shrank. If the allowance for shrinkage was insufficient, the timber might split or severely warp. Subsequent alterations to install fireplaces and chimney stacks, and perhaps the rearrangement of rooms, were often ill-considered and resulted in the cutting of members with little regard to their importance to the frame's overall stability.

All this means that the surveyor needs to look very closely indeed at all visible timber and joints and make a careful assessment of the existing structural situation even if there are no immediate problems of rot, beetle infestation, damp penetration or apparent current movement. A detailed assessment of a timber-framed structure is not something to be taken on lightly; but thankfully the increasing appreciation and understanding of traditional construction methods and techniques seems to be enhancing professional opportunities for specialisation in the appraisal, restoration and maintenance of such buildings.

374 A King Post truss (foreground) and two Queen Post trusses, helpfully illustrated in one photograph at a converted restaurant.

375 The flank wall of a single-storey, with attic, timber-framed and brick-infilled cottage of about 1500, showing the end pair of the crucks on which both the roof and the wall framing are based.

376 The deformation in early timber-framed dwellings often leaves the roof in a very uneven condition for the roof covering. That on the right would appear to have been packed out successfully. That on the left has had some attention but needs more.

Cut timber roof structures

Couple roofs

A 100mm × 50mm common rafter can support, on the slope, a load of typical roof covering comprising slates or tiles, sarking and battens over a length of 2.4m. Joined together at the top, they could span between supporting walls 3.5m to 4.5m apart, depending on the pitch adopted. Such a design – a 'couple roof' – would, however, be severely flawed since the rafters would exert a substantial thrust at the top of the walls. Such thrust could only be resisted if the walls were of massive proportions, such as the stone walls of some medieval dwellings. Consequently, couple roofs are generally found only on comparatively ancient dwellings and their use in more recent times has been limited to covering very small spans, such as entrance porches; even then, they may not be appropriate if the roof covering is heavy (**377**).

377 The collapse of this garage demonstrates the consequences of inadequate lateral restraint between the feet of opposing rafters on a couple roof. The outward lean of the wall by the window opening is due to the weight of the tiled roof covering exerting an outward thrust through the feet of the rafters onto the top of an insubstantial wall.

Close couple roofs

The problem of thrust at the top of the supporting walls is solved by the addition of a tie, usually a ceiling joist, to connect together the feet of the rafters to form a 'close couple roof', illustrated in **Figure 25**. As mentioned, the span between supporting walls with this type of roof can vary between 3.5m and 4.5m depending on the pitch. Over this distance, there is a danger of the ceiling joists sagging if subjected to any load or being walked upon, so they are normally provided with a binder hung from the ridge plate. Properly nailed at each of the connecting points, as circled on **Figure 25**, the load of the roof structure and its covering will be transferred vertically downwards through the supporting walls.

Simple close couple roofs of this nature are occasionally found as the main roof over the smallest of dwellings, only one room in depth when the ridge is parallel to the frontage (**378**). Sometimes, however, two close couple roofs will be found constructed side by side, the two ridges parallel with the street and each other. The resulting M-shaped section, when viewed from the side (**379**), requires a central structural partition between front and rear rooms that is taken to top-floor ceiling level to provide support for the lower end of the central rafters. There needs, then, to be a valley gutter between the two

378 A simple close couple roof is all that is required for the straightforward roofing of the type often found on smaller dwellings.

Figure 25 Cross and longitudinal sections of a close couple roof with hangers and binder to support the ceiling joists. Capable of spanning roughly one room and, accordingly, used either gable end, facing street, or in parallel alongside another, but then needing a structural partition in support.

roofs, discharging generally to a flank wall in the case of detached or semi-detached dwellings (**380**). This is a frequent source of trouble if the gutter becomes blocked and overflows. Even more troublesome are those cases where such parallel roofs are provided to terraced properties involving an internal rainwater pipe through the whole height of the building or a secret gutter to take the rainwater either through the roof space to the front or rear or through the floor (as illustrated at **119** in Chapter 7).

379 This end-of-terrace house with parallel small mansard roofs clearly shows the outlet from the valley gutter on the flank wall. Elsewhere in the terrace, rainwater from the valley gutter is taken to the rear via a secret gutter in the top floor.

380 The parallel close couple roofs of this eighteenth-century detached house drain to a central valley gutter with an outlet at the far end. This is a much better arrangement than using secret gutters, but with terraced properties there may be little alternative unless the terrace is short. The surveyor is still faced with the practical difficulty of determining the condition of the concealed valley.

The principle behind the dual-pitch close couple roof is the creation of a rigid triangle that transfers the weight of the roof covering and its own weight vertically downwards through the supporting walls. It relies on the provision of adequately sized timbers that will resist deflection under load, accurate cutting and fitting of the component parts and secure fixing at the joints. A failure in one or more of those requirements can produce unevenness in the roof covering to the extent that it may not be able to fulfil its function of keeping water at bay. Further damage may also be caused by thrust due to deflection or weak jointing. It may be that when such defects are visible, a cause is readily ascertainable and can be included in the report with appropriate recommendations; but it may be that further investigation will needed.

Close couple purlin roofs

The three-up two-down semi-detached dwelling is the most frequently encountered of the period 1920–1940, each half of the pair being relatively square in plan. Something like 40 per cent of all dwellings in the UK were

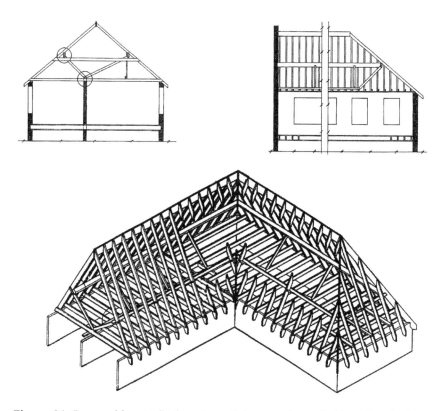

Figure 26 Cross and longitudinal sections of close couple roof with purlins (higher circle) and struts where structural partition with plate on top (lower circle) is parallel to the frontage. The typical square-shaped semi-detached house of the 1930s has a roof of this type. The diagram also shows a similar roof over an L-shaped dwelling, but with collar omitted for clarity.

built during this period (**381**). Usually two rooms from front to back, the roof would consist of a dual-pitch close couple hipped design as in **Figure 26**, comprising common rafters, ridge plate and purlins (shown circled on the diagram). The purlins divide the length of unsupported rafter into approximately 2.4m lengths. To support the purlins, struts are required and sometimes a collar is provided. For a span of about 8m to 10m, with a structural partition separating the front and rear rooms taken up through the first storey and supporting the struts from a plate at the top (also circled on the diagram), it could take the form shown in **Figure 26**. The binder and hangers shown on the cross section would be needed to prevent the ceiling joists from sagging if one room was considerably larger than the other.

Because of the long lengths involved, the 100mm × 50mm ceiling joists were often in two lengths intended to be lapped and nailed together on top of the load-bearing partition to maintain the tie between the feet of the rafters. It is not unknown for the significance of this to be lost and the nailing to be inadequate. Sometimes it is omitted entirely and the two lengths not even lapped, relying instead on skew nailing the ends to the plate on top of the partition. Skew nailing in this position is easily pulled out as the roof starts to spread.

If the dwelling had been designed with a large ground-floor through room, the two-storey structural partition would probably be at right angles to the frontage and the roof arrangement could be as shown on **Figure 27** with binders (as circled) spanning from the partition to the separating wall, supporting both roof joists and roof struts.

Smaller dwellings, particularly those built in the early 1920s in a cottage style but many others going back to the late nineteenth century, were built in the form of long or short terraces. The dual-pitch close couple roof with purlins is also the common form of roof for such dwellings. The purlins are usually strutted from a central structural partition parallel to the frontage (as shown on the half cross section at **Figure 28**).

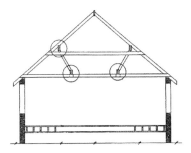
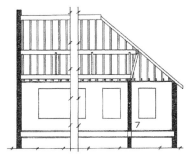

Figure 27 Cross and longitudinal sections of close couple hipped roof with purlins and struts where the structural partition is at right angles to frontage. This form of close couple purlin roof would be required where a large through room was provided.

381 Semi-detached house with typical close couple purlin hipped roof from the period 1920–1940, here covered with plain clay tiles.

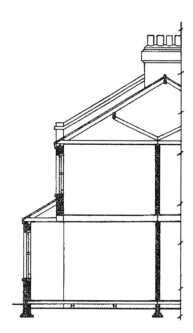

Figure 28
Half cross section of dual-pitch close couple roof with purlins and struts, supported by a structural partition parallel to the frontage – a type of roof much used for late nineteenth-century and early twentieth-century two-storey terrace houses along bylaw streets.

If the frontage of the dwelling was not too wide, the purlins could be arranged to span between the separating walls, in which case there would be no requirement for supporting struts or for the partition to be of a structural nature above first-floor level. Obviously the purlins would need to be of a substantial nature to carry the load without deflection, but such timber would have been available at the time of construction. Later, in the 1950s and 1960s, timbers of the required size were harder to come by for narrow-fronted terraced dwellings; instead, lighter timbers were used, made up in the form of trussed purlins similar to those shown later in **Figure 41**. Existing purlins can sometimes be turned into trussed purlins and this is an option which is often adopted when the struts are removed in older dwellings to provide accommodation in the roof space by way of a loft conversion. Failure to provide a satisfactory alternative to the struts can lead to severe consequences for both roof above and ceiling below.

Collar roofs

A more problematic form of roof can be produced when the tie, instead of connecting the feet of the rafters, is positioned higher as a collar. The higher it is placed the less effective it will be at resisting the outward thrust at the tops of the supporting walls. Nevertheless, a collar roof, as this type of roof is called, will often be found where it is desired to provide rooms half in and half out of the roof space. The arrangement is attractive to designers since the same accommodation can be provided as in a full storey but with a saving in the amount of walling required (as shown in **Figure 29**). Early local authority dwellings favoured such an arrangement, following the examples set earlier by the Garden City and Arts and Crafts movements (**382** and **383**).

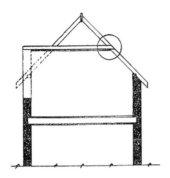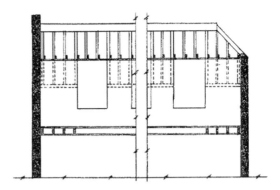

Figure 29 Cross and longitudinal sections of collar roof providing rooms half in and half out of the roof space. There is a saving in the wall material when this type of construction is adopted and, accordingly, it is often encountered on cheaper housing.

To compensate for the less effective nature of the collar – as against a tie at the feet of the rafters – best practice indicates that the size of the rafters and the collars forming the ceiling should be increased to 150mm × 50mm and that the joint between the two (circled in **Figure 29**) be made with a dovetailed halved joint, instead of reliance being placed solely on nailing. Best practice is not always followed, however; and if the joint is merely nailed and these pull out then thrust may push out the tops of the walls, overturning the wall plate in the process. If standard size rafters only are used, these may deflect because they are unrestrained at their feet, possibly disrupting the lie of the roof covering.

382 Rooms half in and half out of the roof space of this pair of early twentieth-century semi-detached houses, necessitating a check on the form of roof construction and whether any roof spread has occurred. None is apparent on the photograph and the distinct hogging of the tiling is due to the separating wall retaining its position while all else has settled slightly.

383 The influence of the Arts and Crafts movement pervaded the dwellings of the Garden Cities, as here, the earliest local authority housing at the beginning of the twentieth century resulting in many with rooms half in and half out of the roof space. Coupled with gables to the dormer windows, they have a cottage-like look but, nevertheless, the construction needs checking.

384 The shape and construction of a mansard type of roof – very common on terraces of the eighteenth and nineteenth centuries since it provided the attic rooms for the servants – can be discerned from this example in process of repair. Stout beams spanning from party wall to party wall at the top of the steeply sloping sections front and rear and at the centre carry the load of either a flat-topped roof, probably covered in lead, or a shallow-pitched, dual-sloped roof possibly covered with large slates or pantiles. The top floor acts as a tie. The front elevation has a cornice and parapet to maintain the classical façade and a parapet gutter, while the rear slope discharges into either an eaves gutter or another parapet gutter. The relationship between the parapet and the parapet gutter is clearly visible here, as is the inaccessible area beneath which is so vulnerable if leakage occurs.

385 The unified façade was not to be disturbed by too much sight of the roof in the eighteenth century yet the servants had to be accommodated somewhere. A parapet or sometimes an ornamental balustrade, as on this terrace, would be used to cap the front elevation along with what could be, at times, troublesome parapet gutters.

386 Because of the gap between the underside of a parapet gutter and the ceiling below, slow leaks can often affect the woodwork, setting up attacks of dry and wet rot before becoming apparent on the ceiling. Nevertheless, the effects are often visible by inspection and it ill behoves a surveyor to miss them. Some replacements have been carried out here but can a clean bill of health be given?

Mansard roofs

Mansard roofs have two slopes on each side with the lower slope, often incorporating dormer windows, being at a steeper angle than the upper. This arrangement has the advantage of allowing the incorporation of additional accommodation within the main roof structure while the topmost section of roof is of close couple design. Properly designed, this overcomes the main disadvantages of a simple collar roof and was ideal for economically providing servant's quarters (**384**).

Apart from the inner-facing portions of the M-section double close couple roofs described above, the slopes of all the roofs described so far are visible from ground level; even those with the very shallow pitches favoured in the periods 1810–1840 and 1960–1980. However, over quite extensive periods and particularly in the eighteenth century, fashion dictated that the roof should be hidden as much as possible from the front of the property. This could be achieved by hiding most of the bulk of a close couple roof or mansard roof behind a parapet (**379**) or sometimes a balustrade (**385**). Unfortunately, the resulting parapet gutters create another area of potentially serious vulnerability where blockages to rainwater run-off and leaks in linings can rapidly lead to extensive damage to structures and internal finishes. These require regular maintenance checks by occupiers if problems are to be avoided and necessitate a thorough examination whenever encountered by surveyors (**386**).

Duopitch butterfly roofs

Later, for smaller houses when there wasn't quite the same need for attic rooms, the height of the parapet could be reduced by the use of a butterfly or double lean-to roof with a V-shaped configuration, as illustrated at **387** and **Figure 30**. The height could be reduced even further if hipped slopes were incorporated at the front.

The top end of the rafters of late nineteenth-century butterfly roofs take their bearing from a plate or corbels or may be built into the flank or separating wall (marked 'C' on **Figure 30**), but the lower end would rely on support from the long beam spanning from front to rear (marked 'A'). Invariably, this beam was intended to have intermediate support from a structural partition dividing front and rear rooms. A frequent cause of problems lies in the construction of this partition when it allows sagging in the beam and consequential overflow from the valley gutter. The partition is often of comparatively flimsy construction on either an inadequate, or practically non-existent, foundation. Timber-framed lath and plaster was frequently used above ground-floor level, and subsequent cutting of principal members to open up through rooms could have serious consequences. In particular, the settlement of the front to rear

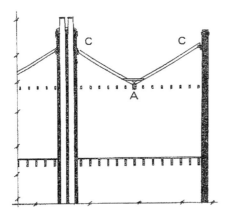
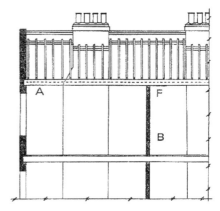

Figure 30 Cross and longitudinal sections of butterfly roof, often employed for the tall, narrow-fronted dwellings built towards the end of the nineteenth century. They came back into fashion for a short while in the period 1960–1980 when monopitch trussed rafters became available.

387 The rear of a terrace of late nineteenth-century dwellings with butterfly roofs. It is not unusual for the defects in one house of a terrace to be repeated elsewhere. A surveyor inspecting adjoining dwellings here would need, therefore, to consider this possibility.

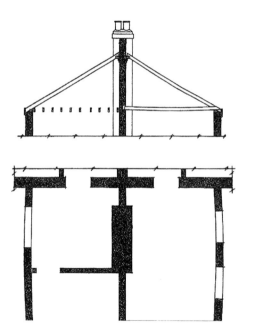

Figure 31 Cross section of typical monopitch back addition offshoot roof with plan of possible accommodation below. These structures are very commonly found attached to small dwellings of the late nineteenth century and early twentieth century.

gutter can cause disruption to the bearing of the top of the rafters, break supporting corbels, disturb flashings and allow water penetration. While this may not reach ceiling level, it can cause rot in the timbers.

At the end of a terrace, the ends of rafters supported at the top of the flank wall can push the wall outwards since there is no balancing thrust from rafters opposite – as there is on a separating wall between two terraced properties. As the flank wall is often inadequately restrained lower down, particularly if the staircase of the dwelling is positioned against it, the movement can be quite dramatic. The result can often be seen in gaps at the edge of floors and ceilings and cracks at the junction of flank with front or rear walls.

Monopitch roofs

Commonly encountered to the rear of many dwellings from all periods up until the 1920s is the back addition or offshoot. This very often has a monopitch lean-to roof. Arranged (as on the right of **Figure 31**) with proper joints, adequately nailed and with suitably sized timbers, all should be well, even though lacking entire compliance with current Building Regulations. However, the span on slope of the 100mm × 50mm rafters is well over the recognised safe span of 2.4m. Accordingly, it is not surprising that the plan shows a bow on the flank wall, particularly as this is broken up by two windows. It could be that deflection in the rafters under the load of roofing could be combined with inadequate nailing, giving rise to roof thrust.

If ceiling joists are provided but there are two rooms in the back addition with a separating partition, the ceiling joists might be arranged as on the left of **Figure 31**. The bow visible in this flank wall could again be caused by roof thrust and manifested to its maximum extent at the centre, away from the restraint provided by the bonding to the rear wall of the main structure and the rear wall of the back addition itself.

In cheaper construction, headroom in the back addition was often increased by omitting the horizontal ceiling joists and simply forming a sloping ceiling directly on the underside of the roof slope. This means that the necessary triangulation to provide lateral restraint at the base of the sloping rafters has also been omitted, increasing the potential for outward thrust and movement at the top of the external wall. Sometimes collar ties create a higher-level horizontal ceiling above a lower sloping section. This is better than having no restraint but, as with collar roofs, can still be problematic and should be properly checked.

Trussed roofs

Trussed roofs are encountered comparatively rarely, but there are a number of variations to this form that the surveyor should be aware of (shown in **Figure 32**); it will immediately be seen that the top and bottom ones very much resemble in shape the roof trusses from earlier times (illustrated in **Figure 24** and **374**). There are structural differences, however. For example, the centre post in a King Post roof truss is a tension member holding up the tie beam whereas the central vertical member in the Crown Post roof is in compression, taking the load from a medial or longitudinal purlin and transferring it down through the tie beam to the side wall framing. Trussed roofs can be found on larger dwellings or where larger spans were required from the eighteenth century up to the earlier part of the 1900s. They do not appear on the cheaper dwellings and were, therefore, usually soundly designed and constructed of well-seasoned timber of ample proportions, put together with joints of long-standing design.

Nevertheless, the bolts, nuts, plates and screws which are used to strap the joints together can rust when exposed to condensation over long periods. Joints can be disrupted if the support to a truss is disturbed, placing the joints under increased strain. This can happen when a pier along a length of wall subsides because the point loading has been miscalculated or the ground has settled. It is usually failures in support of this nature which cause structural problems, not the original design or construction of the truss itself.

Many trussed roofs are provided for dwellings that incorporate parapets. Frequent overflows

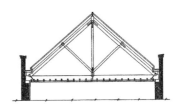

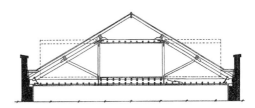

Figure 32 Types of trussed roof. Top: King Post. Centre: a variant of the King Post where the central timber post is replaced by a metal rod, more efficient in tension than the timber, and the purlins support boarding at a shallow pitch, suitable for covering with large slates or alternatively lead, copper or zinc. Bottom: Queen Post, where it was desired to have rooms in the roof space. These will occasionally be found on the more expensive dwellings of the eighteenth, nineteenth or early twentieth centuries where there was often a requirement for roofs of a considerable span to cover large rooms. They are made up of substantial sections of softwood with iron straps to bind the joints together.

from the gutter behind the parapet in heavy storms or at times of snow cause damp penetration. The confined, airless spaces below such gutters are ideal situations for outbreaks of dry and wet rot and are seldom examined when the gutter is repaired and the decorations are made good. Again, truss joints can be disrupted and it is often a repeat of severe flooding that will draw attention to the collapse of gutter boards due to the spread of rot and the consequential need for major repairs after opening up. Traversing the gutters, if the opportunity avails itself, can disclose indications of such problems and the need for 'urgent' warnings to be included in the report.

Prefabricated roof structures

Bolted and connector roof trusses

Over a third of all UK residential accommodation has been built since the early 1960s, and the revolution in roof design and construction which took place around that time will, in the vast majority of cases, be reflected in what is found when properties built since then are inspected. It is now rare to find modern properties with roof timbers cut and fitted on site because computer-designed, factory-manufactured trusses utilise smaller section timbers and are more economical to make as well as faster to erect.

Initially, a partial departure from the traditional form had come about in the 10 years or so before the 1960s with the development of trusses based

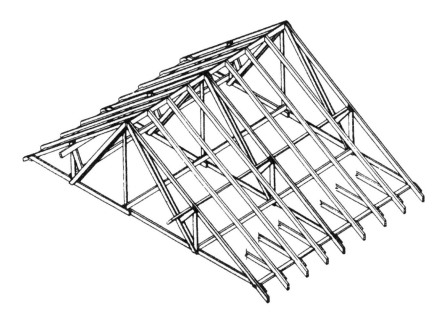

Figure 33 Typical bolted and connector roof truss with purlins. Developed in the 1950s and spaced at 1.8m centres, these freed designers of roofs of the need to rely on intermediate support from structural partitions in small dwellings.

on bolted joints using double-sided toothed plate connectors. Prefabricated, hoisted and spaced at 1.8m centres, site work would be reduced to fixing the purlins and ridge plate and cutting and fitting intermediate 100mm × 50mm common rafters and ceiling joists at 450mm centres (as shown in **Figure 33**).

The use of bolted and connector roof trusses still provided a reasonable amount of storage space between the trusses, a saving in time formerly spent on site, a saving in timber used and, primarily, the freedom to plan space in the building below, unrestricted by the need to include any structural partitions. Depending on the design, sometimes component members would merely consist of two common rafters bolted together and, for some, plywood gussets would be used instead of toothed plate connectors. Their use continues and clear spans of 12m are possible with pitches of between 15° to 45°. Standard details for forming hips are available for some designs.

Trussed rafter roofs

In the early 1960s, with the aim of saving even more timber and yet more time on site by further prefabrication, the idea behind the truss with bolted connections was developed so that, in effect, each 'couple' of rafters could become in themselves a separate truss – a 'trussed rafter' (**Figure 34** and **Figure 35**). These were installed generally at 400mm centres or at the very maximum of 600mm centres. Rafters, ceiling joists and struts together with any other web members became generally thinner at 35mm of stress-graded timber connected together in a factory with thin punched toothed metal plate fasteners of galvanised or, more rarely, stainless steel. On older trusses, nailed or glued plywood connectors may be found.

Considered individually, trussed rafters are very flimsy and should receive considerable care

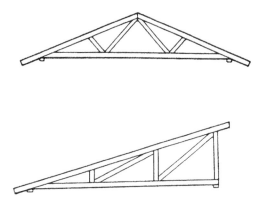

Figure 34 Common types of trussed rafter: fink and monopitch. Generally spaced at 400mm centres, trussed rafters come complete with single-section ceiling joists and web members, joined where they meet by steel-toothed plate connectors, avoiding the need for much further site work once hoisted into position.

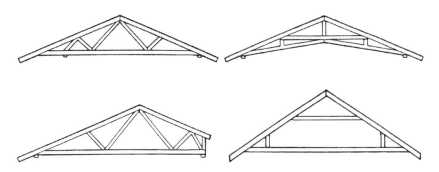

Figure 35 Other trussed rafter designs.

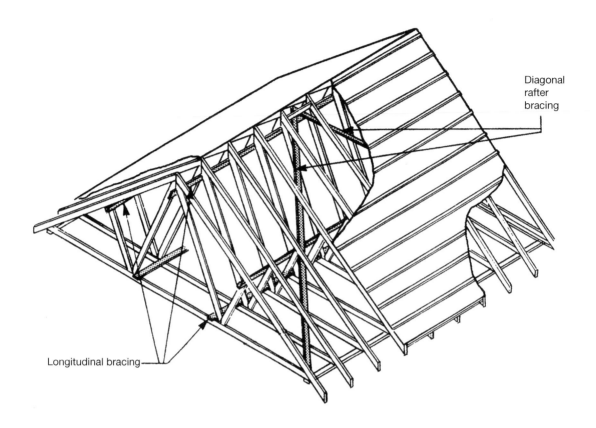

Diagonal
rafter
bracing

Longitudinal bracing

Figure 36 Standard bracing for duopitch trussed rafters (chevron bracing, required on spans over 8m, is omitted).

in storage, handling and loading at the factory and unloading and storage at the site. Fixing of trusses is normally to wall plates, but these need, initially, to be securely held down to the inner face of the external walls to prevent subsequent uplift by strong winds. Hoisting and fixing into position on to the wall plates by either proprietary clips or skew nailing needs further care to avoid damage to the truss joints by distortion. However, when set up and provided with the essential bracing, both diagonally and longitudinally (as shown in **Figure 36**), and secured with straps to the walls at the ends of the roof, they provide a firm base for the roof covering.

Additional chevron bracing is required for duopitch spans over 8m and monopitch spans over 5m unless a sarking underlay of boarding, plywood or chipboard is used, when it can be omitted. Simple trussed rafters suited the then current fashion for clean-lined, minimum-interest roof shapes of low pitch to utilise the availability of new designs of interlocking tiles made of concrete (**388**). The absence of chimneys in this period, following the almost universal inclusion of central heating in new dwellings, often meant that whole roofs could be covered without cutting a single tile. Less helpful to homeowners, however, was that utilisation of space within the roof became virtually impossible because of the close spacing of timbers and the shallow pitch favoured at the time (**389**).

The simplicity of utilising prefabricated trussed rafters encouraged some experimentation in roof forms (**390** and **391**), but these failed to supplant more conventionally shaped roofs. If on occasion – fairly rare in the 1960s and 1970s – a hipped end or pediment with valley gutters was incorporated, these tended to be formed either with truncated trusses or with a girder truss and cut members or, for a pediment, by diminishing trusses fixed on top of the main run (as shown in **Figures 37** and **38**). In either case, their use tended to be solely for cosmetic purposes in the prevailing shallow pitch roofs of the day, perhaps when some individuality was required in the private sector.

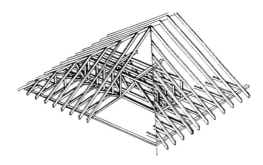

Figure 37 Typical hip construction in trussed rafter roofs using a girder truss.

During the 1980s, when the provision of housing in the public sector was changed from local authorities to housing associations and the styles and fashions of the 1960s and 1970s became anathema in both public and private sectors, hips and featured gables took on much greater importance. Roof pitches increased, often to accommodate rooms in the roof with either roof lights or dormer windows. Increasingly since then, a progressively diverse range of engineered trusses have been designed and manufactured, allowing more interesting and varied roof configurations as well as increased provision of internal space. The trussed rafter roof is now the most common form of roof structure for modern volume construction (**392**).

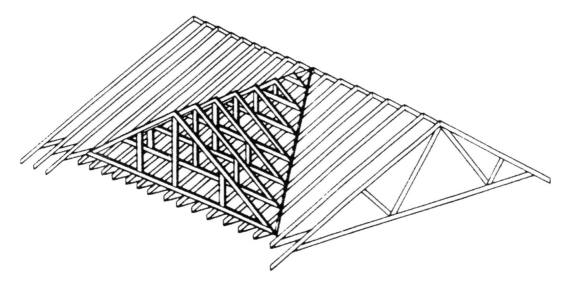

Figure 38 Valley set of diminishing trussed rafters to form a gable. These would be used to purely improve appearance by introducing variety of shape to roofs of the 1960s and 1970s.

388 Typical shallow pitch roof of the 1960s and 1970s. The introduction of trussed rafters and the absence of chimneys made for the simplest of roof designs, a defining characteristic of dwellings built during this period.

389 Some idea of the difficulty of inspecting trussed rafter roofs, dating from the 1960s onwards, can be gained from this example. This one has a comparatively steep pitch but many, particularly in the early days, were provided with pitches as low as 20° to 30° and some even lower.

390 The availability of different shapes of trussed rafters encouraged a variety of nontraditional roof shapes in the 1960s and the 1970s.

391 Experiments with roof shapes during the 1960s and 1970s included butterfly roofs, but problems of drainage, inspection and access inevitably followed.

392 Instructions to carry out surveys on newly constructed dwellings are rare in view of most purchasers' reliance, sometimes unwisely, on NHBC or other warranties. If forthcoming on a dwelling forming part of an estate, the possibility of inspecting other incomplete structures, as here, may present the opportunity to assess the quality of details which may subsequently be hidden from view, such as fixings of and to wall plates and the presence and fixing of bracing straps and connectors.

Attic trussed rafter roofs

A type of trussed rafter design found suitable, after the simple original configurations were further developed, for the purpose of accommodating rooms in the roof space is shown at **Figure 39**. It is generally known as an 'attic' trussed rafter. Because of the increased height required, trans-portation becomes more difficult. The type of single unit shown is normally limited to a height of 4.5m. Intermediate support for the floor tie is normally required when spans exceed 9m, and the use of single-unit attic trussed rafters becomes uneconomic when spans exceed 10m. However, it is possible to design and manufacture larger trusses in sections and for them to be assembled on site, in strict accordance with the designer's instructions – although, of course, this does involve much more site work.

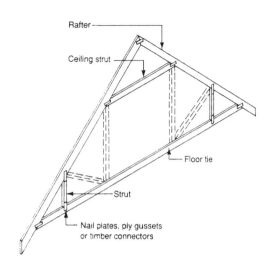

Figure 39 Attic trussed rafter. These were introduced in the 1980s to provide rooms in the roof space.

The maximum spacing for attic trussed rafters is 600mm, but it is possible to form structural openings for dormer windows or roof lights up to three times the spacing adopted. One way to form a hipped slope with this type of roof is to provide a girder truss as shown in **Figure 40** but other ways are possible, depending on span and pitch.

Attic trussed roofs

More freedom in the location and size of dormer windows and staircases can be obtained by the use of more strongly constructed attic trusses. Whereas the lighter attic trussed rafters are spaced at 600mm centres, attic trusses can be spaced at 3.6m centres with only the common rafters needing to be cut and fitted to purlins at the narrower spacing of 600mm. A typical layout where the trusses are spaced at 2.4m is shown at **Figure 41**. Account needs to be taken of the heavier point loads involved with the use of attic trusses in the design of the supporting walls.

Sheathed rafter panel roofs

Where the preference for domestic buildings is still to have simple continuous ridges and separating walls or gable ends, a different form of construction can be used that eliminates all struts, ties or bracing and provides a completely clear triangular roof space. This comprises a structure of sheathed rafter panels as shown in **Figure 42**. The sheathing can be of any number of different materials, including plywood, chipboard, tempered hardboard or bitumen-

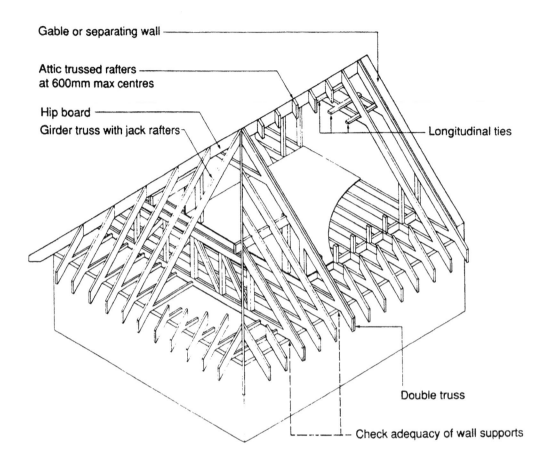

Gable or separating wall

Attic trussed rafters
at 600mm max centres

Hip board

Girder truss with jack rafters

Longitudinal ties

Double truss

Check adequacy of wall supports

Figure 40 Typical hipped roof construction with lightly constructed attic trussed rafters at a maximum spacing of 600mm.

impregnated softboard; in current construction, this is most likely to be of oriented strand board (OSB). Sheathing can be placed either on top of the rafters (as in the traditional manner of boarded sarking in Scotland) or below.

The panels can be factory made (as in **Figure 43**) to include all the necessary noggings, insulation panels and vapour barrier and either hoisted or craned into position. Spans up to 10.5m at roof pitches of 35° to 45° are possible with this method of construction, but it is vital for the floor decking to be taken to connect with a base plate at the outer edge and securely nailed so that it acts to restrain roof spread. Roof lights in the larger panels can be formed with this type of roof construction, but an imperforate panel needs to be located on either side. Panels with dormer windows are also possible, framed up and using short rafter panels for the roofs of the dormers.

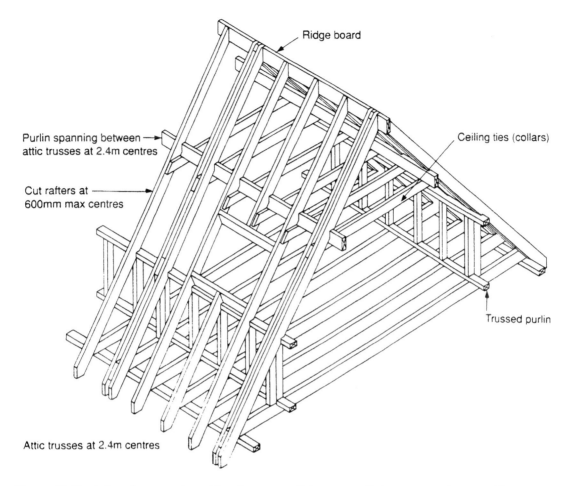

Figure 41 Typical roof construction with attic trusses at 2.4m centres. Being stronger, these provide greater freedom for the arrangement of dormer windows and the staircase, but they do require more site work to provide the cut and fitted common rafters at 600m centres.

Cassette roofs

Taking the process of prefabrication to its logical conclusion, rather than constructing individual trusses or panels for installation on site, it is possible to factory-manufacture a complete roof structure ready to be craned into position once the walls of the building are complete (as illustrated in **306** in Chapter 15). This is usually only carried out as part of the fabrication of a whole-building form of construction, manufactured off-site, because the measurements of the structure on which the finished roof will rest must obviously be exact. The end-to-end quality control possible within a factory environment and the use of dedicated assembly teams on site ensure that all aspects of the manufacture and erection process are properly carried out (as considered in Chapter 15).

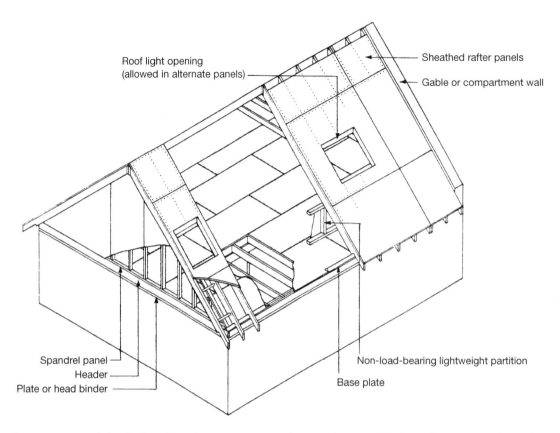

Roof light opening
(allowed in alternate panels)

Sheathed rafter panels

Gable or compartment wall

Spandrel panel
Header
Plate or head binder

Non-load-bearing lightweight partition

Base plate

Figure 42 Typical sheathed roof panel construction. Much strength is gained when rafters are joined together in a panel with a sheathing of plywood or the like. Provided care is taken with jointing to a similarly constructed floor and to walls, a sturdy roof can be produced.

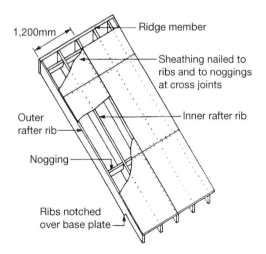

1,200mm

Ridge member

Sheathing nailed to
ribs and to noggings
at cross joints

Outer
rafter rib

Inner rafter rib

Nogging

Ribs notched
over base plate

Figure 43 Typical sheathed roof rafter panel, factory produced and installed on site.

Steel-framed roofs

The prefabricated reinforced concrete framing of the walls of some of the system-built dwellings of the 1940s and 1950s merely provided a base for a conventional cut and fitted roof, typical of the period. By contrast, the framing of some of the steel houses also built around this time, as well as those built during the 1920s and 1930s, was also utilised to form the basis of the roof structure. Experience indicates that where the framing comprises steel sections, visible corrosion is likely to be superficial. However, particular attention needs to be paid to dwellings where thin steel sheet has been folded to form the load-bearing members. Very often, the protection to the steel has gone and corrosion has left little remaining. This necessitates particular attention being paid to dwellings where the metal trussing system is found to be based on fabricated sections such as Trusteel and Hills. Generally, these types are well past their original design life, but if inspection reveals little wrong then they may still last for many more years. Nevertheless, a recommendation for close inspection at regular intervals should be included in any report along with comment on the possible need for eventual replacement when appropriate.

393 Trusteel steel-framed semi-detached houses of the 1950s. These need particular attention because of their extensive use of thin steel fabricated sections which can suffer greatly from corrosion. BRE lists approaching 300 locations throughout England and Wales where they were constructed in both the public and private sectors.

Occasionally, and more often with dwellings originally built by a local authority or the Ministry of Defence than by a private builder, steel trusses or steel purlins will be encountered in combination with timber roof structures. These always warrant a close inspection, but as long as there are no obvious signs of rusting and the joints are sound then there is no necessity for undue alarm. This might, however, be an indication that the main construction is not conventional, so careful investigation elsewhere and judicious enquiries of the owner should be made to ensure that the construction is properly understood and assessed and then accurately described in the final report.

Faults in pitched roof structures

Centuries of carpentry experience lie behind the construction of cut timber roofs built prior to 1960. That is not to suggest that they were always perfectly constructed, but the principles were clearly understood. Bearing in mind they will usually be at least 50 years old and may be several multiples of that age by the time the surveyor arrives to inspect them, any fundamental flaws will already have failed and have been remedied or replaced. The intrinsic over-specification of such roofs will often accommodate a partial failure while allowing the structure to function successfully in its main task of supporting the roof covering and keeping the dwelling dry. Partial failures are continuing points of weakness which must be identified by the surveyor to avoid the risk of complete failure when overstressed by works such as renewing the roof covering with a heavier material. Apart from this, the majority of problems that the surveyor has to recognise and appraise when assessing pre-1960 cut timber roofs are either due to that perennial trio of enemies to building timbers 'dampness, woodworm and rot' or, equally damaging, the consequences of ill-considered actions by homeowners and contractors.

As more innovative roof structures, such as trussed rafter roofs, were introduced from the 1960s, it was inevitable that the earliest versions were found to be wanting in some regards; with experience, refinements such as wind bracing were found to be necessary and have since become standard. When considering post-1960 pitched roof structures, the design and construction of the original structure requires careful consideration; although there is rather less likelihood of timber infestation and rot, dampness can be a problem. Meanwhile, the depredations occasionally wrought by homeowners on the precisely calculated designs of roof trusses provide a consistent source of amazement to property inspectors. The following sections consider some of the more common faults in pitched roof structures which the surveyor is likely to encounter.

Construction faults with cut timber roof structures

Bad cutting during the construction of cut timber roofs will often make jointing difficult to complete satisfactorily since it may reduce the area of overlap and, therefore, the area where nailing can take place. Consequently, the number of nails used could be inadequate. For example, five nails should be used at the joint between ceiling joist and rafter but, more often, three or even two are used with the danger that the joint is weak and could be pulled apart. Cutting square is necessary so that the whole surface of one component is in contact with one to be supported; otherwise, only a proportional part will be effective. Sometimes overcutting to form joints can be damaging as too much material can be removed, thereby reducing strength. Inadequately supported purlins can lead to sagging slopes, and knots in purlin timbers can lead to splits occurring (**395**). Despite these weaknesses, many cut timber roofs continue to perform satisfactorily, but they are points of vulnerability if alterations or renewal of the covering should be required.

The importance of triangulation, which gives close couple roofs such benefits, has already been described, and it is probable that this will be the form of roof structure most frequently encountered on pre-1960 dwellings. The alignment of rafters and ceiling joists should be checked to ensure that the triangulation has been achieved. On terraced properties, this is usually relatively straightforward; but with hipped roofs above flank walls, particular care should be taken that rafters resting on those eaves which are parallel to the ceiling joists have been tied back and properly secured to avoid an outward thrust on the top of the flank wall.

Construction faults with prefabricated roof structures

The requirements for design, fabrication and certification mean that the prefabricated construction of trussed rafters under factory conditions usually assures their individual quality. Provided that the trussed rafters have been correctly selected according to span, pitch and loading required and that they have been carefully transported by the fabricator, they will arrive at the site suitable for their purpose. It is their subsequent treatment on site which invariably causes any problems that arise and, as the standard of site management and erection cannot be taken for granted, a careful inspection is no less important. Storage on site, hoisting into position, installation, bracing and packing require care and are just those operations which are easy to bodge and skimp, compounded perhaps by less than sound adaptations for service

394 A purlin which has unduly deflected. Is this due to inadequate sizing or has the partition supporting the strut undergone settlement? If allowed to continue, the roof covering could be disrupted.

395 This purlin in a close-boarded 1930s roof has split where a knot in the wood has created a point of weakness midway between two supporting struts. The resulting crack, which has penetrated half the width of the timber, has been present for many years without getting any worse or causing any problems. However, making structural changes or increasing loadings on the purlin, perhaps by replacing the original tiles with a heavier replacement covering, could cause a failure if additional support or strengthening is not provided at this point.

396 One way of providing support to a sagging ceiling. A substantial new beam has been inserted and metal hangers provided to support each ceiling joist. It is to be hoped that the inspector heeds what is, in effect, a warning to tread warily.

397 Could this be an entirely DIY effort to provide additional support to ceiling joists by screwing them to steel angles and to the original bearer, which was clearly not fulfilling its function. The lack of insulation at top-floor ceiling level at least aids the inspection but needs commenting upon in the report.

runs, hatchways and possibly the installation of a cold water storage tank. When the roof structure is insulated and enclosed with sarking and the roof covering, the plethora of light timbers and the often shallow pitch makes an inspection of the roof space in post-1960s dwellings much more difficult than in an earlier dwelling with a cut timber roof.

The bracing (as illustrated in **Figure 36**) is necessary in all cases except where a form of boarded sarking has been used, and this is likely to be found only in Scotland where it is traditional. Bracing of this nature prevents racking or lateral movement of trusses, and these were often completely missing with the earliest trusses as the risks of displacement were not fully appreciated. Even when problems became apparent and installation standards increased, diagonal and longitudinal bracing was often minimal and older trussed rafter roofs, in particular, should be carefully checked to ensure that the trusses are adequately braced.

Also necessary but often omitted on earlier trussed rafter roofs, or inadequate even if provided, are straps, packing and noggins that tie the trussed rafters to separating walls between terraced dwellings or to gable walls. Tying the trusses to a separating wall provides the roof structure with an element of restraint. When the roof structure is tied to a gable wall, this additionally provides support against wind pressure to the gable itself by transferring that pressure to the other walls. Instances of gable walls being sucked out by high winds are by no means unknown when such straps, packings and noggings have been omitted. To be effective, straps, packing and noggings need to connect three rafters to the wall in question with adequate screw or nail fixing at each point of connection, as against the one or two often found.

Without both bracing and the straps, cases have been found where all the trussed rafters were leaning over to such an extent that the last one was resting against either a separating or a gable wall. Omission of both or use of only bracing or straps requires urgent attention in view of the danger of collapse at any time, not just when there are exceptionally high winds blowing. To obtain a design for a repair that will be both effective and practical to install, reference to an engineer is essential.

Perfection is, of course, too much to expect from installation in all cases, and a certain amount of lean is considered permissible. The permitted amount is 10mm for the first metre in height with an additional 5mm allowed for every further 1m in height from ceiling tie to apex. Trussed rafters are also permitted to oversail the wall plate to a maximum of 50mm although 50 per cent of the ceiling tie to rafter connecting plate should overlap the wall plate. Any bow between the eaves joint and the apex of the rafter itself should not exceed 10mm at its maximum. A combination of internal and external inspection should establish whether there are any deviations within these permitted tolerances. Out of plumb in excess of the permitted amount will also require reference to a structural engineer although this is probably correctable on site if limited to under twice the permitted amount. Above this, rebuilding is more likely to be required.

If storage facilities on site are inadequate or installation is not carefully carried out then the connecting fasteners which tie together the individual timbers of the trussed rafter can become loosened and trusses distorted, rendering them weaker and less able to carry the applied load. If trusses are cut and altered without taking proper advice, the same effect can be produced. It is not unknown for the pressed toothed fasteners to be prized out, timbers cut and altered and the fasteners nailed back on. Any signs of such nailing needs to be viewed with suspicion and a careful check made for any consequential distortion. Being an engineered product, cutting any component member of a trussed rafter, other than strictly in accordance with the manufacturer's specification and drawings, is hazardous. Despite this, instances are still encountered where trusses have been cut and adapted to form hatchways or for chimney stacks

398 This unusually creative but wholly unauthorised DIY modification of prefabricated timber trusses involved cutting off the upper ends and realigning every pair of struts within each truss, except the final pair which retain the correct arrangement, to create the desired space prior to fixing flooring, linings and decorating. The surveyor was less impressed than the prospective purchasers had been.

or the installation of services during the course of construction. It is also not unusual to come across cases where homeowners, keen to increase the usable storage area in the roof, have happily used a saw to remove and modify whole sections of trusses without appreciating the careful calculations which originally went into maximising the load-bearing capability of an optimal amount of timber in a precisely designed truss (**398**).

Trussed rafters are normally fixed on installation with clips or by skew nailing from both sides to previously secured wall plates. By the time the insulation is in place, the connections are virtually impossible to see and there is, consequently, an opportunity for skimping at this stage of construction. However, recommendation to expose this detail is not usually needed unless there is no other reason ascertainable for a visible defect.

If there is a cold water storage tank in a trussed rafter roof, access must be provided to enable any adjustments or replacements which may become necessary to the float or ball valve. If there are signs of distortion in truss members, the position of the bearers carrying the cistern platform must be checked. To be effectively supported by the roof structure, these need to be as near as possible to the node points (i.e. where the sloping or vertical truss members connect to the ceiling ties) and taken over a sufficient number of trusses – at least three – to avoid distortion. The tank platform itself should be of a suitable material such as softwood, marine plywood or moisture-resistant grades of chipboard or OSB.

Damage caused by dampness, fungal decay or timber-boring insects

Damp penetration can set up attacks of dry rot (*Serpula lacrymans*) or one of the varieties of wet rot (**399**), particularly if the timbers are poorly ventilated. Condensation could be another cause of dampness in the timber, and this has become an increasing problem as insulation levels have increased but ventilation of roof spaces has not always kept pace (**400**). As to insect attack, the common furniture beetle (*Anobium punctatum*) is unlikely to cause undue structural damage in dry timber but persistently damp timber can become badly affected. The ravaging which the deathwatch beetle (*Xestobium rufovillosum*) can cause to structural oak timbers in old church roofs and historic buildings is well known, and it is most essential to indicate the presence of this destructive insect when encountered in the oak framing of both walls and roof of old, wholly framed dwellings. The house longhorn beetle (*Hylotrupes bajulus*), on the other hand, can cause severe damage to softwood timber and is such a serious problem that it is a notifiable pest. Fortunately, this insect is confined mainly to an area of north-west Surrey, though it can also be found to a much lesser extent in London and isolated instances do occur elsewhere, mainly due to insects being transported in infested packing case timbers; so all surveyors should know how to recognise its presence.

The important subjects of fungal decay and wood-boring insects are considered in much more detail in Chapter 21, but in areas where the Building

399 Defective flashings are a frequent source of damp penetration, here affecting the end of a purlin and the rafter nearest the brick wall with wet rot. The absence of roofing felt enables an underside view of the natural slates to be obtained. The signs of discolouration and decomposition near the edges show that they are near the end of their life, while there is also wet rot in the battens. Major replacements and updating are clearly required.

400 The increased requirements for insulation coupled with the provision of roofing felt to the underside of the covering have rendered the timbers of roof structures far more liable to condensation, which can lead to attacks by wet rot and sometimes even dry rot should ventilation levels be inadequate. In the upgrading of this roof, substantial provision for ventilation has been made in the soffit of the generously projecting eaves, but the interior still needs to be checked to make sure that airflow from the vents has not been blocked by insulation pushed into the eaves.

Regulations require the treatment of all roof timbers against the ravages of the house longhorn beetle, it is particularly important to check metal fasteners for signs of corrosion. Use of the wrong type of preservative can in itself cause corrosion. Using the wrong method even with the correct type – if it involves re-wetting the timber and such application is not carefully controlled – can cause distortion of the timbers and this may result in disruption of the toothed metal plate fastener joints.

Alterations to roof structures

The main structural function of a pitched roof structure is to transfer the weight of the roof covering together with the weight of the structure itself vertically downwards on to the solid external walls, or frame, and to provide a level, firm base for the components comprising the roof covering. Less than satisfactory performance in either of these functions can lead to the components of the roof covering being unable to fulfil their purpose of keeping out damp, not necessarily due to any fault in either the material of the covering or the workmanship.

During the external examination of the roof slopes, or when conducting the preliminary external reconnaissance if the roof space is inspected before the exterior, the surveyor will often notice unevenness to the planes of the roof covering. During the inspection of the roof space the construction should be checked to see whether this unevenness is due to an acceptable amount of initial or differential settlement of the roof structure or something which indicates a more serious or ongoing failure.

Many roof structures, particularly those of early timber-framed dwellings, may have moved in the past but then settled into a new position of repose, leaving the base for the roof covering sound but by no means level. Much of this can be corrected by packing out and firring up timbers to leave the roof structure if not exactly flat then at least with undulations despite which, with a bit of skill, the roofing contractor can provide a satisfactory covering.

If the surveyor encounters evidence of movement in the structure and suspects that this is a continuing problem, either causing dampness currently or likely to do so later by way of disruption to the covering, then it will be necessary for the report to include a description of what has been seen from both the exterior and the interior of the roof space along with advice on what needs to be done next. If a remedy is fairly obvious then that can be included; otherwise, it may be necessary to advise further detailed investigation which might include monitoring the situation over time.

Problems often arise, however, not because the original roof structure was defective but because it has been exposed to stresses which had not been anticipated when it was originally designed and constructed. Sometimes this is due to significant alterations, such as the construction of a loft conversion or an extension, which require substantial structural modifications. Frequently, however, the cause is simply the result of replacing the original covering with a different material. Doing this without proper regard to pitch and structure, and without appreciating the implications of changing the loadings on the original structure, can cause problems and two examples are provided next.

Replacement roof coverings

The first example concerns roofs of old timber-framed dwellings. Many of these, particularly on the eastern side of England and Scotland and in some other places such as the area around Bridgwater, were originally roofed in clay pantiles at a shallow pitch of around 30°. Undue movement in the frame renders these unable to fulfil their function and the temptation to re-cover in plain tiles, which fit together much better on undulating slopes, has sometimes been given in to. Unfortunately, plain tiles do not perform satisfactorily at such a low pitch and the presence of torching (pointing beneath the tiles with lime mortar or cement) is often evidence of continual efforts to keep out damp and windblown snow and rain. When either enters, it tends to be soaked up by the torching. This is fine for preventing staining on the ceiling below but not so good for its effect on the oak pegs used to fix the tiles, which will eventually disintegrate due to wet rot. A check on the angle of pitch can be useful corroboration if a change from the original is suspected and there are signs of damp and slipped tiles.

The second example relates to the replacement of slates. Towards the end of the nineteenth century, substantial numbers of the two-storey terrace houses lining the streets, as laid out under the bylaws of the time, had roofs which were covered with the widely available natural Welsh slates, also laid at a pitch of around 30°. The Welsh slates usually lasted well – around 100 years in most cases – but as they increasingly required replacement during the final third of the twentieth century, replacement slates were either very expensive or often said to be unobtainable. Nevertheless, interlocking profiled concrete tiles were then available and said by their makers to give a satisfactory performance at this relatively shallow pitch, unlike plain tiles. Roofing contractors were happy to use them but failed to appreciate, along with the difficulty of cutting and fitting them properly to the awkward shapes that often occurred, that their weight could be around twice that of the slates they were replacing. In consequence, many roofs settled badly and there were reports of some even collapsing during the 1970s and 1980s. There is now a requirement for calculations to be made whenever a roofing material is to be replaced by one which is either significantly heavier or lighter.

In this connection, it is useful to have an awareness of the relative weights of different types of slate and tile roof coverings. As a rule of thumb, the very lightest artificial fibre cement slates are around $20kg/m^2$ with the thinnest Welsh slates being around $27kg/m^2$. Heavier natural slates, clay pantiles, and interlocking concrete tiles are typically around $50kg/m^2$ while double-lapped plain clay tiles are around $70kg/m^2$. The weight of natural stone slates will depend

401 Defects in roof structures or a replacement of lighter-weight slates with heavier components, such as concrete tiles, often require additional strengthening, as here. Additional struts off a load-bearing partition and collars with bolted connections have been provided and the ends of purlins are supported by corbels. As a structural alteration, the work should have been subject to Building Regulation approval.

402 The contrast between natural (on the left of this interwar terrace) and artificial slates (far right) is obvious if the artificial ones have faded after some years in use. It is usually possible to tell them apart though, ironically, some types of natural slates imported during the 1990s are prone to colour loss while artificial slates are now more resistant! Until relatively recently, an equally popular replacement for natural slates has been profiled concrete tiles (centre), which are suitable for the original pitch but they are a great deal heavier and change the appearance. As the standard of artificial slates has improved, they have become more popular when replacement is required, not least because the roof structure is less likely to require additional strengthening.

403 Concrete slates replace the original coloured asbestos cement slates on this 1930s bungalow. They are much heavier and the surveyor will need to check what has been done to strengthen the roof structure.

on the type and thickness of stone but can exceed 100kg/m². On inspection, a surveyor should be able to form an opinion of the material likely to have been used originally according to the dwelling's age and quality. Evidence suggesting a change of material, much more likely to be found on cheaper dwellings, will necessitate a check on whether there have been changes to the roof structure appropriate to the circumstances. If not, there should be a thorough examination for signs of movement and, if necessary, recommendations for additional strengthening should be made.

While on the topic of roof loadings, concerns have been raised more recently about the additionally imposed load of solar PV or solar thermal panels or tubes on existing roof structures. Solar PV panels, which generate electricity, are typically around 14kg/m^2 so will not usually pose difficulties. Solar thermal panels, which have a similar appearance but provide water heating, are significantly heavier, ranging from about 20kg/m^2 to 28kg/m^2, and are, therefore, more likely to require additional strengthening in some cases. Evacuated solar tubes come midway between the two – typically around 20kg/m^2. Approved Document A of the Building Regulations states that where there is a significant change in the roof loading, defined as an increase in loading of more than 15 per cent, the structural integrity of the roof structure and the supporting structure should be checked. Panel providers should have undertaken an appraisal of roof structures prior to installation, including structural calculations if appropriate; but surveyors cannot assume that this has always been done properly and should take nothing for granted when inspecting roof structures where solar panels are present.

Loft conversions

A loft conversion has been a popular improvement to increase the living accommodation of a dwelling for the last 50 years or so. Not all roof structures are either suitable or capable of conversion. For example, the trussed rafter roofs almost universally used since the 1960s, comprising a multitude of light timbers at close centres and often with shallow pitches of 30° or less, do not lend themselves for this purpose. On the other hand, cut and fitted roofs, to be found on the vast majority of older dwellings, can usually be adapted; although it is simpler to do so if the dwelling has gable ends or separating party walls on either side rather than hipped slopes. It is even more practical if the dwelling has a reasonably steep pitched roof and there are internal structural partitions. A proper load-bearing floor for the new living accommodation will always be required and, of course, Building Regulation approval for the necessary structural work.

It is unlikely that there would be any difficulty in identifying a property where a loft conversion has been carried out. A comparison with similar, neighbouring properties can provide a clue, as can the presence of roof lights of modern design or unsympathetic dormer windows or what seems to be a curiously arranged internal staircase. Once the presence of a loft conversion and its date of construction have been established, depending on the circumstances, a request to the owner for a sight of the Building Regulations approval and the submitted plans and details on which it was based would be a sensible first step. If produced and there are no signs of defects, checking the details may have to be taken on trust as much may be hidden from view.

Although not the principal cause of complaint in the case of *Cross and Another* v *David Martin & Mortimer* 1989, alterations to roof trusses to form a room in the loft featured prominently. Whenever a conversion with substandard work of a DIY nature, an inadequately supported floor (**404**), missing struts or other defects are seen, the absence of appropriate consents should always be suspected and reported.

If no Building Regulations approval was obtained – and such instances are by no means uncommon, particularly if no alterations were made to the front of the dwelling – then the report should recommend that steps be taken prior to purchase to regularise the situation. At the very least, the extent and cost of any works likely to be required before retrospective consent would be granted should be determined. In an attempt to be helpful and to reduce delays in the purchase process, a purchaser may be offered the opportunity to take out insurance against the cost of any works that may be found to be necessary after completion. Such insurance policies are primarily intended to indemnify against the cost of works which might be imposed by the Building Inspector if enforcement action should be notified to remedy a breach in the Regulations. This circumstance occurs only relatively rarely but, irrespective of whether or not enforcement action is anticipated, the actual work needed to improve the substandard conversion should still be carried out by the new owner to bring the conversion up to the appropriate standard. Policies obviously vary but, in most cases, the work itself is unlikely to be covered by indemnity insurance in the absence of enforcement action by the local authority. While purchasers may understandably be tempted to take out indemnity insurance to avoid delaying a transaction, it is usually far wiser to investigate and consider the full implications of any works – not least the likely disruption and time that may be involved. Most would find delaying their purchase preferable to discovering the ghastly reality of living on a building site after moving in with high but misguided hopes that 'not much work' will be required.

404 The trapdoor access to this loft conversion is a strong indicator that it does not meet appropriate standards for residential accommodation. The depth of the floor, measured by the length of a pen at the edge of the opening, shows that the flooring has been laid directly onto the ceiling joists. Further points of interest await the surveyor, but the tone of the forthcoming report to the client is already clear.

One might have thought that an owner would find it virtually impossible to sell without regularising the work because a prospective purchaser would be warned not to proceed unless retrospective approval was obtained. Surprisingly, however, despite knowing that works have been carried out without the necessary consents, many purchasers do decide to go ahead with their purchase, but often to their later cost and regret.

Groundwork

Those features which are included in pitched roofs below the covering itself but are not part of the supporting structure are sometimes referred to by the term 'groundwork'. Changes to the traditional arrangements during the last half-century or so have had serious consequences, most not anticipated at the time they were made.

For long, groundwork was limited to either close boarding or battens. Close boarding could either be feather-edged for hanging tiles with nibs or plain and battened. Battens, whether with or without close boarding, could support not only tiles with nibs but also those without as well as slates, both of which require nailing. Boarding is more expensive but while it is the traditional form of groundwork in Scotland, it is more rarely found in England and Wales. It is thought to be present on only a small percentage of dwellings south of the border, almost exclusively ones pre-dating the Second World War.

In roofs without boarding, the penetration of windblown rain and snow was dealt with, in many cases, by the provision of torching with lime and hair plaster around the nibs, nails and battens. The torching had a secondary function of helping to secure the tiles and slates. While successful at keeping damp away from ceilings, not always possible if omitted, torching nevertheless tended to cause wet rot in the battens by retaining moisture. Fortunately, this can be checked by the surveyor in a roof of this type along with the actual condition of the torching.

Before the introduction and general use of bituminised roofing felt as a moisture – but not vapour – barrier since the middle of the twentieth century, wet rot could also be a problem in boarded roofs. Torching is not possible with boarded roofs, and windblown rain and snow could creep under both tiles and slates. Even though there was usually sufficient ventilation to prevent outbreaks of dry rot, as there usually was in the case of unboarded and unfelted batten roofs that were provided with torching, wet rot could affect close boarding. This was particularly a possibility if tiles were hung on the feather-

405 *(left)* Profiled clay tiles hung at a shallow pitch entirely by their nibs on battens, viewed from below where there is an absence of roofing felt.

406 *(middle)* Torching of lime and hair plaster to the underside of clay tiles and battens, which helps to keep windblown rain and snow at bay and adds to the security of the fixing, whether by pegs, nails or nibs. The downside is a possibility of wet rot in the battens as the torching soaks up the rain and snow to remain damp for some time.

407 *(right)* The sight of boarding below the roof covering provides the surveyor with food for thought. If the covering is of tiles or slates, has underfelt been provided and have counter battens been laid to prevent wet rot setting in to the transverse battens on which the tiles are hung or tiles and slates nailed? Damp staining to the edges of the boards usually indicates an absence of underfelt and probably counter battens, but it is not possible to be certain unless there is a gap somewhere in the boarding, more likely to occur at abutments.

edged variety or slates were nailed direct to flat boarding – each of which is detailing that tends to limit ventilation. The provision of battens and counter battens would improve conditions, but the omission of the counter battens could substantially increase the risk of wet rot in the cross battens.

Boarding fixed to the top of the rafters allows the surveyor only a glimpse, between cracks, of what arrangement has been provided above. If the dwelling is from the end of the nineteenth century or the earlier part of the twentieth century and of good quality, as were many built at that time, and still appears to have its original roof covering, it is highly probable that there is no roofing felt. Confirmation of this may well be obtained by indications of damp staining on the boards, principally on the edges, as the boards are probably of sufficient thickness and quality as to prevent moisture penetrating the total depth. It would be entirely appropriate in these circumstances, and indeed essential in the event of the roof needing to be re-covered, to provide a warning as to the likelihood of considerable renewal being required to the boarding and total renewal if the boarding was found to be feather-edged – although complete removal and replacement of boards with felt and battens could prove to be more economical. Such a warning would be even more essential if there were signs of slipped or dislodged tiles or slates indicating that battens or boards were no longer sound enough to hold nibs or nails.

As indicated, there is a danger of wet rot in both boarded and unboarded roofs of earlier dwellings but little danger of dry rot, generally, above the level of the boarding. Below boarding, however, in the roof space and in the boarding itself, there could be a danger of dry rot if ventilation is limited because of tightly closed eaves, particularly in confined parts of the structure – for

408 *(left)* Underfelt to roof coverings, fixed through battens to the rafters and which prevents the ingress of wind-driven rain or snow, must be fixed to sag slightly between each rafter – not too little, not too much (and consistently) but just enough to allow any water to run away below the battens into the gutter; otherwise, there is a danger of wet rot in the battens eventually depriving slates or tiles of support. When a section of roofing felt is allowed to sag too much and there is insufficient overlap between sections, as here, there is a danger of water flowing back into the roof space.

409 *(middle)* Early types of roofing felt often deteriorated rather rapidly and would split under the weight of any substantial amount of water. Here, an old friable felt has been replaced with new and given a fairly generous sag between fixings. It would have been better to have removed the old felt.

410 *(right)* Roofing felt clumsily torn away to allow fitting of a tube to ventilate the roof space. Such tearing can be useful at times to the surveyor by permitting a sight of the underside of the roof covering but, obviously, can allow damp penetration.

example, towards the foot of a valley gutter or at the corners of hipped slopes. The surveyor, therefore, needs to pay particular attention to these areas, even at the risk of getting a little dusty, and all the more so if there is evidence of damp penetration nearby.

The development of roof lining felts enabled both boarded and unboarded roofs to be provided with protection against windblown rain and snow and rendered the provision of torching unnecessary for unboarded roofs. Lining felt should always be laid loosely enough to hang down slightly between each rafter – a point always to be checked. If it is fitted too tightly across the rafters and below the battens, it can cause wet rot in the battens if water is retained along the top edge, especially as accumulations of dust and debris on the surface of the lining material can gradually build up against the battens and help to keep them damp. If felt has been fitted above boarding then counter battens are needed; otherwise, there is again a serious risk of wet rot developing in the battens. As before, the presence of boarding usually prevents the surveyor from seeing what has been provided above and it may, therefore, be necessary to provide appropriate warnings of what might happen in the future if there are not already signs of dislodged tiles or slates providing strong evidence that the fault has already materialised.

Insulation

Up to the 1960s it was rare for top-storey rooms to be heated. In consequence, roof spaces remained cold, with the air at a temperature and water vapour content not too dissimilar to that of the exterior. This meant that no problems of condensation occurred although there were often plenty of other problems – for example, if the dwelling was provided with an unlagged cold water storage cistern. The advent of widespread central heating and the use of plasterboard and a skim coat of plaster, rather than lath and plaster, to top-floor ceilings for new dwellings in the 1960s and 1970s led to roof spaces being almost as warm as the rest of the house. This could lead to a high level of localised condensation when the warm air within the roof space, able to support a much greater level of water vapour, condensed overnight on the cold underside of the pitched roof structure, the covering and any metal fastenings forming part of the roof structure. Usually, this caused few problems because roofs were still relatively well ventilated. There might be some damp staining on the timbers but there would only be problems if the roof was constructed in such a way that ventilation was very restricted, in which case there was an increased possibility of wet rot or even dry rot in the timbers, corrosion in metal components and perhaps even damage to the electrical installation through dampness.

Considerable increases in fuel costs from the mid 1960s led to the introduction of energy conservation measures – as a requirement under the Building Regulations for new dwellings – with most owners adding a layer of insulating material at top-floor ceiling level as an improvement. The extremely modest degree of insulation initially required has been progressively increased as national commitments have been made to reduce energy consumption and to limit greenhouse gas emissions. Around the same time as insulation became a requirement for new dwellings so too did a requirement for ventilation to roof spaces to counteract the risk of condensation. The attempt to provide a vapour control barrier at top-floor ceiling level by the provision of foil-backed plasterboard, while practical from the constructional point of view, proved totally unsuitable as water vapour condensed on the room side of the foil, ruining the plasterboard.

As insulation standards have increased and homeowners have been encouraged to insulate their roof spaces, surveyors' lives have become more difficult. Moving from joist to joist across a roof space always called for careful steps and a steady balance but with insulation quilt now laid across as well as between the joists venturing into many roof spaces has become a seriously hazardous exercise. A false step risks causing not only damage to the ceiling beneath but also serious injury to the surveyor. Where a clear and negotiable path across the roof space is not apparent, the surveyor should think very seriously before venturing beyond the point of access, especially if working alone in the building.

Installing insulation may conserve energy but it can cause problems of its own. An inspection of a roof space might bring to light insulation quilts stuffed into the eaves restricting ventilation, insulation missing from areas where access is difficult, or the creation of cold bridging and heavy condensation in small areas causing damp stains on ceilings. Further problems are the top of access trap covers not being insulated and the edges inadequately sealed, dormer windows not ventilated, insulation carried under cold water storage cisterns, a lack of cross ventilation where there are separating walls and poor ventilation to monopitch roofs. In roof spaces where condensation occurs regularly, it is sometimes possible to detect numerous small indentations or 'pits' in the surface of the insulation where drops of moisture have consistently dripped from exactly the same points of the roof lining fabric above. Condensation at these levels can cause corrosion damage to the punched toothed plate metal fasteners of prefabricated trusses so these should be closely checked along with the possibility of rot in the timbers from this cause.

When insulation has been laid, it is important that ventilation of the colder upper part of the roof space is adequate to ensure that condensation does not occur. This is usually achieved by providing vents in gable ends and around the eaves. For projecting eaves, forming ventilation apertures in the soffit may not present too much of a problem, but complications arise with flush closed eaves where the guttering is virtually at the wall head. Ridge ventilation or specially designed ventilation tiles may then be the answer and may also be used to provide cross ventilation air currents to dead areas around roof lights and dormers. Even with satisfactory ventilators in the eaves, maintaining an airflow from the main area of the space can lead to a band of thin insulation quilt around the edge of the roof below the base of the roof

411 *(left)* When insulation at top-floor ceiling level is encountered, a depth of around 100mm, roughly to the top of the ceiling joists – as here in fibreglass and no longer considered adequate – it does at least enable the surveyor to step gingerly, if safe to do so, on to joists and move around to inspect storage cisterns, wiring, etc. as well as the underside of the roof covering or boarding.

412 *(middle)* Surveyors, of course, are not the only persons needing to be able to move around in the roof space. Plumbers, electricians and, not least, the occupier may need access, sometimes in a hurry. Partial flooring or a defined boarded pathway is helpful. The boarding on the right might previously have served for the purpose, though if not fixed in place then this would be positively dangerous for anyone stepping onto the ends. When the minimal insulation, something like only 50mm of exfoliated vermiculate, is upgraded, the boarding will need to be removed and a walkway then provided above the much thicker depth of insulation.

413 *(right)* Full access to a roof space can often be frustrated by thick layers of insulation. It is not safe to venture into such areas. The surveyor should not disturb the insulation in any way but report the limitation on the inspection and advise that something needs to be done to improve the situation.

slope, potentially creating a cold bridge, localised condensation and mould growth at the edges of the rooms beneath. Ridge ventilators and ventilation tiles again may provide a better solution.

Separating walls

Where the inspection is of a terraced or semi-detached dwelling, a sight should be obtained in the roof space of the wall or walls separating the dwelling under consideration from its neighbours. Their condition should be noted as they often provide an indication of the quality of construction to be found elsewhere in the dwelling. Although not seen from outside, they should be of reasonable construction without gaps or broken bricks and be well jointed.

Occasionally in older properties, such separating walls, or party walls as they are also called, will not be found as there were times and circumstances when they were not thought necessary. The principal danger when separating walls are absent is, of course, from fire. A fire next door will soon vent itself by breaching the flimsy nature of most top-floor ceilings. It will take longer to penetrate the roof structure and covering but, in this time, especially if doors and windows are open below, it will spread fiercely into the roof spaces of adjoining dwellings immediately above top-floor bedrooms and can spread along whole terraces of buildings. Accordingly, for safety's sake, surveyors should recommend the building up of a barrier between roof spaces that is equal in fire resistance to that pertaining in the party walls lower down. When incomplete separating walls or holes through separating walls are found, there should be a recommendation to carry out work to create complete fire separation between adjoining buildings. While the main danger to a dwelling without separating walls is from fire, there is also a security risk as their absence potentially allows access from one dwelling to its neighbours and, with it, the potential for burglary. It is essential to include comments in the report to this effect if no separating walls are found.

414 Defective brickwork with loose and missing pointing to a flue incorporated in a separating wall, constituting a hazard from fume leakage should any form of space heating appliance, open fire, stove or gas fire utilise it. The remaining brickwork and pointing to the separating wall is also poor, suggesting a fairly low standard of construction for the dwelling originally.

Chimney stacks should have been rendered originally as a precaution against any untoward leakage of gases from a flue, but this was not always done. With the increasing popularity of open fires and stoves, this aspect is assuming more importance and flues running through the roof space of older houses and flats, whether individual or as part of a separating wall, should not be assumed redundant. If unrendered and in poor condition with defective brickwork and pointing then the hazard of fume leakage should be indicated in the report. It is also important to note their position so that they can be checked against the presence of chimney breasts in the rooms below. Sometimes to increase the floor area, chimney breasts are removed from rooms without providing support to the remaining brickwork above – a failure which can have serious consequences should the brickwork collapse. The presence of support must be checked in these circumstances. Further comments on this matter are made in Chapter 23.

Intruders

Dwellings provide an appropriate habitat not only for human beings but also for other creatures from the natural world. Those that are welcome become pets and usually, apart from the odd cat, move when the owner moves, although leaving their aroma behind for a while. The unwelcome ones remain and since many new owners have individual dislikes – sometimes amounting to 'a horror of . . . and I wouldn't have bought the place if I had been told' – the surveyor needs to keep a wary eye open for any visible evidence of intruders so that the client's attention can be drawn to any implications.

415 Not to be disturbed. The white, paper-like enclosure of a wasps' nest in the eaves of a slated roof.

The roof space, being an area of the dwelling infrequently visited by the human occupants, offers the ideal area for roosting and nesting, along with ready access to other parts of the dwelling, so that it is appropriate to deal with the unwelcome intruders here even though they may also be encountered elsewhere. The surveyor, accordingly, needs to be on the lookout for any visible evidence during the roof space inspection.

Bats

Bats can be found in all kinds of buildings – usually in the roof but also in other areas. Many are dependent on buildings for roosting though they may not be present throughout the year. It is unwise to assume that bats only favour very old buildings with large roof spaces. Many do, of course, but the pipistrelle species – the most common – will happily take up residence in modern dwellings given the opportunity. Because many species of bat are endangered or threatened, legislation gives them very full protection and it is illegal to disturb them or to recklessly damage, destroy or block up their roosts. This places a special responsibility, not to say burden, on the residential surveyor as one of a select group whose work routinely puts them in exactly the right place to detect evidence of bat activity. A significant consequence following on from this is that surveyors making recommendations for maintenance or remedial works following an inspection may put themselves, their clients and their client's contractors at risk of prosecution if such evidence was present but ignored.

Bats are small harmless mammals which feed only on insects. They do not damage property and there is no known health risk associated with them in the UK. Because they tend to return to the same roosts each year, these sites are protected whether the bats are present or not. If evidence of bats is found in a dwelling and any action is proposed which would affect bats or their roosts, the relevant Statutory Nature Conservation Organisation (SNCO) – Natural England in England, for example – must be notified before any work is started. Amongst the works likely to affect bats are exactly the sorts that might follow a surveyor's inspection: blocking holes, renovation, reroofing, conversion or any application of pesticides, such as for remedial timber treatment. The notification to the SNCO must be adequate to allow them time to advise on whether the operation should be carried out and, if so, the method and timing of the work and any special precautions that may be necessary.

In view of this, surveyors should make it an automatic part of their standard inspection process to check for the presence of bats. Only rarely will bats actually be seen during the day, although the greatest likelihood of doing so is on hot days between June and September when breeding colonies may be present and the numbers are likely to be highest. The key identification feature is the presence of droppings. These are dark brown or black and vary between 4mm and 8mm in length. They look very similar to mouse droppings but with one identifying difference – bat droppings crumble into a fine powder when crushed because they are made up of fragments of insects. Mouse droppings are initially smooth and plastic but quickly become hard and retain their shape.

During the normal course of an inspection, the surveyor should remain alert for bat droppings scattered anywhere across the floor of the roof space; but particular attention should be paid to the areas beneath the ridge, junctions between two ridges, below hip rafters and bays, around the base of chimneys and gables, and all around the eaves. Only a small number of droppings may be visible but accumulations into piles beneath ridge boards or hips and around chimneys and gable ends are typical of bats whereas mice and rats rarely produce such an accumulation in these places. Examining the point immediately above an accumulation may give the fortunate surveyor the pleasure of a rare personal sighting, whilst taking care to ensure that the object of the examination is not disturbed. The quantity of droppings readily visible in the roof space does not necessarily provide a guide to the number of bats using the roost.

If evidence of bats is found then the surveyor must obviously report it, and the implications, to the client. In view of the legal implications, it is also appropriate to mention it to the owner if present at the time or to the agent if a key was used for access. Because of their endangered status, the presence of bats in a dwelling ought to be, and sometimes is, greeted by owners and clients as a privilege, but some will have concerns and others may not be at all happy at the news. Usually it is sufficient to give some words of encouragement, reassurance about the lack of health risk and directions to the helpful publications of the Bat Conservation Trust (www.bats.org.uk) along with a word of caution about the legal position, just to reduce the risk of any rash actions.

Birds

The provision of sturdy roof linings beneath roof slopes and the increasing prevalence of uPVC cladding to eaves, soffits and gables during recent decades have severely reduced the number of roofs to which birds can readily gain access. When entering a roof space in the spring, it used to be common to hear a brood of nestlings calling noisily from the eaves, but this has now become a comparative rarity. Nevertheless, if access to a roof space is available, birds of many different species will take the opportunity of making a home there. Evidence of their presence are the nests and the birds themselves, fouling by droppings, feathers and the remains of their food. They can cause considerable disturbance to occupants by their movements and even, should cold water storage cisterns be left uncovered, contamination to the water. The sight of a dead bird floating on the water is not unknown.

In older properties without the ubiquitous plastic cladding, it can be a constant battle to keep birds at bay; but it is necessary because, quite apart from the mess from droppings, they carry parasitic mites that can cause dermatological problems for human occupants. Seeking out and closing up all openings likely to provide access is the solution, taking care not to eliminate ventilation to the roof space in the process.

Even when this is done and the space cleaned and fumigated, it may be found that others, particularly those that visit for the summer months, will nest below the eaves or in hopper heads or the tops of ventilators or flues. Any blockage of ventilators or flues can be potentially dangerous and must be looked out for and reported; at one time there were a number of deaths caused by asphyxiation through birds' nests blocking the flue outlets from instantaneous gas water heaters.

In coastal areas, and increasingly further inland, gulls have been known to peck at and damage the gaskets to plastic pipes and have a habit of congregating tiresomely, as do pigeons, which are a particular problem where they tend to cluster as a group and are hard to dislodge. While many owners will enjoy the sight of house martins nesting under the eaves and be willing to put up with the droppings on the ground beneath, no surveyor who has been in a roof space or attic room where pigeons have been in occupation for an extended period will forget the experience or wish to repeat it. Specialist methods are needed both in terms of dispersal and prevention by means of small strips of spikes on ledges and by scaring devices. One surveyor, called back to an 1890s house, was at a loss for words when the owner pointed out a seemingly innocuous hole in an outer wall which provided access for pigeons to an enclosed roof void: the purchaser had bought not only a house but an aviary!

Insects

The most important insects for the surveyor to watch out for and recognise are obviously signs of the different types of wood-boring beetle, and these are considered separately in Chapter 21. Older dwellings with large roof spaces may, of course, provide space for other types of insects such as cockroaches and pharaoh ants, which may also be found elsewhere. Any sightings by the surveyor should be mentioned in the report, though the presence of such will probably merely provide further evidence that a major infestation exists, perhaps extending to lice and bed bugs, which will need drastic action. The surveyor may, with good reason, think twice about continuing an inspection beyond an initial reconnaissance until such work is carried out.

The presence of other insects such as wasps, which can form substantial nests in the cleanest of dwellings, need commenting on in the report but, obviously, for safety reasons on no account should these be disturbed. Wasp nests usually take the form of a white paper-like enclosure down towards the eaves and although the surveyor is not likely to trip over one, prodding with a folding measuring rod is not to be recommended.

Although wasps can be a nuisance in the garden to owners and their families due to their coming and going during the day and returning in numbers, probably tired and irritable, on a summer's evening towards dusk and even more so when drunk in the autumn, they tend to occupy a nest for 1 year only. As they do not usually return to the same nest the following year, it can be more sensible to put up with the nuisance for one season rather than hazard an attempt to exterminate them one evening when they are all, hopefully, back in the nest (**415**). If the problem persists the next season with another nest then it is often possible to identify the access point by watching the comings and goings so it can be blocked up or covered with a 3mm to 4mm wire mesh during the following winter.

Rodents

Along with other parts of a dwelling, rats and mice can use the roof space for nesting purposes and as runs to give access to other areas. Fortunately the rat most commonly found is not the Black Rat, the Plague carrier, but the prolific grey common Norway rat which seems to have arrived in the early 1700s and which produces about 40 young a year. Even so, all rats and mice are vermin and present an infestation hazard, causing a nuisance by fouling of surfaces which has health and hygiene implications as this is responsible for leptospirosis (Weil's disease, which can be fatal), salmonella and tapeworms. They also gnaw at the insulation to plumbing and even wiring for electricity and telephones, although this is comparatively rare.

Droppings provide evidence of their presence, comprising smooth, hard, cylindrical, black pellets about 20mm long for rats and about 5mm to 8mm long for mice (**416**). Eradication can prove difficult, requiring the blocking of access and runs, baiting and poisoning. It can be expensive, depending on the number of visits required and on the attitude of the local authority. Some still accept responsibility to deal with infestations but increasingly they merely list contractors. Even if there is a local authority service, it is rare nowadays for it to be provided free. The British Pest Control Association also lists members on their website (www.bpca.org.uk). Ultrasonic devices which emit a high frequency sound that is beyond the range of the human ear but sufficiently distressful to encourage mice and other vermin to pack their bags and move elsewhere have proved to be successful in some circumstances.

BRE has reported problems from infestations by edible dormice in an area to the north-west of London in parts of Berkshire, Oxfordshire, Buckinghamshire, Bedfordshire and Hertfordshire. These dormice are a protected species so they cannot be exterminated in the same way as rats and mice; by law, only licensed pest control contractors may be employed to remove them. The affected area seems to be extending gradually outwards so surveyors in bordering areas, as well as those directly affected, should ensure they are familiar with the issue. Dormice are much smaller than ordinary mice and the sight of clusters of very small droppings to the order of 2mm to 3mm long should put the surveyor on guard.

Like edible dormice, grey squirrels are attractive to look at; but unfortunately their population increases apace and, as intruders to roof spaces,

416 These black pellets are too large for mouse or bat droppings so the suspects are rats or squirrels, probably the latter as the shape is more variable and the colour of some is lighter. Both are rodents and it is the work of a specialist to get rid of them if they are still in occupation.

they are probably the most troublesome and difficult to remove. They are energetic and resourceful and are able to shin up rainwater pipes and jump from nearby trees to gain access to vents in fascias and soffits and gaps in tiling or slates. Sealing up the gaps does not always keep them at bay as they are quite capable of lifting flashings or removing grilles and the odd tile to get back inside. Once inside, they can cause extensive damage because, like all rodents, they have front teeth which grow continually and need to be constantly trimmed and sharpened. They do this on the woodwork and metalwork in the roof space and on the insulation to electric wiring. A group of them larking around at night is enough to keep the soundest of sleepers awake, and during the day, the odd mysterious clatter will cause occupants to jump. It is debatable whether the squirrel or the surveyor will be the more surprised when, as the hatchway opens, their eyes meet. Even if squirrels are outside when the surveyor calls, the remains of food and excrement, to say nothing of the visible damage, should leave no doubt about their intrusion.

Also like edible dormice, red squirrels are a protected species, and in the unlikely event that they are found to be causing problems in those parts of the country where they live, special measures will be required. In the majority of the country, however, the invaders of the roof space will be grey squirrels and the most reliable way to deal with them is to employ a specialist pest control contractor, but this must also be coupled with effective prevention of re-entry. Points of access must be blocked with strong wire netting that is properly secured in place, and any loose tiles or rotten timber must be repaired. Any weak points in the defences will be ruthlessly exploited as squirrels are not easily deterred. One successful method at the eaves is to remove the lower two or three courses of tiles or slates and fix and fold over the fascia and soffit, lengths of expanded metal lathing, subsequently refixing the tiles or slates. It ill behoves a surveyor to miss the intrusion of squirrels. It has been known to cause owners to move without necessarily giving their presence as one of the reasons.

19

Internal walls, partitions and ceilings

The inspection

It is important that the structure of a dwelling is considered as a whole and not just as a series of individual components. A sound and satisfactory roof and well-constructed walls are, obviously, vital elements but more often than not they in turn are dependent on the internal structural features of partitions and floors to provide the necessary support and restraint. Without these, external walls can develop leans, bows and bulges and roofs can sag. Even when the necessary support and restraint are provided, inadequacies of design and construction can undermine what would otherwise be satisfactory construction elsewhere. For example, an internal partition wall could well be carrying a substantial load from floors and roof. Failure of its foundation, causing it to settle, can lead to sloping floors and a distorted roof and possibly disturb the stability of the external walls. It is vitally necessary, therefore, to establish during the internal inspection which partitions provide support to the structure above.

It is often overlooked that internal partitions can also be crucial in providing vertical restraint to assist the prevention of outward movement in the external walls. Floors at first storey and above should be examined to determine the nature of the bearings providing support in order to ensure that intended lateral restraint to the external walls has not been altered or rendered ineffective by decay. The internal inspection needs also, therefore, to take the aspects of construction, support and restraint of floors and walls, both external and internal, into account in every room on every floor.

Partitions

Until lightweight timber roof trusses began to become available to span large areas of floor space from about the 1960s, most dwellings had at least one internal structural partition to support the floor of the first and any further upper storeys as well as providing support for the struts forming an essential part of the on-site cut and fitted close couple roofs of the time. Even modern properties built with lightweight truss roofs may require internal load-bearing partitions to support upper floors and partitions.

Whether the dwelling is of the common two-storey semi-detached or detached type built between 1920 and 1960 or the average-sized three- to four-storey terraced dwelling of the eighteenth, nineteenth or early twentieth centuries, the main internal load-bearing partition will generally be found to be that separating the front and rear principal rooms on the ground floor. This partition not only supports the floor joists of the first storey but also those of any floors above and, furthermore, provided the most convenient position from which to support the struts forming part of the roof structure, being roughly half the distance of the total roof span.

In a typical property, therefore, this intermediate partition wall shares half of the load of the front of the property with the front elevation wall and the same proportion of the rear half with the rear elevation wall. Consequently, the internal load-bearing partition of the average-sized dwelling built over a period of some 250 years will be supporting a load in excess of twice that of any of the external walls. While it is true that for domestic construction, most of the external walls, because of the need for weatherproofing, are built far stronger than technically necessary, it has not always been the case that the requirement for this extra degree of strength by the internal load-bearing partition has in all cases been fully appreciated and accommodated. Thus, it will sometimes be found that while the external walls may be substantial, the partition can be of comparatively flimsy construction. Recognition where this is found to be the case often provides a simple explanation for many of the internal defects found in dwellings.

Of course, typical average-sized dwellings are being considered here, but this is not to overlook the fact that it has always been possible to rearrange the structural elements of dwellings to provide the types of accommodation required. For the more expensive dwelling of the early to middle twentieth century with wider frontage where large areas were needed at ground-floor level, structural partitions could run from front to rear, dividing a basically rectangular dwelling into separate, though integrated, structural units. In the eighteenth and nineteenth centuries, on the other hand, large spaces were more often needed at first-storey level, seemingly interrupting a transverse structural partition, which is apparent at basement, ground- and second-storey levels in a large terraced dwelling but not in the first storey. This could cause the surveyor a degree of puzzlement; accordingly, he or she needs to be aware that trussed partitions, based on similar joints with the metal straps and tension rods used for the elaborate roof trusses of the period, could be arranged so that a floor could be suspended below the partition which would also, if required, be supporting a floor above.

Partitions are often considered a source of annoyance to owners who wish to make improvements. Combining the front and rear rooms into one is a popular measure. It can be done at ground-storey level, always assuming that an adequate beam is provided to replace the structural partition. A smaller opening in the form of an arch or a door still needs care, and an adequate spread of the load from above on to either side of the opening is essential. Where the partition is of timber, and perhaps to the uninitiated thought less important, unwise cutting of sill plates, struts or braces can lead to a serious loss of support. A case in 1996, *Gardner* v *Marsh and Parsons*, concerned openings formed by a developer in the central spine partition during the conversion to flats of a listed mid nineteenth-century five-storey terraced dwelling. The surveyor missed the significance of rucking wallpaper, indicating developing cracks which later necessitated the insertion of steel beams. The Court of Appeal upheld the view of the lower court that the surveyor had been negligent.

When ascertaining the construction and significance of internal partitions, the age of the dwelling is an important factor, as always, for providing clues. Partitions invariably present a face of plaster on both sides so that, in addition to a knowledge of or an assessment of age, reliance needs to be placed on the partition's thickness, the sound it makes when tapped plus information derived from surrounding features. The adjacent floors on either side may reveal floorboards running parallel – a good indication that the partition is supporting floors joists. Boards running at right angles suggest joists are running parallel to the partition. The layout of rooms above and below may need consideration, such as whether the partition under review continues in the same form above and below and in the same position.

Measurements of partition thickness are often most conveniently taken on leaving a room. This should be part of a routine sequence in combination with an examination of the door and frame. If the latter are out of square, this can be an indication of movement in the partition itself. Solid-sounding partitions in dwellings built before 1920 will almost certainly be based on brick and have a width of 260mm to 270mm if of one brick thickness and around 130mm to 140mm for a half brick thickness with plaster on both sides. Comparatively rarely, solid load-bearing partitions in large houses of the eighteenth, nineteenth and early twentieth centuries will be found to be of one and a half and two brick thickness, 370mm to 380mm and 480mm to 490mm thick respectively, and if the dwelling is of stone construction, upwards of 450mm. For houses built after 1920, the load-bearing partitions are more likely to be of cheaper concrete blocks or fired clay units, and thicknesses would usually be some 50mm to 60mm less than would be the case if they were of brick. Unless a sight of a partition is obtained where it is not plastered, only reasonable assumptions can be made based on the evidence and these should be reported as such.

It should not be assumed that a hollow-sounding partition is of no structural significance, even if it is of comparatively slender thickness. With good-quality timber, careful design and construction and adequate ties and bracing, softwood timber is perfectly capable of supporting floors and a degree of roof loading in two-storey domestic construction. With lath and plaster both sides, it is unlikely that the thickness would normally be less than 140mm to 150mm and could possibly be much more. Once again, only assumptions can be made of construction, but information derived from adjacent floors and from rooms above and below should enable the status of the partition to be clarified as structural or non-structural.

The important factor about a timber and plaster load-bearing partition is the treatment at the base. In dwellings built before 1900, very often the partition would be built off a plate fixed to the top of a sleeper wall at the same height as other plates supporting the timber joists of a hollow suspended ground floor. Such plates should, of course, be laid on a damp-proof membrane, which along with the ventilation provision through the sleeper walls is intended to help prevent any rising damp causing an outbreak of rot. However, the damp-proof membrane was often omitted, and even though the ventilation may be adequate to prevent an outbreak of dry rot, wet rot in the plate can affect the soundness of the timber at the base over time, causing it to settle under the imposed load.

A cautionary note needs to be introduced here in regard to the transverse load-bearing partition in the speculative housing of the eighteenth and nineteenth centuries. This will often sound solid and provide a thickness measurement indicating brick construction. A common practice of the time, however, was not to build the partition with sound bricks, proper bonding and adequate mortar but to use over-burnt, inferior, misshapen or under-burnt, soft bricks along with broken bats and inferior mortar, all put together within a rough and ready framework of timber. Ground leases in the early eighteenth century were short – as little as 33 years at times – and the builder's intention was that the construction should last very little longer than the length of the lease. Even when leases became more standardised at 66 or 99 years, the practice continued and was common until the late nineteenth century. As with the hollow timber and plaster partitions noted above, the important factor was the performance of the base plate in relation to rot and degradation.

Partitions are not necessarily of the same construction or of the same thickness throughout their height. If the floor for the topmost storey has been provided with adequate support and there is no load to be carried from the roof, the uppermost portion of the partition can be constructed to a lesser standard, sufficient merely to separate rooms.

Generally, the foundations of structural partitions should have followed the practice at the time for the construction of the foundations for the external walls; and where this has been carried out, there is perhaps less likelihood of problems than there could be with the external walls. The dwelling itself protects the ground in which they are placed from the effects of undue weather changes. However, shortcuts and economies may have been taken, leaving the partition less able to perform its function than would otherwise be the case. The inspector will not be able to see the form of the original construction, but sloping floors, binding doors and cracks in the plasterwork, where they occur and are serious, need to be reported and coupled with advice that further investigation will need to be undertaken.

In particular, the founding of internal structural partitions on the oversite construction below hollow timber-joisted ground floors can lead to problems of settlement where no allowance for the loading on the partition has been taken. Even founding on the solid concrete ground floor slabs that were frequently used because of timber shortages in the 1940s and 1950s produced an array of problems due to insufficient preparation of the site, inadequate compaction and often the omission of appropriate thickening of the slab where the partition is located.

The quality of the dwelling may well have been a determining factor in the construction of non-load-bearing partitions. Expensive dwellings are likely to have brick and plaster partitions in anticipation of the need to support large, glazed pictures and other heavy objects or fittings on the wall, for superior sound insulation and for a more impressive effect. The occupants can find the thin timber and lath and plaster or lightweight modern plasterboard partitions found frequently in cheaper dwellings less than ideal in these regards.

Non-structural partitions at any storey level which are not a continuation of a partition in the storey below need some form of support unless they are of very lightweight construction and the partition is running at right angles to the floor joists below. If comparatively heavy, the partition should have been provided with support from a beam. Otherwise, support is normally obtained where the partition is running parallel to the floor joists from a doubled up joist. In the conversion of larger dwellings into flats, support for the new partitions, which are normally needed, does not always receive the attention it deserves with the result that cracks can appear in the ceiling below as deflection in the floor occurs.

The ubiquitous breeze blocks with which solid partitions were often formed in houses built between 1918 and 1960 were of dismal quality and difficult to cut to form angles or openings. If intact, they are best left as they are; but cracks and or bulges suggest instability and this may entail renewal on a more extensive scale than might, at first, be thought. The introduction of a range of concrete and lightweight blocks has meant a vast improvement, but examination of cracks and distortion is just as necessary, in both structural and detailing terms, as with timber partitions to determine whether there are major or local faults.

417 In the process of refurbishment, the ceiling of the front ground-floor room with cornice (seen on the left) is higher than that of the rear room beyond, providing a sight of the inner end of the joists of the floor above and the timbers of the braced stud partition carrying the ends of the joists and separating the rooms. The removal of a stud to form a later door opening has caused some deflection in the top plate.

Ceilings

The ceiling to the underside of a hollow suspended timber floor in the upper storey, or storeys, of a dwelling can be considered an integral part of the structure in that it reflects the structural condition of the floor. Sometimes that condition may be revealed in a more pronounced fashion in the ceiling below than in the flooring above where, for example, timber floorboarding will provide more resilience to accommodate movement than the homogeneous plaster of the ceiling.

Irrespective of whether the ceiling is formed of lath and plaster, 19mm to 25mm thick, or plasterboard with a skim coat of plaster amounting to about 12mm to 15mm in thickness, if the structure is inherently weak or has been weakened by undue cutting, the ceiling may well show signs of bowing or cracking. If the ends of joists built in to external walls are beginning to take up dampness and expand, it could well be that the first sign of this will be cracks in the ceiling plaster. Should any rot develop as a result of dampness, the laths are likely to be the first affected.

Undue deflection in the floor of an older dwelling can cause the plaster to lose its key to the laths and become in danger of a collapse. In time, the simple effects of age can also lead to a lath and plaster ceiling becoming separated from its key and sagging. As suggested when lath and plaster ceilings were mentioned in the first section of Chapter 5 in the context of inspecting flat roofs, it may be possible with the owner's consent, having first made the owner aware of the possible risks, to gently test the sagging section with the palm of the hand to see whether the plaster is loose – and to assess the extent of the area affected – or whether, in fact, the plaster is still firmly attached and the ceiling structure itself is sagging. Care must be exercised to avoid bringing the whole of a loose section of plaster down on the hapless surveyor, but if the plaster is loose then it should be attended to before the effects of gravity take their natural course in such circumstances.

With floors, however, it is possible for the surveyor to advise that the floorboards above be lifted to enable an examination of the plaster key to be made before any other steps are taken, such as gentle probing or pressure from the hand, which might cause premature collapse. Depending on the outcome of the inspection from above, it may be necessary to provide advice to erect an internal scaffold to support loose plaster until repairs are undertaken.

Penetrating damp can be lethal to plaster ceilings. Leakage from showers, generally obvious, can completely destroy the characteristics of plaster; and even where staining seems light, the plaster can be reduced to the point where it is no longer fit to take a decorative surface. Less obvious, however, is gradual but slight leakage from a pinhole-type leak in an old lead pipe or a defective connection in a copper pipe which allows gradual seepage to occur. This can result in the sudden and unexpected collapse of heavy lath and plaster ceilings, often added to in layers over the years to compensate for unevenness. Heavy decorative cornice sections also pose a particular risk of sudden collapse when affected by dampness. The inspector should be aware of the possible implications of small as well as large stains.

Decorations

While considering the internal walls and ceilings, it is appropriate to touch briefly on their decorative finishes. The materials, style and colours for internal decorations are very much a matter of individual, subjective taste. What suits one person will not be liked by another. Comment on the specifics of the decorations should usually, therefore, be restricted to general observations on their overall condition and the likely extent of redecoration that the new purchaser will probably need to contemplate after taking occupation. Occasionally, some aspect of the decorations is so striking that it justifies particular comment, especially if it has a detrimental effect on the property or dealing with it will involve unusual effort or expenditure.

418 The single-storey bay window, favoured for many of the smaller houses built in the latter half of the nineteenth century and the early part of the twentieth century, needed a beam above to support the remainder of the front wall. Frequently, a crack will become apparent in the ceiling of the bay on the line of the front wall, as here, even though there is no sign of any really serious movement.

In older dwellings, plasterwork may have deteriorated. Areas affected by cracks may have been made good, parts may have become soft, loose and bulging due to damp, and large areas may actually be held together by wall lining materials. If these are stripped, the plaster may, unsurprisingly, part company from the backing, adding greatly to the cost of redecoration. The condition of plaster and decorations to walls and ceilings needs therefore to be noted in each room on the internal inspection together with notes on any circumstances which might affect the ease or otherwise of redecoration.

If there are damp patches, the moisture may have brought with it salts from the ground or from the bricks themselves into the plaster and thence to the surface. These salts are invariably of a hygroscopic nature, which may not show in dry weather but take up moisture from the atmosphere at other times and mysteriously show up as damp patches for no apparent reason. Staining to chimney breasts from condensation in disused flues causes the same problem. The only way to prevent this happening repeatedly, probably after an initial spoiling of new decorations, is to strip back to the brickwork and render with cement and sand before replastering. A buyer will be unlikely to know this unless told by the surveyor. Again, a buyer may not realise that it takes so long for a wall to dry out following any form of damp penetration or damp treatment works. The rule of thumb is to allow at least a month for every inch (25mm) of brickwork so, depending on weather conditions, it can take anything from nine months to a year for a wall of one brick thickness to dry out, and proportionally longer for thicker walls.

While the decorations normally require little comment in general, the surveyor should make a point of mentioning, when appropriate, that some textured decorative finishes which were popular earlier in the second half of the last century may contain asbestos. While these remain in place and in sound condition, there is no risk to occupants and they can be redecorated without worry. If, however, there is a desire to have the textured finish removed, or should there be a need to drill or cut into the finish, this would risk releasing any asbestos fibres that may be present. In such cases, the material should be properly assessed by an appropriate expert before being disturbed to see whether asbestos is present and, if so, what measures need to be taken to eliminate the risk. More recent products do not contain asbestos fibres but it is not possible to differentiate between them without proper testing.

The risk of lead in older paintwork is now widely known, but warnings about breathing in dust from paint while sanding down or about young children chewing on painted surfaces should still be included when reporting on older properties. Depending on the particular format in which the report is being completed, it may be appropriate to cross-reference warnings about such risks as lead in paint or asbestos in textured finishes between the part of the report dealing specifically with decorations and a more general section dealing with risks.

The decorations may include not only paint and paper but also other applied finishes such as decorative tiling. This may range from modest areas of splashback tiling behind handbasins to whole kitchens and bathrooms fully and expensively tiled from floor to ceiling. Shower cubicles are usually tiled and the increasing fashion for wet rooms means that tiled floors will also need to be examined and assessed. Whenever tiling is encountered, it should be tapped for indications of hollowness, which may indicate that the hollow-sounding area has lost its adhesion to the surface behind. Small areas may not cause concern but if more than a couple of adjoining tiles are affected then there is a risk that whole sections may be loose. Gently pressing the surface of the affected area may reveal an ominous slight movement that betrays the separation of the tiles from their backing and the inevitability that they will, at some stage, become completely detached.

The unexpected need to repair or replace a large area of defective decorative tiling is not a matter to be taken lightly by most homeowners, especially if it involves one of the more expensive tiling materials. The location of the affected area – whether in a shower or above a kitchen worktop, for example – will determine the risk that water will be getting behind the loose tiles and spreading to other parts of the structure. Ceilings below showers and especially wet rooms should always be checked very carefully for signs of damp staining; and if any is noted then very strong warnings about the risk of deterioration and rot should be sounded along with firm recommendations that any concealed areas which may have been affected by the dampness should be exposed and thoroughly checked.

20

Floors

Suspended timber floor construction

In domestic construction, the vast majority of floors are of wood – almost universally so above the lowest storey. Along with partitions, they have a contribution to make to the overall structure of a dwelling above ground level by reason of their self-weight and imposed loads from furniture and occupants, being supported on some of the external walls that, in turn, benefit from the very necessary restraint they provide. Little attention, however, was ever paid to the aspect of providing restraint to the other walls of a dwelling until the 1960s. Accordingly, it was only those walls that supported the ends of floor joists which received the benefit of restraint. Walls parallel to the run of floor joists in many dwellings built before that time were left without any restraint at all. This often resulted in those walls developing the leans and bulges, even in two-storey dwellings, described in Chapter 11.

Similarly, there was little attention paid by way of regulation to the requirements of strength in the construction of upper suspended floors until the introduction of bylaws in the late nineteenth century. In earlier times, but particularly in the eighteenth and nineteenth centuries, reliance would be placed on rules of thumb or pattern books to determine the depth required for floor joists to span the largest room in the proposed dwelling. The rule would assume usage of the common width of 50mm for softwood structural timber, available from the merchants at the generally adopted spacing for joists of 350mm. Once selected, this would set the depth of the floor at that level throughout the dwelling. The margin of safety provided by the rule of thumb and the quality of timber obtainable at the time meant that, in the main, a satisfactory result was produced. Difficulties could arise, however, when, a lesser standard was adopted to save money – say, a depth 50mm less than that suggested by the rule of thumb. Furthermore, in practical terms, all the timber floor joists of the eighteenth and nineteenth centuries have subsequently been notched and drilled for the installation of the service pipes to supply water, gas and then electricity as well as the improvements found to be necessary to the systems when new bathrooms and kitchens have been installed during the twentieth century.

Guidance on the sizing of floor timbers for new buildings in England was removed from the Approved Documents under the Building Regulations in 2004; the relevant Timber Tables for sizing roof and floor timbers are now published by TRADA (Timber Research and Development Association). It is to this document, or its equivalent for Scotland, Wales or Northern Ireland, that the surveyor needs to turn if the opportunity arises to examine the specification for a new or near new dwelling to confirm that the specified sizes are in compliance with the regulations or standards. Should a specification not be available then the surveyor is in no better position for checking the adequacy of an upper-storey floor than for any other dwelling.

In the absence of permission to open up, the information for checking the adequacy of floors will be limited; though it can be greater in some circumstances than others. Fitted carpets – which cannot be lifted at any point and which prevent a sight of the floorboards and, therefore, the direction of run – deprive the surveyor of even information on the span distance. The presence of large flooring panels, whether chipboard or oriented strand board (OSB), can also seriously inhibit the ability to determine which way the joists are spanning, although it is sometimes possible to identify where lines of nailing correspond with joist runs if a large enough area of the floor surface is exposed or accessible. In circumstances where there is doubt about the ability of the floor to support domestic loading because of undue deflection, the only advice that can be given is to expose the structure for further investigation.

In cases where floorboards are visible, the run of joists will be evident, enabling the span to be measured. The frequency of nailing will probably also be apparent, giving the spacing of the joists. It should be possible to measure the thickness of the floor at some point on the staircase; deducting 40mm from this measurement for ceiling and floor finishes in pre-1940 dwellings, this provides the depth of the joists. The only item on which information is then missing will be the width of the floor joists. Where plain-edged boards are present, the tops of joists can sometimes be seen and the width measured. Otherwise, for a dwelling built up to 1940, a reasonable assumption for the width would be the 50mm already mentioned; and with this amount of information, it is possible to compare the size as assessed with the tables in the regulations and standards to check whether the correct size for the span was originally used. A shortfall in the assessed size in comparison with the correct size from the tables based on the lowest strength of timber would indicate that any undue deflection is due, at least partly, to an inherently wrong choice of joist size. If the comparison suggests, however, that the joists ought to be satisfactory then the cause of the inadequacy could lie in the floor's subsequent history – for example, excessive notching and drilling for the installation of services as described previously.

For houses built after 1940, however, it would not be reasonable to make the assumptions suggested because of the much wider range of sizes of timber and forms of construction that could have been used. Equally, there is a greater likelihood of tongue-and-groove boarding so no joist tops will be visible. If loading is a concern, advice on opening up for a close examination would then need to be given.

Some of the information that it is necessary to assemble before an opinion can be given on the inherent strength of an upper floor should have been collected on the room by room inspection of the interior, particularly the run of floor joists. Other information, more related to the development of subsequent faults, includes whether the floor is sloping, whether as a whole or in part and in which direction, whether it sags towards the centre, and whether there is a gap between the floorboards and the skirting or if a moulding has been fixed at that point to disguise a gap which exists. It is, of course, in connection with ascertaining these faults that the surveyor's spirit level is put to good use.

Not quite as obvious as the above defects, which are visible, are those which may be hidden but which are possibly contributing to the overall appearance of a defective floor. Among these could be decay and rot in the ends of joists where they are built into external walls for support and rot in the timber plate on which they bear. Building in to external walls was the practice until the 1940s, and as no ventilation was ever provided to the timbers in upper-floor construction, persistent driving rain could penetrate through what was, in the case of walls of one brick thickness, little more than 120mm or so of brickwork. Floors of dwellings built after 1940 can also have problems where joists, instead of being built in, are supported on joist hangers. It is important that joists are properly seated and fitted tightly to the back of the hanger and that the hangers are securely built in to the brickwork and prevented from twisting by the fixing of a batten between them, attached firmly to the adjacent wall.

It is unlikely that solid construction will be found for the floors at upper levels except in the case of purpose-built blocks of flats. Such floors are likely to be of reinforced concrete construction if built from the 1920s onwards or of rather specialised types of construction in the mansion blocks of flats built from the 1880s to around 1920. This is not so, however, in the case of floors constructed at ground level, and it may need a sharp stamp on the floor or a 'heel drop' with the feet together to distinguish solid from hollow construction. The sound of a hollow suspended timber floor can be produced at various degrees of intensity: a bright and clear sound to signify a tight, sound, well-constructed floor; or a deep, rumbly, dead sound with a distinct degree of vibration to indicate a rather shaky floor, suggesting either less than adequate construction or the development of faults.

Generally, suspended timber floors at ground-storey level are formed on sleeper walls built off such site ground cover as there may be – for instance, rammed earth, ashes or weak concrete, depending on the era of construction and the rigor, or otherwise, of the enforcement of the regulations in existence at the time. Early versions in the eighteenth century could be built off the compacted earth, after the topsoil had been removed, with a baulk of timber or a plate on a course of bricks, the ends of the joists built into the external walls and with minimal through ventilation beneath the floor. In the absence of any damp-proof membranes, the ends of the joists would probably degrade, first from wet rot and then dry rot. Later versions, instead of being built in, would have an additional sleeper wall close to the external wall and a greater depth above an ash-covered site with more airbricks for underfloor through ventilation. At the end of the nineteenth century and into the early twentieth century, site concrete and damp-proof courses became common, bringing the standard up to that which generally pertained up to the 1940s and indeed, in many instances, applies right up to this day.

The trouble was that it was not until the period 1920–1940 that the typical house became more or less square in shape; those of the later nineteenth century and early twentieth century were very far from square. With their back additions or 'offshoots' usually provided with solid floors, the suspended timber floors of the main part would be left with quite large areas of subfloor space inadequately ventilated, particularly if there were comparatively few airbricks positioned in the external walls. On a damp site or where damp-proof courses had been omitted or had failed, there could be the ideal conditions for the development of dry rot.

Many of the typical terrace houses of the eighteenth, nineteenth and early twentieth centuries were, of course, provided with cellars, either below the whole of the ground-floor areas of the dwelling or limited to that area beneath the entrance hall. These were primarily for the storage of coal, which provided the only form of heating up to around 1900 and even up to the 1940s was still the cheapest. Cellars, now often full of old or unused household appliances, provide inspectors with useful access from which to view the underside of the floor at ground level. If a cellar is provided across the whole site area then, of course, the floor above will as likely as not be totally suspended in the same way as the floors to the storeys above. Assuming there is no ceiling, the surveyor in these circumstances can clearly see the size and spacing of the joists. It will probably also be possible to identify and see where the main internal structural partition is placed, establish its construction and measure its thickness along with any other partition that may, for example, be supporting the staircase and landings. Entrance hall floors are often tiled on top of joists and floorboards, and the surveyor will be able to see how this form of construction is arranged. Without a cellar, this is impossible to ascertain without opening up; in the presence of tiling or fitted floor covering, permission to do this is unlikely to be forthcoming unless some serious defect, obvious to all eyes, is present.

Even though the cellar may only extend to the area below the entrance hall, a sight of the construction of the sleeper walls supporting the bulk of the floor at ground level is usually obtainable. This enables a check to be made on the position, the thickness and the presence, or otherwise, of the damp-proof membrane below the wall plates and also the adequacy of the openings

419 A section of suspended timber floor construction at ground level has been taken up here for renewal. A new sleeper wall of three courses of brick with wood plate has been provided to carry the new joists. The surveyor will need to check that a damp-proof membrane has been provided below the plate and that there is adequate underfloor through ventilation. None is to be seen on the section visible here. The old joists (one can be seen on the right) are carried on the partition, the bottom of which can be viewed beyond the new sleeper wall.

420 Generally, hollow timber suspended floors at approximately ground level are single – i.e. the joists are supported directly off the plates on the sleeper walls. In a somewhat unorthodox fashion, this floor is double. Substantial cross beams support the floor joists, which in turn are carried on brick piers, approximately 850mm high.

left for through ventilation in the sleeper walls. In many ways, these partial cellars can be considered the saving grace of such houses since not only do they provide useful space and access for inspection but the extra degree of ventilation can also keep the area free of the problems which could arise if the space below the hallway flooring was more confined and deficient in ventilation. The not uncommon burst or pinhole leak from an old lead mains water pipe is a case in point. It will usually be spotted early on and dealt with if there is a cellar. If not and it occurs below a floor with tiling or fitted carpet, it may go unnoticed for a long time until an outbreak of dry rot causes the floor to collapse.

Higher up the scale, so to speak, from 'cellars' are 'basements' with, usually, their better light, ventilation, a proper ceiling and plastered walls. The durability of any hollow suspended timber floor at this level, well below the level of the surrounding ground, will be very much dependent on the degree of dampness present and any provisions for its prevention or amelioration in the enclosing walls but, in particular, on the availability of through ventilation to the underside and the avoidance of pockets of damp air which are liable to accumulate. This can be achieved through the provision of external areas front and back for the full width of the dwelling, extending to a depth of around 400mm below the internal finished floor level, and the installation of airbricks at each extremity and at intervals in accordance with current requirements – 1,500mm^2 for every metre run of wall.

These low-level external areas should extend, front to back, from the walls enclosing the basement, free of obstruction, to at least the minimum current requirements wherever possible. These ideal requirements will be found to be present in very few old dwellings, even apart from the factors of damp-proofing which are often difficult to achieve in the basements of the older terraced dwellings, limiting the use of such accommodation to utility purposes. The surveyor's routine and the requirements of the Conditions of Engagement, Practice Notes and Guidance Notes will necessitate the use of the damp meter at frequent intervals, and it is more than likely that considerable areas of dampness will be found. The shortcomings of a basement need identifying and describing in the report.

Solid floor construction

It was not until the period 1940–1960 that solid construction for floors at ground level began to take on the structural significance that they were to assume for some years, particularly in many local authority and New Town Corporation housing schemes and those schemes at the lower end of the private sector price range. Once introduced, however, they became very popular, and it was estimated by BRE that in one English New Town, 97 per cent of houses had solid floors at ground level.

It was not a good period for the propagation of such construction on a major scale even though the reasons seemed sound at the time – urgent need for new housing, shortage of timber and economy. The policy of retaining green belts around the major cities meant that sites not previously considered suitable for house construction were being brought into use, some steeply sloping, some with drainage problems and others where household waste had previously been dumped or minerals extracted. Skills were lacking and less experienced developers were in the market. It was thought that sites could be quickly cleared and levelled, a layer of hardcore put down, a slab of concrete laid and the floor would be ready for its finish. As a bonus, the slab provided support for partitions, structural or otherwise, together with all the fittings in the kitchen and support for the staircase.

Poor sites, poor site preparation, inappropriate materials in the hardcore such as colliery shale and pyrites, and indifferent compaction of the concrete resulted in overall and differential settlement in many cases (**421**). In others, there was substantial heave – as much as 150mm in some – due to sulphate attack on the concrete. In one New Town, BRE recorded 600 houses with major problems of heave in the ground-floor concrete slabs in the late 1970s.

NHBC was paying out over £1 million a year in the 1970s and 1980s on claims in respect of defective floors slabs – in some cases, at the rate of £7,500 per house at prices then ruling. It had to raise its standards in 1974 to require a suspended floor at ground level where the depth of fill exceeded 600mm. Much of the subsequent advice and the requirements for the construction of floors at ground level tends to reinforce the long-held traditional practice in Scotland where, for house construction, the site was cleared, covered with pitch and the lowest floor suspended from the walls – in the same way as upper floors – with a suitable space beneath for ventilation but with no reliance placed on oversite concrete.

Of course, a suspended floor does not necessarily have to be of timber. Suspended solid floors of precast concrete beam and block construction have become increasingly popular in housing developments since the 1980s. With a new house, the surveyor may be able to obtain details about the type used from the developer; but once a development has finished, research can be much more difficult unless a visit to the local authority proves productive. The requirements for a new dwelling in England with a floor at ground level of solid construction are set out in Approved Document C of the Building Regulations.

Unless a floor surface has been deliberately exposed, perhaps stripped and varnished to show off the original boarding, the prevailing fashions dictate that the surveyor is likely to find the original floor surface concealed by either fitted carpets or a fitted laminate strip floor covering and, effectively, inaccessible. In scanning the walls and ceilings and in examining windows, doors and other fittings, there is a danger that the surveyor will skim over the carpeting or applied surface, giving the floors less than adequate attention. A few cases from the 1980s will stand as a reminder of what needs consideration. In one case, the surveyor on a mortgage valuation missed a crack in a floor. Damages of £29,000 were awarded although the house only cost £14,850 (*London and South of England Building Society* v *Stone* 1983). In another case, damages of £5,000 were awarded where a surveyor missed a 20mm gap below the skirting to rooms on the ground floor (*Westlake* v *Bracknell District Council* 1987). In a third case (*Cross* v *David Martin and Mortimer* 1989), the surveyor failed to relate the evidence of sloping floors and doors out of alignment to the degree of settlement in a house built on a sloping site. The award in this case included amounts in respect of other items where the surveyor was also found negligent.

Should solid construction for the floor at ground level be encountered in a dwelling built within the last 40 to 50 years and where the floor is covered with fitted carpet, it will not be possible to see whether there are cracks in the surface and even, perhaps, whether there is a gap below the skirting. Nevertheless, the surveyor's spirit level will show whether the floor is sloping; if it is and should there be displaced framing and door openings out of true alignment in partitions then there should be careful consideration about whether the two defects are related. A conclusion that they are related could well provide the reason for other defects such as roof and staircase distortions.

As already mentioned, solid construction for floors at upper levels in domestic buildings is unlikely to be found except in the case of flats. Although multi-storey blocks of flats were common on the continent and in Scotland, producing apartments and tenements, they were comparatively rare in England and Wales until the charitable trusts began to build them from the 1860s onwards. These were followed from the late nineteenth century and early twentieth century by private developers of mansion blocks and local authorities, all producing flats to let and now also sold on long leases. The flats in the mansion blocks could be quite large, often built to a high standard that could extend to an expensive finish to the floor structure like superior hardwood boarding, parquet or woodblock. There could also be accommodation for servants with separate access and the provision of services to include centralised heating and hot water systems, lifts and porterage.

The form of construction for the external walls and internal structural partitions would traditionally be of brick or stone as although a patent for reinforced concrete had been taken out here in the 1850s, its development and use was mainly on the continent; its employment for the walls of commercial buildings in the UK did not commence until the early 1900s – examples being the Royal Liver building in Liverpool and Kings College Hospital in London, both of 1909. For domestic construction, its use came even later in the 1920s and 1930s for the structural frames of blocks of flats. It had not been until 1915 that the London County Council introduced regulations governing the use of reinforced concrete and the first British Standard Code of Practice did not appear until 1934. This lapse of time, however, did not occur in the case of floors which, as long as they had continued to be formed of timber, represented a hazard by allowing a fairly rapid spread of fire and resulting in substantial conflagrations and loss of life. Such fires led to the earlier development of floors with a degree of fire resistance.

Fire-resisting floors became possible following the Industrial Revolution from the late eighteenth century onwards. Cast iron had been an early outcome and was used as beams of various shapes from inverted T-sections to those not too dissimilar from the familiar rolled steel joist of today. From the lower flange on either side would be sprung a shallow brick 'jack' arch with the top brought level with an infilling of mass concrete, rediscovered in the late eighteenth century after a lapse of over 1,000 years from the time the Romans departed. Early rediscovered mass concrete was first based on lime and then on Parker's Roman cement from 1796 until the patenting of Portland cement in 1824. Cast iron was brittle but strong in compression and, therefore, satisfactory for columns; but it was weak in tension and following the collapse of a number of bridges in the mid nineteenth century, it was replaced for beams of wrought iron, which was much stronger in tension and had been used since medieval times, in small quantities, for cramps and tie bars.

Combinations of steel beams and mass concrete were known as 'filler joist floors' but, during the 120 years or so from about 1800 to 1920, many designs were patented for floors with a degree of fire resistance based on cast or wrought iron and, from the 1880s, mild steel beams of various configurations and mass concrete of differing formulations. Some included types of the ribbed and hollow terracotta clay sections that became more commonly familiar from the 1920s onwards as the reinforced concrete hollow pot floors. Names such as Holmans, Frazzi, Monolithic, King and Fawcett became known for their patent floors.

Problems arose, however, from the formulation of the mass concrete. BRE quote a writer commenting on Holman's patent filler joist floors as follows:

> the use of fine coke in the aggregate may have seemed at the time a useful way of disposing of a plentiful waste material but it contained sulphur that slowly oxidised and many floors of the period later suffered grave damage through the corrosion of the ironwork with which the coke or breeze made contact.

This problem was by no means restricted to Holman's patent floors and it was much aggravated in the presence of prolonged dampness from plumbing leaks. In cases where corrosion occurred, the concrete infill would expand and crack along the line of the beams, pushing up the top surface.

Mild steel was much more pliable than cast or wrought iron for forming into different shapes, including bars and strips, and has greater and more reliable all-round strength; after its introduction in the 1880s, it gradually replaced both cast and wrought iron for structural purposes and, with the standardisation of sections and the publication of stress tables, supplanted both by the 1920s. The ingredients, comprising Portland cement and mild steel, and regulations governing the construction were accordingly available from the 1920s for the development by designers of cast *in situ* reinforced concrete fire-resisting floors in flat slab, hollow pot or waffle form for use in blocks of flats. These could be used either separately in an otherwise masonry-constructed block or as an integral part of a reinforced concrete frame. The patented proprietary systems gradually fell out of use and the alternative forms of cast *in situ* slabs served well for the purpose until the high-rise developments of the 1950s and 1960s. At this time, various new manufacturers' systems were introduced involving, along with other components, floors of prefabricated prestressed concrete sections for hoisting into place on site and subsequent bolted connection to wall slabs. The requirement, at the time, for speed brought into use High Alumina Cement and additives such as calcium chloride for rapid hardening, which along with the development of lightweight aggregates, in turn, produced new problems of carbonation and corrosion of steel reinforcement.

The significant challenges presented when inspecting and reporting on a flat in a block built since the 1950s where concrete floors are encountered and where the construction is otherwise in nontraditional form have already been discussed at some length in Chapter 13 under the section 'Precast concrete panel walls'. It is not necessary to repeat here the difficulties of striking the right balance when inspecting and reporting on such properties, including the need to take into account the requirement for an engineer's inspection of such buildings at regular intervals, but they should not be underestimated. For blocks of flats with reinforced concrete floors built between 1920 and the 1950s, whether with walls constructed of masonry or with a reinforced concrete frame, and for blocks with solid floors built before 1920, the advice is that if there are no visible defects then there is no need to recommend the engagement of a structural engineer unless substantial alterations are proposed. Such advice, however, can only follow the inspection of those visible parts both above and, whenever possible, below. This should seek not only to ascertain the mode of construction but also to detect any disturbance or damp staining to finishes, in particular to the floors – such as disturbed woodblock or parquet – or to tiling in kitchens, bathrooms and hallways, and also possible disturbance to other features such as partitions where defects may have had a 'knock-on' effect. The inspection must, obviously, include the likes of cupboards or storerooms where floors may originally not have been provided with a finish, and it could, most usefully, be extended to the common parts of landings, staircases and lift shafts. If there is a chance to inspect basement rooms and perhaps old boiler compartments, fuel stores, cleaner's stores and the like, the opportunity should be taken as the underside of the ground floor may reveal not only the form of construction but also give a clearer view of any sign of movement. It is permissible to conclude that the type of solid floor in a block of flats cannot be ascertained but that there are no visible defects. What is inexcusable is to miss those defects if they are visible and fail to advise further investigation by an engineer. With an outline knowledge of the types of construction which could be present according to the age of the block, the surveyor will be in a better position to know where to look.

421 The wood block surface to this solid floor helps to show how far it has dropped, the threshold of the French window on the left being the original level. Poorly consolidated hardcore is the most likely cause in this particular 1950s house rather than anything more sinister.

422 A diagonal crack in a partition separating these two rooms, running from the top corner of the connecting door, coupled with distortion in the frame, is a sign of subsidence in the solid floor on which it is supported.

Floor finishes

For hollow suspended timber flooring, the traditional top layer of the structure has for long been softwood floorboards nailed to the joists, almost universally plain edged or tongued and grooved for dwellings of fine quality and, in the main, between 19mm and 25mm thick. It would then be up to the occupant to decide whether to leave the boards bare, to have them painted or varnished, or to cover them with some form of surface material such as linoleum, plastic tiles, some other type of wood in the form of parquet or strip, or a woven material in the form of loose or fitted carpet. An intermediate layer may be needed before a wearing surface can be applied. Plywood or hardwood are most commonly used, fixed to the floorboards to even out any discrepancies in the original boards, before laying tiles or hardwood strip flooring. For carpet, an underlay of some material is usually laid.

Accordingly, it may not be possible to see the original floorboards. Depending on the age and the history of the dwelling, floorboards can be in good condition; but more often than not they will have been taken up repeatedly, cut for the installation of services and frequently renailed, and they are likely to be heavily worn. This is the reason why most old floors are covered over; if they are left bare, the unevenness, loose sections and, in particular, any sections which have been attacked by woodworm could not only be extremely unsightly but could also present a hazard.

If possible, the surveyor should – depending on the level of inspection being undertaken – attempt to peel back at least a small section of any fitted carpet or other potentially 'liftable' covering just to see what the floor surface material is and perhaps form an opinion on its likely condition. The corners of rooms, the points where radiator pipes pass through the carpet, and the backs of toilets and washbasin pedestals are often productive points for such access. It may not be very much but simple knowledge such as whether the floorboards are plain edged or tongue and grooved, the direction they run or whether chipboard sheet flooring is present can give confidence to the dating assessment of the property, add to the overall analysis of construction and perhaps provide reassurance about the source of the creaks and squeaks originating from the floor while walking around the property.

Where a chipboard-surfaced floor is found, it may not be possible to determine whether a moisture-resistant grade has been used in the areas likely to be affected by dampness, such as the kitchen and bathroom. With a modern property, it is reasonable to assume that it complies with the requirements imposed in 1990; but with dwellings built earlier than that, the report should include a warning that in the event of any areas of chipboard flooring being affected by moisture they will probably require complete replacement.

Where laminate strip flooring has been laid, the surveyor should consider whether the standard of installation can be ascertained. If the floor surface is level and firm and the edge detailing is tidy then all may be well; but frequently work will have been carried out by a well-intentioned but not especially skilled

423 A natural stone flag floor can add much to the character of a historic dwelling and, despite some unevenness and the inevitable absence of a damp-proof membrane, such floors should be retained as far as possible. Lifting and replacing the slabs is rarely satisfactory and can often introduce more problems than it solves. In a listed building, permission for such work is unlikely to be forthcoming and guidance should be sought from a specialist in historic buildings if it should be deemed essential.

DIY enthusiast and the floor surface will be uneven, possibly with spongy or distinctly raised sections. The edges of such a floor are likely to be poorly finished. Most purchasers anticipate a laminate-surfaced floor to be durable and, while not exactly adding value, to at least be an asset in their new home. Although such a surface is not part of the structure of the dwelling, it is arguably something more that a decorative finish, so a warning that it has been poorly laid and is unlikely to prove satisfactory in the long term should be included in the report when appropriate.

What the surveyor can see and discern from the inspection should be noted, but whether it falls to be included in the report will depend on the type of report for which the inspection is being carried out – with the exception that where a hazard is noted, it should be included whatever the type. If such is present in one room but others are covered over then it is necessary to include a warning that similar conditions could become apparent when the owner's carpets or the like are removed.

The same comments, in general, apply to the top surface finish applied to floors of solid construction. However, in the process of application many of the treatments acquire a degree of permanence that renders them non-removable on a change of ownership – woodblocks, for example. Most of the defects in the finishes to floors of solid construction are related either to the absence or the lack of effectiveness of the damp-proof membrane or the use of unsuitable wood or adhesives. If the finish is of wood then changes in moisture content will cause movement and buckling due to expansion if no allowance has been made. Alternatively, rot will eventually set in from the underside. Unsuitable woods can be prone to inconsistent colour change, uneven wear and, in some cases, a degree of dangerous splintering. Coverings of materials that are impervious to moisture, while keeping damp at bay, will not necessarily resist upward pressure or the effects of an accumulation of water below and, whether in tile or sheet form, will arch, lift or billow upwards. All porous ceramic tiles run the risk of expansion and it is this which causes crazing in the glaze. Movement joints are necessary and many problems are caused by differential rates of expansion in the tiles, the adhesive material and the base. As with tiled wall finishes, the surveyor needs to note what has been seen and to provide warnings if necessary.

21

Timber insects and rot

Timber insects

The modern centrally heated house with fitted floor coverings and heavily insulated roof space, and often incorporating pretreated timbers, is no longer the hospitable environment provided by houses over the centuries for wood-boring beetles. This is just as well for surveyors because the same carpets and roof insulation, by concealing so much of the timber within a building, make life very much more difficult for the detection of beetle infestation than it was in the days of loose-laid carpets and uninsulated roofs. It is not common to find infestations in houses built during the last 50 or so years, but with older houses – indeed, any built prior to 1960 – a significant proportion have at least some areas of timber where beetles have been active.

During the years of steadily rising home ownership from the middle of the twentieth century, the presence of wood-boring beetles was identified in a high proportion of the associated mortgage valuation inspections for prospective borrowers. An industry of timber treatment companies quickly became established to meet the need to remedy this apparent plague, the same contractors also offering to cure the rising damp which mortgage valuers frequently reported at the same time. While dampness has occasionally proved rather problematic to remedy, the generous application of lethal chemicals throughout many of the nation's houses has significantly reduced the overall scale of beetle infestation. The preferred habitat of timber beetles, cool and ideally tending towards damp, used to be provided by the previously ubiquitous uninsulated roof spaces with rough-surfaced timbers and unlined slopes (**424**) or in damp and draughty subfloors. Modern building practices, with planed timbers, lined roof slopes, full central heating and properly screeded subfloor areas – where the ground floor is not some form of solid construction – have reduced the attractiveness of many homes for potential invaders. While the scale of the problem may have lessened, however, it has not disappeared; the surveyor needs to be just as diligent as ever in seeking evidence of beetle infestation even though the amount of accessible timber for inspection has reduced, making detection correspondingly harder.

Wood-boring beetles are commonly known as woodworm – probably because the larvae might seem like a worm burrowing through the ground. They are flying insects of varying sizes, depending on species, which lay eggs and whose larvae tunnel and consume the structural timbers, emerging later through flight holes to mate, lay eggs on more timber and continue the life cycle. It is the larvae which cause damage to the timber but the emerging insect that causes the distinctive 'woodworm' holes, by which time the damage has been done. The surveyor needs to ascertain whether flight holes are present and, if so, their size, shape and positions along with the extent of any damage already caused. Some species pose very serious threats to buildings but most will cause little serious structural damage to dry wood, although even these are capable of devastating damp wood.

While possible, it is unlikely that the surveyor will see any beetles flying around in roof spaces. Most beetles are small, extending only to about 3mm to 6mm in length. The exception is the house longhorn beetle which is about 12mm long but which is only found in the northern parts of Surrey and parts of Hampshire and Berkshire – the reason why the Building Regulations in these areas require pretreatment of new roof timbers. Furthermore, beetle flight is quiet and not rapid, so unless the surveyor peers intently and has good reason to look at a particular spot – in an area of a damp patch perhaps – beetles which have alighted on timbers are unlikely to be glimpsed since their appearance and colour tend to blend in with the rougher and dustier surfaces of the unplaned timber usually found in roof spaces of older houses. It is also a fact that most look very similar – small beetles with heads and wing cases – so it is not very likely that the non-expert surveyor would be able to distinguish one type from another if one was seen. For practical purposes, the appearance of the adult beetle can be discounted as a means of identification. Having said this, however, the sight of a number of little brown beetles – usually between May and August but, with warmer winters, now at other times of the year as well – should set alarm bells ringing and renew the search for clean, fresh dust and new-looking holes in the timber.

What helps to identify the type of beetle is the flight hole the adult makes when biting its way out of the wood. Although the holes left by wood-boring insects which can cause structural damage can look very similar in size and shape to others which cause little or no damage, it is the type and condition of the wood from which they emerge and possibly the bore dust from within the hole that pinpoint the different species. **Table 5** should clarify the deduction process.

As will be seen from **Table 5**, consideration of the size and shape of the exit hole, from what wood and from where the adult beetle has emerged should point towards a possible identification. Knowing, or assessing, the age of the dwelling could help to eliminate six of the possible types of beetle. Some that come in or are brought in from the forest die out within a year or so of construction while another only infests wood which is already damp or decayed, probably due to other defects in the structure.

That leaves four types that can cause structural damage if an active infestation is left untreated. Surveyors practising to the south-west of London will no doubt be well aware of possible infestation by the house longhorn beetle, and indeed the sight of its exit holes is enough to bring anyone up sharply. However, its activities are confined, to date, to a comparatively small area. In practice, infestations by the powderpost beetle are also comparatively rare since attacks are confined to hardwood timbers, which are more likely to be a feature of the more expensive dwellings. Since it is unusual, and indeed wasteful, to hide attractive and expensive hardwood, the effect of any infestation is often readily apparent.

In practical terms, damage by wood-boring beetles generally narrows down to infestations by two types of beetle – both fairly common and both of which can cause structural damage if left to their own devices, but which are easy to differentiate. In regard to one type, the common furniture beetle, it is believed that between a third and a half of all dwellings with softwood roof structures, floors and fittings, which comprise the vast majority, are or have been infested – a higher proportion if built before 1900, rather lower if built after that date and a much lower proportion of those built since 1960. The other type, the deathwatch beetle, favours old hardwood timbers, particularly oak and elm. These are usually found in buildings going back in time before the 1700s, either wholly timber framed or with masonry walls and an oak-framed roof and floors. There are very few such buildings that have not been infested by deathwatch beetle at some time or other and many that still are. However, the exit holes of this beetle are about twice the size of those made by the common furniture beetle, making recognition easier than it might otherwise be.

The importance of the surveyor pointing out the difference in the report between the two types of beetle if one or the other is found to be present, or even both as that is by no means uncommon, was brought out in the case of *Oswald* v *Countrywide Surveyors Ltd* in 1994. The claimants had bought a timber-framed Essex farmhouse for £225,000 even though the survey report had indicated that the old timbers were much attacked by woodworm. However, the surveyors did not state the type of insect that had caused the attack, which the judge considered vital in the circumstances. It transpired that one of the claimants had a horror of deathwatch beetle. The plaintiffs suffered a plague of the insects after moving in, some of which got into their bedding. The judge awarded around £50,000 in damages and a subsequent appeal by the surveyors in 1996 was dismissed.

With deep layers of insulation quilt now routinely laid across roof spaces, surveyors often face a balancing act – both literally and metaphorically – when deciding the extent to which physical entry is practical and safe. Even before stepping off the surveyor's ladder into the roof space, however, it is good practice to stand at the top of the ladder and look in detail around the access hatch surround and then at all the timbers visible from the access point. The access hatch timbers are, for some reason, a prime location for infestation and

Table 5 Identification of wood-boring insects

Size of exit holes in ascending order (mm)	Shape of exit holes	Wood affected	Condition of wood and whether infestation likely to be active or inactive	Bore dust resulting from active infestation	Likely type of insect and whether liable to cause structural damage	If active, whether treatment is required
1.5	Circular	Softwoods and hardwoods, but mainly sapwood	For active infestation, cool and damp conditions preferred; tends to die out with central heating	Cream-coloured pellets; fine, granular, gritty texture	Common furniture beetle (*Anobium punctatum*); yes, liable because of successive infestations to structural timbers	Yes
1.5	Circular	Hardwoods only – particularly sapwood of those with open pores such as oak, ash, walnut and elm	Dry and warm; wood generally employed for framing, flooring and decorative features	Flour-like texture; colour depends on wood	Powderpost beetle (*Lyctus* sp.); yes, liable in flooring and framing where such woods used	Yes
1.5	Circular but ragged	Softwoods and hardwoods in damp areas	Only infested when damp and decaying	Fine, granular pellets	Wood-boring weevil (*Euophryum confine*); not liable, but earlier damage from other causes would probably require renewal	No
2	Circular	Softwoods, but only in the bark or sapwood	Only found adjacent to bark; dies out naturally due to limited amount	Brown or cream; bun-shaped pellets	Waney edge borer beetle (*Ernobius mollis*); not liable as does not affect heartwood where strength lies	No
3	Circular	Hardwoods only – usually oak or elm when old; both sapwood and heartwood	Damp and poorly ventilated wood preferred, but attacks extend into relatively dry areas; if active, tapping of mating call may be heard	Oval or bun-shaped pellets in dust, visible to the naked eye	Deathwatch beetle (*Xestobium rufovillosum*); scourge of old timber-framed dwellings and oak-framed roofs; very liable to cause damage	Yes

Size of exit holes in ascending order (mm)	Shape of exit holes	Wood affected	Condition of wood and whether infestation likely to be active or inactive	Bore dust resulting from active infestation	Likely type of insect and whether liable to cause structural damage	If active, whether treatment is required
3	Circular	All woods; sapwood and heartwood, but only when growing in forest	Infestation dies out on conversion and seasoning into usable building timber; blue or black stain may be left around flight holes	None	Pinhole or shothole ambrosia beetle (SCOLYTIDAE and PLATYPODIDAE species); not liable	No
6	Circular	Imported hardwoods, but only in the sapwood	Dies out within a year of importation	Flour-like texture	Bostrichid beetles (BOSTRICHIDAE species of powderpost beetle); not liable	No
6	Circular	Sapwood and heartwood of softwoods only	Dies out within a year of construction as normally lives only in trees and logs	Chips of wood, tightly packed	Wood wasp (SIRICIDAE species); not liable	No
6–10	Oval	Sapwoods only of softwoods and hardwoods	Originates and normally lives only in trees and logs; dies out	Mixture of pellets and wood splinters	Forest longhorn beetle (*Phymatodes testaceus*); not liable	No
6–10	Oval but may be ragged	Generally only the sapwood of softwoods is attacked first	Dry, warm timbers favoured, but attacks will spread; larvae may tunnel for years leaving a mere shell without much external sign of the galleries formed by tunnelling, except for possible blistering; to date, attacks have been limited to areas mentioned in Building Regulations	Mixture of fine wood dust and short, compact cylinders of excrement fill the galleries bored by the tunnelling larvae	House longhorn beetle (*Hylotrupes bajulus*); very liable to cause structural damage, necessitating replacement and treatment	Yes

424 It is now unusual to have such an uninterrupted view of the roof timbers, but where piles of frass like this can be seen, they are unmistakeable evidence of an active infestation of wood-boring beetles.

finding activity here reduces the necessity to crawl into all of the eaves, which provide a further popular location for beetles to take up residence. The second reason for diligent search at this point is for reasons of professional self-preservation. If called to account at a later date for not spotting an infestation in a roof space, there can be few satisfactory responses to an opposing expert or, worse, a non-expert homeowner who states that: 'It wasn't even necessary to go into the roof space because affected timber could be seen from the access hatch'. Making it an invariable rule to conduct a thorough search before entering the roof space will ensure this embarrassment is avoided. Having made a decision to proceed, the roof space timbers then need to be checked until either signs of infestation are found or the surveyor is satisfied that the inspection has been as full as possible.

With roof spaces and floor surfaces frequently now relatively inaccessible, the surveyor, while obviously keeping an eye on any areas of timber, must check particularly the other areas where beetles tend to congregate – namely in the understairs cupboard and, in older houses where the meters are still fitted internally, beside the electricity meter where meter mounting boards seem to have a propensity to attract beetles. Other areas to make a point of checking for infestation are any which might be affected by persistent dampness from minor plumbing leaks, especially the floor at the back of toilets – helpfully a location where floor coverings are often missing or can be peeled back.

On finding exit holes, further diligence is needed to establish whether there is evidence that the infestation is active. Activity can be deduced, and accordingly reported, if the exit holes are clean and freshly made and by the presence of bore dust in the holes, on surfaces underneath and, perhaps, by the beetles themselves. Advice to have the infestation treated must, therefore, follow.

Unfortunately, the sight of old exit holes and the absence of bore dust or beetles is not conclusive evidence that activity has ceased and that no infestation is present. Once the beetles have emerged, they will be looking around for somewhere to lay their eggs. The nearest and most convenient timber for this purpose is right at hand and, accordingly, re-infestation is a common phenomenon, although less likely in properties where all timbers have been sprayed or painted during a previous timber treatment process. An infestation by woodworm is a natural hazard for dwellings, and if there is visible evidence of exit holes then the surveyor must report their presence and should give an opinion on whether they seem recent and active or of old origin. However, it is unwise for a surveyor to state categorically that all activity has ceased as it could be going on in a concealed area without anyone knowing. If the holes seem old and there is no evidence of active infestation, the facts should be reported and the owner warned to keep a watch for any indications of further activity in the future. There is little merit in recommending an inspection by a timber treatment specialist simply as a precaution. On the other hand, if there are solid grounds for suspecting that an infestation may be active or if significant areas of high-risk timbers were not accessible or not inspected for some reason, a specialist inspection by a reputable timber treatment company – such as a member of the Property Care Association – should be recommended.

Timber rots

Wood is an organic material so in addition to being attacked by insects it is also subject to attack by other organisms that can derive sustenance from the cellulose material in its composition. It can provide the source of food for fungi, which develop and grow from spores present in the air at all times and which are just waiting for the right conditions to arise. In the countryside, this occurs simply as the natural process of dead trees being broken down by decaying; but in a dwelling, it only happens when something has gone wrong. The first requirement is for timber to be subjected to persistent dampness, and should those conditions not be remedied then it is usually only a matter of time before fungus attack takes place and the timber starts to deteriorate. One particular type of fungus, *Serpula lacrymans* or 'dry rot' as it is commonly called, is a serious problem, and once it becomes established, the damage can be considerable and the cost of eradication and repair substantial. It is important for the surveyor, therefore, to be aware of the places in a dwelling where those conditions are likely to already exist and perhaps where an attack might already be flourishing or where, if some aspect of the dwelling is not corrected, conditions suitable for an attack could develop.

Dry rot

There are three essential conditions required in the wood for it to be attractive to the spores and susceptible to the depredations of the dry rot fungus.

1   *Damp:* Wood in a dwelling settles down after construction to a level of approximately 15 to 18 per cent moisture content of the oven dry weight. This reduces to less than 12 per cent with central heating. However, if it rises to 20 per cent then the wood becomes susceptible to fungal growth. Once established, dry rot is able to transport the necessary moisture from the existing outbreak to areas of adjoining dry timber and thus facilitate spreading. Other rots do not have this facility and this ability to spread is one of the things that makes dry rot such a problem.

2   *Oxygen:* At 40 per cent moisture content, there is ample oxygen in the cells of the wood to support growth at its optimum level and maximise

the availability of the food supply. The availability of oxygen declines above this level of moisture content, and at saturation level, the wood is immune from attack.

3   *Warmth:* The ideal temperature for vigorous fungal growth is 18.5°C (65°F), but the normal range of temperature found in most dwellings will support growth as long as the air is still and not draughty. Dry rot does not like moving air; hence the provision of ventilation to certain areas of a dwelling likely to contain moist air – under suspended floor construction at ground level, for example.

Two relevant points can be made here. One is that wood-based products are even more susceptible to attack than the structural softwood timber for floors, roofs and partitions and the timber used for joinery. Plywood, chipboard and hardboard are more easily consumed as are related products such as books and papers. The second point is that following a fire or a flood, the drying out process after wood has been saturated brings it down to ideal levels for fungal growth. With the addition of warmth to help the drying out, then very dangerous conditions indeed are created and these are not always appreciated by the occupants after such calamities have occurred.

The most likely places in a dwelling where dry rot is a possibility if warm, damp and still conditions prevail can be summarised as follows:

*   below parapet or valley gutters in pitched roofs that are poorly ventilated; boarded roofs with closed eaves are more prone to be affected from slow leaks left unattended and which may not give away their presence on ceilings below;
*   where timbers are built in to brickwork and subsequently concealed either as plates, floor joists, lintels, fixing blocks or bonding timbers as part of the construction of walls – practices universally adopted in domestic construction from the seventeenth century to the early twentieth century;
*   in lined wall construction – for example, in panelling or matchboarding to walls and, in particular, around window and door openings where there are shutter boxes, panelled heads and aprons; also where the interiors of external walls are battened out and covered with lath and plaster;
*   in basements or floor construction at ground level where there is inadequate damp-proofing and a lack of ventilation;
*   in floors at upper-storey levels and below flat roofs where there have been leaks and there is an absence of ventilation;
*   in wood floors generally if there have been plumbing leaks.

It is often the case that in the early stages of an outbreak of dry rot there will be either no indications or very few signs of anything amiss. In cases where a dwelling has been neglected, the surveyor needs to pay the closest attention to the likely places set out above, most particularly where there are signs of leaks or overflows, past or present, in rainwater or waste pipes externally (**425**) and, internally, around bathroom and kitchen fittings. These latter positions also require close attention in dwellings that have apparently been well looked after. Intermittent or slow leaks from plumbing can remain unnoticed for sufficient time to allow attacks to develop, more rapidly if fittings are boxed in. The surveyor's moisture meter will be calibrated to show moisture content in plaster and wood in excess of safe levels and should be used freely, but judiciously, in the likely places where damp staining will not necessarily be unmistakably visible.

On painted wood panelling, the only visible evidence of something amiss may be slight opening of the joints and slight cracking to the paintwork as the wood behind initially starts to shrink (**426**). On skirtings and other features fixed to walls that are damp, the meter could indicate levels of moisture content in the wood beneath the paint sufficient to support growth. At a later stage, painted wood surfaces might warp or appear wavy before bigger cracks start to appear, while unpainted wood may darken in colour.

It is only later that the surface of the wood will provide evidence of characteristic cuboid cross-cracking as the fungal growth from behind consumes all the goodness in the wood, causing it to completely dry out and shrink in all directions. Just occasionally, a pancake-shaped, soft, fleshy fruiting body will grow out from behind an architrave or skirting. This will normally have whitish margins to the top surface and, on the underside, a ridged and furrowed spore-bearing surface, rusty red in colour (**427**). If the architrave or other feature is given a thump, the reddish spores might drop out from behind. Occasionally, the spores can form a brick-red dust on surfaces, betraying the presence of a fruiting body in some concealed location.

If the affected feature is removed or falls off or if a floor can be seen from below then strands coloured white to grey, sometimes up to the thickness of a pencil, may be seen. It is these strands or strings, known as mycelium, which can travel through brickwork and other material for some considerable distance to seek out more timber for attack. In the case of dry rot, the mycelium can transport a supply of moisture sufficient to encourage further growth in adjoining timber; this means that newly affected areas need not be unduly damp, unlike the timber where the outbreak originated. In damp and dark places such as cellars, soft, white cushion-like patches – reminiscent of childhood candyfloss – may appear, sometimes also lemon yellow coloured and with tinges of lilac (**428**). It is at this later stage that a mushroom-like smell may become apparent when cellar or cupboard doors are opened after having been left shut for a while.

There is no way of telling how far the dry rot fungus has spread in a dark, damp, unventilated habitat, and the surveyor can only report what has been seen and that full opening up will be required to ascertain the extent before an estimate can be obtained for eradication along with the necessary renewal of timber and plasterwork and any improvements necessary to prevent it happening again. The estimate should be from a well-known, experienced firm that is a member of the Property Care Association, which will provide

425 The need to repair this defect to the rainwater drainage arrangements has obviously been ignored for a considerable period. Not only is the balcony becoming dangerous but the extent of staining beneath the balcony also suggests that the wall may have become saturated, resulting in a serious risk that the adjacent floor timbers and the lintels of windows below will be affected by dry rot. The strongest warnings must be sounded by the surveyor and exploratory opening up to check the condition of vulnerable areas should be strongly recommended.

426 With the development of a dry rot attack, splitting across the grain begins to appear as the wood shrinks from behind with the loss of moisture. The attack will be extending into the adjacent floor at this stage, but crafty placing of furniture can sometimes effectively hide what is happening. In this case, an attempt to conceal the effects of the damp and this attack of dry rot would be difficult to achieve and it could be that the carpet is effectively spanning a hole in the floor below. It would be inadvisable to tread too heavily in this area!

427 A particularly dramatic example of dry rot with fruiting bodies spreading from the far side of the staircase and every surrounding surface coated with the distinctive fine, brick-red dust of the spores.

428 *(right)* The white, cushion-like, fluffy growths which the dry rot fungus can produce in dark, damp and unventilated areas, such as cellars and vaults. Apart from the tinges of colour – here, lemon yellow and grey although lilac can also be featured – such growths can be likened to the candyfloss of childhood memory. Droplets of moisture are sometimes present at this stage – and might just be made out in the centre of the photograph; hence the Latin name for the fungus, *Serpula lacrymans* – *lacrymans* meaning weeping or crying.

a guarantee and operate under a Code of Practice. This should provide for the correct identification of the fungal growth and whether it is active or not as well as advice on appropriate treatment.

Dry rot will not continue to thrive in the total absence of moisture so eliminating dampness is a primary objective in treatment. Upon a reappearance of dampness, however, any areas of dried out growth will reawaken from their dormant state and happily resume their destructive activity; so a second objective of treatment is to completely remove all traces of fungal growth. Unfortunately, the leading edge of the spreading growth is microscopic and can extend up to a metre in any direction from areas of visible growth, so the traditional treatment regime involves removing all timber and chemically treating all areas extending 1m beyond the last visible signs of an outbreak. It will be appreciated that even a small outbreak can involve significant disruption and expenditure and dealing with a large and well-established outbreak can involve major structural works and substantial expense. With this in mind, some specialists will now offer somewhat less drastic remedial treatments, while still guaranteeing the outcome as long as they are confident that any dampness issues can be effectively dealt with. The sceptical surveyor will always seek to ensure that the client only deals with an organisation of the highest repute but this applies particularly when dealing with dry rot.

Wet rot

Much visible evidence of wood rot will no doubt be seen during the course of the surveyor's work, but there are many different types of fungi which attack wood. Most of this will not be caused by the dry rot fungus *Serpula lacrymans*, but by others commonly known as 'wet rot' fungi. These are not so virulent, quite so damaging or so expensive to eradicate since, while the primary measures of eradication are the same, the secondary measures are not necessarily so comprehensive.

The term 'wet rot' is a general term for different types of fungus that prefer more moist conditions than dry rot. Probably the most prevalent of the wet rot fungi is *Conifora puteana*, known as 'cellar fungus', which is usually found in very damp situations where there has been a leakage of water and, as its name implies, in vaults and cellars. Darkening of the wood occurs and longitudinal cracking develops. Cracking across the grain is less frequent but it is common enough for it to cause confusion with dry rot where cuboidal cracking is often mentioned as a distinguishing characteristic; but it should not be regarded as conclusive proof of identification (**430**). Deep brown or black, thin strings distinguish *Conifora puteana* from dry rot, the strings of which are white or grey and normally much thicker. Even more helpfully, dry rot mycelia are brittle when dry but those of *Conifora puteana* remain flexible. The underside of the sheet-like fruiting bodies of *Conifora puteana* are greenish to olive brown as against the russet red of dry rot. Other wet rots have different types of fruiting bodies: some have yellow- to amber-coloured strings; most require a higher

430 Darkening of wood and longitudinal cracking are present in this window frame, badly affected by wet rot and in the course of being removed. Some cracking across the grain is also apparent, and this sometimes causes confusion with dry rot, so it should not be regarded as conclusive identification.

429 Softwood windows and doors, and their frames, are prone to attacks of wet rot if the timber contains a lot of sapwood and the paintwork is neglected, allowing damp to penetrate the slightest opening of the joints. Continual overflows from a hopper head have allowed damp to penetrate to the walls and the woodwork in this area so that total renewal needs to be contemplated.

degree of moisture than that required for dry rot but can flourish in the open air and in well-ventilated areas; and one, *Lentinus lepideus*, sometimes found in damp skirting boards, does not produce strings, has fruiting bodies which can be cylindrical in shape, and gives off a sweet, aromatic smell from both the fungus and the wood affected.

Wet rots had a field day in the 1960s and 1970s when economic necessity required the inclusion of much sapwood in the joinery of windows and doors. The building up of sills and other components from smaller sections allowed damp to penetrate on the slightest of differential movement and, consequently, rot to develop. Much renewal had to be carried out after as little as 4 to 6 years from the date of construction – a circumstance which led directly to the creation of the replacement window industry and, eventually, to the double glazing and conservatory specialists of today.

The elimination of moisture will effectively end an outbreak of wet rot, although badly affected timber will obviously need to be replaced (**360** in Chapter 17); but the remedial measures for treatment are nothing like as extensive as with dry rot. Depending on the circumstances, and the risk that dry rot may be present elsewhere but concealed, it may be prudent to recommend an inspection by a timber treatment specialist. But at times this may not be needed if the cause and full extent of the problem and the necessary remedy are clear.

There is much helpful information available online from numerous specialist sources, including many illustrations of insects and rots, and these can be used to assist familiarisation and identification. BRE publish the useful booklet *Recognising Wood Rot and Insect Damage in Buildings*, and every surveyor inspecting dwellings would benefit from owning a copy. The essential requirement for a surveyor, however, is to be aware of where the dry rot fungus is likely to attack timber and to actively seek it out if the conditions suggest the possibility, committing to memory the requirements for its development and the visible indications of its activity. It is also important to realise that because dampness encourages attacks by both insect and rot, they are both often found together. The softening of the wood by fungal growth encourages the insect attack by making it easier to digest and, between them, they can lead to its total disintegration.

22

Internal joinery

Staircases

The structural condition of the staircase in a dwelling is more often bound up with the condition of its support at the lowest level from the floor and the further support, if necessary, from the strings fixed to partitions. Furthermore, the satisfactory trimming to the openings in the floors above at landing level plays an important role in maintaining stability. Failures in these components will disrupt the staircase, opening up the joints, loosening carriages and treads from risers, and disturbing the support to handrails and balustrades. The staircase is an area of hazard in any dwelling. BRE estimate that there are roughly a quarter of a million accidents a year on stairs and these result, consistently, in around 600 deaths. Even if the structure of a staircase itself and its supports are sound, the original design may leave a lot to be desired from the aspect of safety, and the inspector needs to draw out particular failings in this respect in the report. Approved Document K of the Building Regulations for England and Wales lists the standard required for new dwellings as distinct from institutional, assembly or other types of buildings south of the border. It is to these requirements that the staircases in older buildings can be related and where many will be found to be well below current standards – one of the reasons why there are so many accidents.

Among the requirements are that the rise should be between 150mm and 220mm and the going, between 220mm and 300mm, but the pitch for the stairs in a dwelling should not exceed 42°. If the staircase has open risers – a feature popular in dwellings of the 1960s and 1970s and in some modern conversions – the treads should overlap each other by 16mm and the open space between the treads should not allow a sphere of 100mm diameter to pass through. Small children can fall through open space in excess of 100mm and this amount is also the maximum for the space between balusters on the staircase and the balusters forming part of any open balustrade around landings and stairwells. Gaps larger than this are frequently found on staircases and balustrades, commonly in dwellings built during the 1960s and 1970s (**432** and

433) but also often following 'improvements' by DIY enthusiasts, so this is always worth checking.

Handrails are a necessary requirement for all staircases, only on one side if the width is less than 1m but on both sides if in excess of this amount. The height of the handrail should be between 900mm and 1,000mm. Balustrades at the sides to enclose staircases in the absence of a wall are needed not only to provide a handrail of the appropriate height but also to provide sufficient strength for the purpose. This, obviously, also applies to the balustrades around open stairwells so that they do not collapse in the spectacular fashion seen in so many classic Western films. In older dwellings, they are often much lower in height than current requirements permit, constituting a hazard which should be mentioned in the report.

A headroom of no less than 2m is considered necessary for the stairs between floors, although on the top flight leading to an attic, a minimum height measured at the centre of the stairs of 1.9m will suffice, and this can even be reduced to 1.8m at the sides for an arched bulkhead – a concession useful in the case of loft conversions where space is likely to be restricted.

The layout on plan of staircases is also an important factor in reducing the risk of accidents. Long flights should be broken up into sections. Landings

431 The underside of the landing to a staircase where the ceiling has been removed to provide a view of the substantial joists and a glimpse of the landing floorboards (between the joists near the top of the image). Beyond and to the lower right is the exposed underside of the staircase. The carriage is clearly visible, as is the base of the top landing newel post. The curving dado rail on the wall beyond (lower left) follows the line of the staircase curving downwards to the left (out of shot).

432 The common part staircase in a 1960s block where gaps in the handrail will allow the passage of a 100mm sphere and, therefore, not complying with current standards. Improvement of an existing fitting is not essential but the risks, especially to children and the elderly, should be pointed out by the surveyor.

should be provided at the top and bottom of every flight at least to the width and length of the narrowest width of the flight, although this may include part of the floor. Doors to rooms at the top of a flight should not open directly onto the staircase as this can be a particularly dangerous feature. Tapered treads are also a hazard and should have a minimum width of 50mm at the narrow end (**434** and **435**). Their safety can be improved in use when well lit, if not naturally then by good artificial lighting.

Any dwelling constructed since 1990 should comply with these standards, but in older properties – especially in period properties and some of the smaller dwellings at the lower end of the price range built before 1900 – the staircase will often be found to be so steep as to be a hazard to the elderly and the very young. There is no legal requirement for owners to improve such arrangements in the properties they purchase; in some cases, it would be totally impractical do so, and it may even be illegal in listed buildings. It falls to the surveyor to report any shortcomings in comparison with current standards and then to highlight the associated risks. Altering the width of tapered steps or reducing the angle of a staircase may not be realistic options, but an additional handrail attached to a wall might make a steep staircase safer to use. While the danger of a child falling through a large gap in a landing balustrade might seem obvious

433 Balustrade fittings with broad parallel rails were popular in 1960s houses but, as safety features, they are sadly lacking. Not only are there large gaps for children to crawl and fall through but the parallel rails create a welcoming ladder for them to climb up. The gaps in the improvised timber grid fitted as a safety measure by this owner are larger than 100mm, but the result is far better than leaving the original untouched.

434 Winders on a staircase, particularly when they are as steep as here, present a hazard to children and the elderly – an aspect which should be mentioned in the report, strenuously if they are worn and uneven.

435 Spiral staircases can be attractive features, even if they are somewhat lacking in practicality, but this one meets the necessary standards for the maximum sizes of the openings and the minimum width at the narrow end of the tread.

to the surveyor, it still needs to be pointed out in the report so that such important safety considerations are not overlooked by the purchaser and appropriate measures can be taken to prevent an accident.

Dwellings with more than one upper floor, including properties where a loft conversion has been added by converting an existing roof space, should have a fire-protected staircase leading to a final exit as a means of escape in case of fire. Approved Document B of the Building Regulations sets out the relevant requirements. The surveyor should report when a property doesn't meet the necessary standards and in addition to reporting any significant hazards in this part of the report they should be repeated in the section summarising risks, if one is included separately within the report format.

Doors, skirting boards and other joinery

It is good practice for the surveyor to close the door of each room on entering it during the course of the internal inspection. Doing so allows the simple functioning of the door to be checked and immediately reveals any which will not close or will do so only with difficulty. In this way, the handle, door latch mechanism and hinges are all tested in normal use. Once closed, the relationship of the door to its frame should be examined to see whether they are square with each other; if not, the degree and direction of any differential movement which has taken place around the door opening can be assessed from the gaps, cutting or filling which are evident (**436**). This simple procedure not only ensures that every door is routinely examined during the course of the inspection, but it can be very revealing about wider patterns of internal movement within the property. Checking the individual doors also indicates their quality and whether or not they are original or later replacements.

During the 1960s, it was fashionable for internal doors to be glazed but it was not a requirement to use safety glass (**437**). Safety glass is now required in such locations but older buildings may well have the original glass still fitted and the surveyor should recommend replacement in the interests of safety, especially when properties are likely to be used by children or the elderly.

As moisture meter checks for dampness are carried out around the property, the skirting boards should also be checked for signs of rot and deterioration due to dampness. As with the doors, a note should be made as to whether or not the

436 The wedge-shaped gap between the top of this door and the door frame above is wider over the handle side on the left than over the hinge side on the right. From this, it can be concluded that the slight distortion of the door frame is due to downward movement of the wall facing the camera in relation to the wall on the left of the picture. A hairline crack in the plasterwork above the door, just visible, indicates that movement is continuing slightly. Routinely checking every door means that all such minor observations contribute towards developing a complete picture of the building.

437 Glazed doors were popular during the 1960s but safety glass was not used. The surveyor should point out the risk they pose unless there are markings on the glass indicating that safety glass has replaced the original glazing.

skirting boards are original. If the skirtings are original then they will provide a good indication about the quality of the original construction. Skirting board joinery is a detail which would not have been skimped on a high-quality dwelling and, equally, it is somewhere money would not be wasted if economy was required. If skirtings are not original, it begs the question: why not? Replacing doors is often a matter of fashion but skirtings are very rarely removed unless the originals have deteriorated or been damaged for some reason. Replacement at ground-floor level suggests past deterioration, possible damp treatment work or even replacement of an original floor, all of which require further investigation or enquiries.

Other joinery features may be present depending on the age of the property, such as window shutters, picture rails and dado rails. Built-in cupboards, such as meter cupboards, should be inspected and, again depending on age, the surveyor should be aware that there may be asbestos-based materials in some places, such as in cupboard linings or on older fire doors. Unless particular issues are found, it may not be necessary to do more than make a general comment describing the nature and overall condition of the joinery.

Kitchen fittings

There was a time when kitchens were fitted to only a very basic standard and little needed to be said about the units; but with some fully fitted kitchens now costing tens of thousands of pounds to install, purchasers rightly expect to be given some indication as to whether the actual standard of fittings is what it superficially purports to be. The inspection of the kitchen should include consideration of the built-in cupboards, storage units and worktops but not an examination of any of the built-in appliances such as hobs, ovens, grills, refrigerators, dishwashers or laundry appliances.

Most surveyors will not be able to make fine judgements between different top of the market ranges, but by checking inside cupboards, opening and closing drawers and looking at the quality of the fittings, joints and finishes, it will not normally take very long to decide the relative standard of the units, whether they have been fitted well or badly and how well they have weathered the day-to-day use to which they have been exposed since being installed.

Any significant defects in hinges, drawer runners or damage to finishes should be noted so that appropriate comments can be made to the client. Depending on the level of inspection and report, this may range from a very general observation about the overall standard of the fittings and their condition to detailed reporting about individual defects (**438**).

In addition to checking the fittings themselves, the backs of cupboards at lower levels and areas in proximity to plumbing should also be checked as far as is practicable for evidence of dampness which might have affected the fittings and that might provide clues to possible problems elsewhere.

438 While there are specialists available to repair kitchen worktops, damage such as this caused by a hot pan is usually regarded as unrepairable and the remedy is to replace the whole section. The person who caused it may be covered by accidental damage insurance but a new owner will not have that benefit. Depending on the complexity of the remedial work, the cost could be significant, so it is the sort of defect which should be reported by the surveyor.

23

Fireplaces and chimney breasts

Fireplaces

Until about the 1940s, fireplaces were provided as a matter of course to practically every room in a dwelling since the burning of wood or coal had been the most common form of domestic heating until then. This was so even though other forms of heating had been developed from the latter part of the nineteenth century onwards. Gas and electric fires were available from the turn of the century and came into more widespread use after the First World War. Central heating using solid fuel with the boiler in the basement and heavy cast iron wide-bore pipes and radiators was viable for public and commercial buildings but remained expensive and inefficient for the average two-storey dwelling. This situation lasted until the 1960s when the development of the electric pump, time controls, small-bore copper tubing and steel radiators enabled cheaply available gas to be used cleanly and efficiently to provide heating and a supply of constant hot water to the average flat as well as house.

Fireplaces for bedrooms had already started to be left out of designers' plans, but from the 1960s, they began to be left out elsewhere as well. However, it wasn't very long before their absence as the focal point to the main room began to be felt, and in many new homes, the provision of at least one fireplace resumed after the 1980s. However, while one fireplace is considered desirable, that still leaves others in older dwellings – especially those in bedrooms – which most occupants regard as redundant and, therefore, candidates for blocking up or removal. The first alternative is fine and causes no problem provided that the flue is swept first, a vent is left in the panel used to block the opening to allow ventilation through the disused flue, and the top of the flue is suitably capped to prevent damp penetration while allowing ventilation to take place. Without this being maintained, there is a danger of staining along the line of the flue from condensation.

The surveyor will obviously need to note during the internal inspection what is present in the way of fireplaces, mantelpieces and hearths. Some will not have been blocked up and the fire parts may still remain. Their type and condition need to be noted. While fireplaces were being used as originally intended or as a simple framework for decorative effect 'fires', there was no real reason for the surveyor to trouble too greatly about the technicalities of the hearth and flue details, other than to recommend that flues should be checked prior to use (**440**). In recent years, however, there has been an increasing interest in reusing old fireplaces as a focal point in the main reception room and, particularly, in installing wood-burning room heaters. Consequently homeowners may have a greater expectation that more specific advice will be given about the suitability of existing fireplaces for modification. The current requirements for England are set out in Approved Document J of the Building Regulations.

In every case where potentially usable flues are found, the client should be recommended to have the flue checked and possibly swept by a 'competent person' if this has not been done during the preceding 12 months. Registered members of HETAS (Heating Equipment and Testing Scheme) are 'approved persons' for biomass and solid fuel domestic heating installations and the HETAS website (www.hetas.co.uk) lists a number of trade bodies whose members sweep chimneys. If a recently fitted appliance is found when inspecting an older property then it is important to ensure that it has been installed in accordance with the requirements for reusing flues as set out in Approved Document J of the Building Regulations. This is likely to have required relining of the flue, which is 'notifiable work' under the Building Regulations. As 'approved persons', HETAS engineers can undertake and self-approve such work in the same way that Gas Safe engineers do for gas appliances and FENSA members do for glazing. It is not easy to determine by inspection whether or how a flue has been relined though there may sometimes be a modern terminal visible externally. After such installation work, however, there should be a 'notice plate' – typically a plastic-covered card – left at the property.

439 An elegant marble fireplace surround in classic style with Ionic columns and moulded frieze enclosing a cast iron register grate with Dutch-style tiled hearth and oak curb. Possibly earlier than the mid 1800s house in which it is installed and still providing a fine focal point to a living room.

440 All original fireplace fittings have been completely removed in this 1930s reception room, leaving the opening very much as the original bricklayer did. It is being used with a basic firedog – a thermally inefficient arrangement more suitable for decoration than as a room warmer. The client should be advised to have the flue checked before using it. Blocking the fireplace would only necessitate provision of a ventilator and an external cap to the flue, but if installation of a new working fireplace or a wood burner should be proposed then specialist advice should be sought from an appropriate 'approved person', such as a HETAS member.

The notice plate can be secured beside the appliance or next to the chimney or hearth or, alternatively, in an unobtrusive but obvious position in the building such as next to the electricity consumer unit or the water supply stopcock. The notice plate should detail information specified by the Building Regulations, typically the location and type of appliance, the type and size of the flue or the liner if it was relined, the types of appliance that can be used with the flue which has been installed and the date of the installation.

One particular hazard that has emerged relatively recently with the increased reuse of open fireplaces is where old back boiler installations are still present. In some houses, usually ones built after the Second World War before central heating became commonplace, a direct heating system was sometimes installed to heat hot water and one or two radiators from a back boiler behind the main reception room solid fuel open fire. Where these have been superseded by full heating, the original installation has not always been removed. If an open fire is lit against the old back boiler, any water remaining in the back boiler will be heated and if it is not properly vented then the resulting build-up in pressure can cause an explosion. Several injuries and one death have occurred in this way, and in 2009 a housing association was fined £50,000 following one incident. If an original open fire is found in a property of this age, it may be advisable to recommend a check to assess the safety of the installation.

Chimney breasts

The redundancy of the fireplaces frequently tempts owners to go to the trouble of removing chimney breasts to create additional space. The inevitable expense and disruption is deemed worthwhile by many, but to do it properly involves cutting away the brickwork and making good all the way to the top (**442**). If the whole flue and chimney stack above are completely removed then the work can be relatively straightforward but complications arise if part of a flue or chimney needs to be retained. This might occur where, say, the chimney breasts in rear rooms are being removed and those in front rooms are being retained but the front and rear flues combine within the roof space beneath a single stack. Similar complications arise when a chimney serves two properties on either side of a party wall and the owner of one property wants to remove their chimney breasts.

For many years, this sort of work was done without a great deal of thought; but experience has shown that the eccentric loadings of the retained

441 Flues within a party, separating or flank wall, soundly rendered and complementing the good standard of brickwork and pointing elsewhere in the roof space. There should be no fume hazard should the flue be utilised for a space heating appliance, but the surveyor should advise a thorough check in the interest of safety prior to installing one at any time.

442 Having seen that chimney breasts have been removed from the rear ground and first-floor rooms, the surveyor will be pleased to see, on entering the roof space, that the top part of the breast has also been removed, leaving no unsupported remainder above – by no means always the case. The external stack has also been demolished. The front fireplaces and breasts have been retained but the flues are ventilated so no further work seems to be required.

upper sections above the party wall can create significant problems, and in many cases, the improvised supports beneath retained upper sections have been completely inadequate. Many surveyors have encountered the residual upper section of a chimney breast within roof spaces 'supported' by pieces of wood or metal resting on the ceiling joists (**443**). Even trusty gallows brackets, long recommended for the task, are no longer considered suitable for supporting what is usually a massive piece of masonry, precariously poised above the bedrooms of the oblivious homeowner. The means of support now considered necessary in most cases is likely to be a steel joist spanning from two load-bearing walls, frequently at two opposite elevations of the building. This is not an undertaking to be contemplated by most DIY'ers. The practical implications, not to mention the likely cost of work on this scale, should be reflected in the recommendations in the report and in any allowances being made for the cost of remedial works. If the chimney flue is on a party wall, notices need to be given; and if work is not being carried out by an 'approved person' then the Building Inspector's approval should be obtained. Inadequately supported chimneys following removal of chimney breasts have featured in several cases – notably *Smith* v *Bush* 1990 and *Sneesby* v *Goldings* 1995, both of which related to mortgage valuations.

It should be immediately apparent to every surveyor entering a dwelling built before the 1960s whether there is something 'wrong' and that where a chimney breast or fireplace should be, there is nothing. A mental note should, therefore, be triggered to see whether sections of brickwork have been left and if they have, whether they have been properly supported. The roof space will no doubt provide further evidence of whether work has or has not been carried out. If a ground-floor fireplace and chimney breast have been removed and the first-floor ones retained, it will frequently be unclear how the necessary support to the upper part has been provided; so recommendations should be made to undertake investigations to see what, if any, additional support may be needed.

443 This is a novel and unduly complicated but still completely inadequate way to try to support to the retained upper section of a chimney breast after removal of the lower part from the rooms below; it certainly makes a change from the popular but equally unsatisfactory gallows brackets. Structural calculations are required to determine what support ought to be provided, but a steel joist spanning between two load-bearing walls is almost certain to be necessary.

444 Pipes such as this running through the roof space in a comparatively new dwelling should be checked out for their purpose and support and the soundness of any jointing and taping. Manufacturers' instructions are not always followed. Do they comply with any relevant Building Regulations or bylaws?

445 These pipes in an older dwelling could be the work of a DIY enthusiast and need checking out, especially as the dark staining probably indicates deterioration and it is not unknown for such staining to be caused by flue gases leaking from small holes in the lining material, with obvious dangers.

PART 4

The services

24

Service provision

Introduction – reporting on the services

Everyone understands that venerable historic houses, the Elizabethan and Georgian mansions which are the staple of National Trust visits, were built before the days of electric lighting, mains water and internal toilets, still less central heating. It is more of a surprise to realise that a significant proportion of the nation's domestic housing stock was also built without such services, and it wasn't even until the interwar housing boom of the 1920s and 1930s that electrical wiring was routinely installed in new housing. As services became more widely available and homeowners began to demand their installation, the existing housing stock has continually adapted to new demands. The same process continues in the twenty-first century with an ever wider range of expectations from homeowners, most of whom now consider a fast internet service, for example, to be essential.

All of these services now comprise an important part of the domestic environment for the owner so it is understandable that the prospective purchaser, when thinking about commissioning a survey, makes no automatic distinction between the fabric of the building and its services. For the property inspector, however, there is an important, if rather blurred, distinction between inspecting and reporting on the basic structure – the roof, walls, floors and windows of a dwelling – and what can be seen and said about its services. It is likely that, in their hearts, some inspectors would like to exclude any comment at all on the services but they recognise that their clients would consider this very remiss. At the same time, inspectors are not experts in the provision, installation and maintenance of services; any comments which they do make must either be of the most general nature or run the risk of being negligent if later found to be incorrect or inadequate in some respect. Finding a satisfactory means of resolving this tension between what the client would like and what the inspector can legitimately provide is a significant challenge to those undertaking residential surveys.

In many ways, the surveyor's role in reporting on services has been simplified over the last few years. The Guidance Notes supporting the RICS

Home Survey suite make it clear that services are only visually inspected in order to form an overall opinion on the type and age of each installation, its visible condition and whether further investigation is needed. It is stressed that no tests whatsoever are carried out for the Condition or HomeBuyer Reports; and for the Building Survey, only everyday operations such as turning on taps and flushing toilets are undertaken. The RICS Guidance Note *Surveys of Residential Property* (third edition, pp. 15–16) states that, irrespective of the level of service being offered:

> The surveyor does not perform or comment on design calculations, or test service installations or appliances in any way. . . . In all cases, the surveyor will advise the client that further tests and inspections will be required if the owner/occupier does not provide evidence of appropriate installation and/or maintenance, or the client requires assurance as to their condition, capability and safety.

For the first time, the requirements for surveyors' inspections of services have been formally defined and, as discussed in the first pages of this book, it is inevitable that these Guidance Notes will become the benchmark against which surveyors will be measured.

Following directly from this is the matter of how the surveyor should report on the services to the client. Once again, the introduction of the RICS Home Survey suite with its traffic light style of reporting has been of assistance because it has led to fundamental change in the approach to reporting, especially on gas and electricity installations. In the 2011 edition of *A Surveyor's Guide to RICS Home Surveys* and more fully in his excellent and comprehensive *Assessing Building Services*, published in 2012 also by RICS, Phil Parnham outlines an assessment strategy for building services which was somewhat controversial when first mooted but which has since become widely regarded as a wholly appropriate way of approaching this vexed topic. Describing gas, oil, solid fuel and electricity emotively but accurately as '*services that can kill*', Parnham acknowledges that surveyors don't have the specialist knowledge that is required to make sophisticated judgements so they need to combine caution with clarity. The suggested approach is simply for the surveyor to decide whether or not there is sufficient evidence to state categorically that a service installation is satisfactory and requires no action. Under this approach, a system can only be considered satisfactory when it appears to meet current standards, there are no indications of any problems or DIY alterations, and there is documentary evidence that the installation has been properly installed, maintained and/or recently tested. Consequently, when there is any uncertainty whatsoever or if works of any nature are identified as necessary, there can be no middle ground and the only safe course is for the surveyor to conclude that the system is or may be faulty until proved otherwise – and to report accordingly.

This black or white approach has been criticised as reducing the surveyor's role to that of box-ticking. Many experienced surveyors have been accustomed to reporting systems in terms such as: 'being in reasonable condition and requiring only minor works to be brought up to a satisfactory standard'. They might then specify individual matters requiring attention, such as improving earth bonding to pipework, updating the consumer unit or renewing an isolated section of older wiring, perhaps retained in an outhouse when the remainder of the installation was renewed, while pointing out that in all other respects the installation appears acceptable. At one level, this seems helpful to the client in that it quantifies relatively modest levels of work and expenditure, but it begs the question: Is there such a thing as a 'partly safe' electrical installation? Any system containing a single fault that might kill surely means that the whole installation should be regarded as potentially unsafe until the fault is remedied. And how confident can the surveyor be that other matters would not be revealed by a full electrical test?

For many, the clinching argument about the merit of this approach has been the comparison of privately occupied properties with rented ones where residential landlords now have legal obligations to provide current gas and electrical certificates at the start of new tenancies and to regularly service gas appliances – normally annually. Would the private purchaser of a dwelling expect to expose the lives of family members to greater risks than a landlord would if the same property was being let to tenants? Clearly not, the argument goes; so the default position with property purchases now ought to be that recent test certificates should be provided by the vendor or estate agent prior to marketing. If they are not available at the time of inspection then the surveyor should be recommending that the client obtain them. With this comparison in mind, the logic of a black or white assessment is clear.

Alterations to gas or electrical installations, which once might have been undertaken by a jobbing builder or DIY enthusiast, can now only be carried out legally by a 'competent person' who has been trained and approved by a related supervisory organisation. Approved individuals and organisations can self-certify that their work complies with Building Regulation requirements, so if work of dubious quality is encountered then it should be questioned. Electrical faults may kill but they also cause thousands of domestic fires each year with inevitable impact on the lives of those affected.

When considering who should carry out tests in the meantime, the introduction of the Competent Persons Scheme (CPS) has made life more complicated for surveyors because there are now as many as six organisations which can carry out electrical work and while the expression 'competent person' has a specific meaning under the Building Regulations, some services – for example, drainage – are not covered by the CPS. Rather than go into lengthy explanations in the report about the options and definitions, it is probably better to simply state that tests should be carried out by 'an appropriately qualified person' or 'an appropriately experienced person'. If the client wants further clarification, that can be given afterwards; and if the client requires assistance with appointing contractors and arranging tests then that can be offered as an additional service. There are benefits in surveyors cultivating an association with suitable local specialists so that recommendations can be made with confidence and clients are not left to flounder on their own.

So much then for the 'services that can kill', but what does this mean for the surveyor in respect of the potentially less dangerous parts of the services such as the basic plumbing system, the sanitary fittings or the drainage installation? Because the risks to life are not the same, there is more leeway available to the surveyor in these areas. In categorising them, there is the opportunity to discriminate between defects or shortcomings which are serious and warrant, to use the HomeBuyer traffic light terminology, a red Condition Rating 3 because they are 'serious and/or need to be repaired, replaced or investigated urgently' (*RICS HomeBuyer Report* Practice Note, p. 8) and other matters which are less significant but still require attention. These can be reported as defects that 'need repairing or replacing but are not considered to be either serious or urgent' or, if using the traffic light format, an amber Condition Rating 2 (*RICS HomeBuyer Report* Practice Note, p. 8). No such distinction is appropriate with the 'services that can kill'.

Finally, although the surveyor should resist the temptation to offer observations on subjects where the averagely competent surveyor is not expected to have an expert level of knowledge, the fact remains that residential inspections continually pose questions of those who undertake them and it is often a combination of small clues noted during the course of the inspection which add up to an important conclusion. In the matter of service installations, the surveyor must have a certain level of knowledge to be in a position to recognise these clues for what they are rather than proceeding in a state of blissful ignorance. In the context of the services, probably more than any other aspect of an inspection, it is important to be able to recognise the 'known unknowns' – that is, the ability to realise that something is not as it should be even if not understanding exactly what that something might be. Inevitably in a book of this type, it is only possible to skim the surface of this subject and point the surveyor in the right general direction. To fully describe the level of knowledge which the residential surveyor should aspire to in order to properly advise clients about services would require a full further volume. Fortunately, one such volume already exists – that mentioned previously, *Assessing Building Services* by Phil Parnham, which deals comprehensively yet readably with the whole topic. Every residential surveyor should have a copy on the bookshelf, having first read and absorbed the contents. This and the following two chapters dealing with the services highlight only briefly the main points that the surveyor should take into account in this area.

Electrical installation

Clearly a test is not necessary where an electrical installation patently requires renewal. Unlike the components making up the other service installations, some of which have a degree of permanence about them, those of an electrical installation have a distinctly limited life as standards are constantly raised. Dwellings of 100 years of age may already have had their installations renewed twice because the covering to the wires carrying the current can degrade with possibly dangerous results. The sight of ancient metal-clad light switches, two-pin lighting points and a few ancient power points with round pins during the surveyor's internal inspection will leave no doubts about the antiquity of the installation, and it will be no surprise to find that the incoming main, switchgear and fuses are equally ancient with, perhaps, a few examples of newer wiring tacked on (**446**). There is no point in advising a test since a completely new installation will be required and the surveyor's report should say so without equivocation, albeit with a preamble giving the reasons.

It is where fittings and wiring in PVC-covered cable are superficially in reasonable condition that the evidence is not so clear-cut. The Electrical Safety Council (www.esc.org.uk) recommends a Periodic Inspection Report (PIR) as the standard test which should be carried out at least every 10 years or on the change of occupancy. This is a much more thorough examination than the simple visual inspection sometimes offered at no cost by electricity suppliers. Ideally, vendors should offer their properties for sale with the benefit of a recent visual inspection as a minimum, but preferably with a recent PIR. Although the qualifier 'recent' has not been formally defined in this context, it would seem realistic to consider anything less than a year to be acceptable and more than that to be questionable. Any changes or damage to the installation after the date of the report will invalidate it and necessitate an update.

Since 1 January 2005, all electrical installation work, whether installing new or altering an existing installation, has come within the scope of the Building Regulations Part P for England and Wales, although this had been a requirement in Scotland since 1990. Until then, it was up to owners and contractors to carry out the work in any way they wished. Some owners 'know' about electricity, though most do not, but fortunately most contractors were aware of the IET Wiring Regulations, which are regularly updated. Until 2005, it was not mandatory to follow them so there has been plenty of opportunity over the years for divergence from the Regulations not only by the less responsible contractors but also by the DIY handyman.

Although there will be no attempt to carry out any test, the surveyor must still inspect what there is to be seen of the electrical installation and its condition and make notes of those observations. Even if it is rapidly concluded that all is not well and a test will be recommended, it is still necessary to continue to make and record observations throughout the inspection. Omitting to do so will inevitably, at some time or another, mean that some crucial point is overlooked and the final report will not properly reflect the visible condition of this important element of the property. The surveyor's notes should include such matters as the location of the meter and consumer unit/fuse board; whether a residual current device or miniature circuit breakers are present; the condition of the visible wiring and the type of cable to lighting pendants; whether supplementary bonding is present in the usual places; the visible condition of a

446 This is obviously not a modern installation. The differing ages of wiring connected, ranging from rubber encased to modern PVC, indicate that alterations and additions have been carried out in several stages so it is likely that complete rewiring will be required throughout the property, regardless of any modern outlets and fittings elsewhere.

447 It is increasingly less common to find properties with twin-core lighting cable, a wall-mounted two-bar electric fire (in working condition!) and, just visible on the skirting board near the corner, two round pin sockets to serve bedside or standard lamps. These all date from before the 1960s and any one of them would indicate an outdated electrical installation in need of replacement. Together they almost amount to a heritage piece.

sample of the power sockets, light switches and fittings; and the nature and condition of any fixed electrical appliances such as heaters, electric showers or towel rails. The notes should include external electrical installations to garages, sheds, and other features such as garden lighting or pond pumps. There is no need to count the numbers of sockets but it is helpful to gain an impression of whether there seems to be an adequate provision bearing in mind the size of the property. If the owner or occupier is present, questions should be asked about repairs, alterations or additions to the electrical system. If any have been carried out since 2005, a Part P certificate should have been issued on completion unless the works were only minor repairs or replacements, or very minor works such as adding a fused spur to an existing circuit.

One alteration that is increasingly being encountered is the presence of some form of microgeneration installation, usually solar photovoltaic (PV) panels on one or more of the roof slopes or, less frequently, a wind turbine. The visible condition of these installations and any implications for the part of the structure to which they are attached should be noted during the external inspection. There are comments regarding the external assessment of solar PV

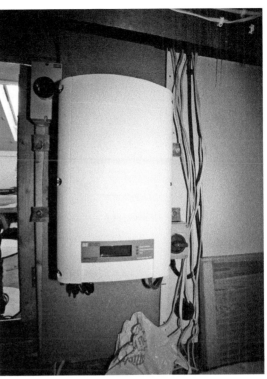

448 There is no apparent reason why this double 13 amp socket outlet has come away from the wall, but it represents a distinct danger and the surveyor must warn the owner or the estate agent as soon as possible of its presence as well as including it in the report. Unevenness to the skirting below looks as though it might have been caused by rot and should certainly be checked.

449 If solar PV panels have been fitted, there will also be an inverter unit within the property to convert the direct current coming from the panels to alternating current for the mains supply. The untidy wiring by this unit would usually be a warning sign about the quality of workmanship but this photograph was taken while the system was still being fitted.

and thermal panels on roof slopes in Chapter 4 in the section 'Solar generation installations'. During the internal inspection, the various components of PV electrical systems should be considered to see whether they appear to be adequately and neatly fixed or whether they have more of a DIY look about them (**449**). Not all installers provide Part P certificates but to benefit from payments under the government Feed-in Tariff (FIT) scheme, both system and installer must be registered under the Microgeneration Certification Scheme. If the installation was not provided under the terms of this scheme then the client should be warned that the financial benefits from the installation will be much lower without the FIT subsidy and the surveyor must recommend an assessment of the installation by an appropriately qualified person. In all cases, the client's legal adviser should be advised that a microgeneration installation is present so that appropriate checks can be made about the terms and conditions of any related contracts or leases (see Chapter 30).

At the end of the inspection, the surveyor must reach a conclusion about the electrical installation and make an appropriate recommendation in the report, supported by the notes made on site. If in any doubt, the surveyor must advise in the firmest of possible words that a new installation is required or that an inspection and test should be carried out, coupled with a recommendation that any part of the installation not in accordance with the Regulations should be brought up to standard. Any other advice is not only dangerously misleading but could rebound with serious consequences on the surveyor. A surveyor is just not qualified to assess the significance of what can be seen of the installation as far as safety is concerned.

Gas and oil installations

Mains gas is available for most properties in the UK although not everyone chooses to be connected, but some 3.6 million homes, mostly in rural areas, do not have access to a mains supply. One alternative fuel source is oil, but if residents in these locations want to use gas appliances then an alternative is liquid petroleum gas (LPG), which requires the provision of either a bulk tank or smaller separate cylinders on site. The Office of Fair Trading estimated in 2011 that about 150,000 homes use LPG as their main source of heating while 1.5 million use fuel oil.

With oil or LPG installations, the surveyor's responsibilities are to describe the installation and general condition along with evidence of certification of the installation or alteration and annual inspections. Where mains gas is connected, the location and condition of the meter and any visible pipework should be checked and any alterations noted. Gas meters should ideally be located externally and should not be in an area that is the sole means of escape, such an entrance hall, unless contained within a fire-resisting box or cupboard with self-closing doors. The gas meter should be at least 150mm away from an electricity meter but if it is nearer then there should be a fire-resisting partition between them (**450**). Gas pipes, especially if they are of lead, should be properly supported along their length to prevent undue strain on joints, which might lead to leaks.

450 These gas and electricity meters are an adequate distance from each other to avoid the need for a fire-resisting partition between them. Apart from a query about the curious switch apparently dangling down just right of centre, the electrical installation looks modern enough to tempt a surveyor into giving it a 'qualified' OK, except that there can be no such thing as a qualification with 'services that can kill'. Either they are demonstrably completely satisfactory or a test must be recommended.

With LPG installations, there are specific requirements about the locations and arrangements of the cylinders or bulk storage tanks. Most LPG suppliers have details of their requirements on their websites and Approved Document J of the Building Regulations sets out the minimum standards in England. These specify matters such as the minimum distances from boundaries and dwellings, whether or not firewalls are required to protect bulk tanks from a fire in the dwelling, and the requirements for ensuring bulk tanks can be safely resupplied by delivery vehicles. The requirements for oil storage are similarly outlined but with important differences, such as whether the tank is internally or externally bunded to prevent contamination of watercourses, drains or sources of drinking water in the event of spillage or leaks from the tank. Close proximity to any of these features may impose a higher standard on the installation so their relative location to the oil storage facility should be noted in order that compliance can be checked.

Water supply

A plentiful supply of clean, wholesome drinking water is a prerequisite for satisfactory household management and the water companies in urban locations provide this through their underground mains to which the dwelling's installation is connected by a stop valve. This is turned on and off by the company dependent on payment of its charges. Within the curtilage of a dwelling, the responsibility for the condition of the mains supply pipe is that of the owner.

In more remote rural areas, the surveyor will occasionally come across a property with a private supply (**451**). Legislation under the Private Water Supplies Regulations 2009 for England and the Private Water Supplies (Wales) Regulations 2010 increased the obligations on owners with a private supply in England and Wales. Where more than 50 people are served by a supply, there is a requirement for regular testing and monitoring by the local authority. Supplies to single domestic premises with fewer than 50 people just need to be 'risk assessed'. This can be done by the owner, but it is more likely to be effective if a suitably qualified person such as an Environmental Health Officer is requested to assist. The current owner should be able to advise where the private supply comes from, whether the supply has been risk assessed and/or tested and what the outcome was, and whether there is any water treatment equipment. Documentary evidence supporting the answers should be available and the client's legal advisers should check all matters, including the extent of the owner's rights and responsibilities relating to the supply. More detailed advice is available from local Environmental Health offices and the Private Water Supplies website (www.privatewatersupplies.gov.uk), where a lengthy technical manual is available, including guidance specific to the differing UK jurisdictions.

If a mains supply is installed, the surveyor should try to identify the location of the external stop valve and, as far as possible, determine whether there may be any sections of supply pipe which pass across other properties or whether there are other properties which rely on a supply which passes across the property that is the subject of the inspection. If either of these possibilities is suspected then this should be referred to the client's legal advisers for guidance. The company stop valve is usually at the location of a meter, if one is present.

The point where the supply enters the building should be located if possible, and hopefully there will be an internal stop valve at that point which will enable the domestic supply to be turned off whenever necessary. From that point, it is a question of seeing as much of the internal installation as possible during the course of the inspection and noting such matters as whether the pipes are run in dissimilar metals (risking electrolytic action) and whether lead pipework is present, the nature and condition of any water storage tanks, and whether tanks and pipes, including overflow pipes, in unheated areas are insulated. If lead pipework is found then the client should have the various disadvantages pointed out and, especially if in a soft water area, consideration should be given to replacing them. There is no obligation for owners to undertake replacement but the Water Supply (Water Fittings) Regulations 1999 control their repair and alteration.

Steel pipes are found less frequently nowadays. They were widely used up to the 1930s and briefly came back into popularity during the 1960s and 1970s when there was a shortage of copper pipes. They suffer from rusting and if water is left undisturbed in lengths of steel pipe for an extended period, it can emerge from the taps with a gruesome-looking rust-red discolouration until fresh water arrives. Rust staining beneath basin or bath taps has the same source. Unlike thicker lead pipes or bendy plastic ones, it can sometimes be difficult to tell whether painted pipes are made of copper or steel. The application of a small magnet will instantly provide the answer, attaching firmly to steel pipework and showing no interest in copper pipes.

To understand the plumbing arrangements at a property, it is helpful to determine whether the system is direct or indirect. The main difference is that direct systems draw cold water straight from the mains supply while indirect systems only draw cold water straight from the mains in the kitchen (and possibly the utility room and an outside tap) but the remaining cold taps are

451 In urban settings, features such as this hand pump are usually only decorative but if it is connected to a private supply, as may be the case in more remote locations, it should be ensured that there is complete compliance with the increasingly stringent regulations.

452 A badly rusted cold water storage cistern in need of urgent renewal before it bursts.

supplied from a large cold water storage tank, usually located in the roof. The easiest way to tell whether a cold tap has a direct supply is to turn it on while trying to stop the flow with a finger or thumb over the outlet. The mains pressure of a direct supply means it is not possible to stop the flow whereas the pressure of an indirect supply is limited because it depends on the location of the tank and, therefore, blocking the flow is easy. Experimenting with this simple but effective procedure will rapidly impress the need for caution to ensure water is not sprayed across the room at waist height, soaking the surveyor and any others in the vicinity. This, incidentally, is also useful when travelling as it is a means of determining whether water from taps in hotel bathrooms, for example, is direct from the mains and therefore suitable for drinking or indirect, from a storage tank, in which case drinking would be inadvisable in case it contains unsavoury matter, such as the dead pigeon encountered by the author at one property.

Indirect systems used to be the standard installation so, consequently, many still remain; but direct systems tend to be more straightforward and economical to install so they have become increasingly popular over the last couple of decades and are usually installed in new homes. Each system has its advantages but the main difference for homeowners is that direct systems provide powerful showers automatically because they utilise mains pressure; showers from indirect systems produce the feeble flow which so appals our international visitors. Indirect systems, meanwhile, have the advantage of an inbuilt reservoir available whenever the water supply is interrupted whereas direct systems lose their supply completely. For the surveyor, the most significant difference will be the presence of a large cold water storage tank, usually in the roof space, which will need to be checked. Irrespective of the cold water supply arrangement, there may be a smaller central heating header tank – because this is determined by the central heating system – but usually homes with a direct supply will have an unvented central heating system, which does not require a header tank.

Modern cold water storage tanks are usually made of plastic or PVC, but older asbestos cement and galvanised steel tanks can still be found in use. Galvanised tanks rust and will almost inevitably be in need of replacement, so care should be taken when checking them because the corroded metal can be very fragile and easily damaged (**452**). Surveyors' examinations have even been known to inadvertently cause galvanised tanks to start leaking, ironically before they have even had the opportunity to leave the roof space and report the risk that a leak might develop. Asbestos cement tanks pose no danger to health in respect of asbestos fibres in water, but the outer surface

453 The water pipes here are insulated against frost damage, but the water storage tank should be covered to prevent contamination and insulated. The need to insulate and provide adequate support for overflow pipes is often overlooked. Insulation to the roof space should also be upgraded.

of the tank may pose a safety hazard if it has deteriorated or if any maintenance is required. Disused asbestos tanks are often seen in roof spaces, and while they are undisturbed they pose little risk but if they should need to be moved – to facilitate a loft conversion, for example – then all appropriate precautions for removing asbestos will be necessary – along with the associated expense.

Cold water storage tanks impose a significant load so it should be ensured that their supports are adequate and properly arranged, depending on whether the roof is of traditional cut rafters or modern trussed rafters design. Tanks and all pipework should be properly insulated, except beneath the tank, and there should be tightly fitting lids (**453**). Each tank should have a separate overflow pipe, which should be insulated and supported along its whole length.

Surveyors should be aware of the degree of hardness of water in the area in which they practise. For clients moving from outside the area, this can be helpful information. Hard water areas suffer more from furring of kettles and hot water pipes, while lead pipework is more of a risk in soft water areas. In hard water areas, a water softener may be encountered during the inspection. There is no expectation that the surveyor should make any observations about the merits of such installations, but the client should be advised to ensure that it has been properly installed and, most importantly, that a drinking tap – usually serving the kitchen sink – has 'unsoftened' water for mainly health reasons.

Other services

As the reach and influence of information technology creeps relentlessly into every corner of our lives, many homes are being networked by their owners to provide wired or wireless connectivity to multiple devices. These are clearly beyond the scope of the survey, but a fast broadband service is considered a necessity by many people, especially the many who now work from home. There are some parts of the country, and not only in more remote rural areas, where the provision of broadband and even a mobile phone service falls far below what many now consider essential – to the extent where it can even have an adverse impact on a sale. If the surveyor knows that the subject property is affected in this way, it would be appropriate to report this, especially if the client is moving from outside the area and may not be aware of the issue. For some, it will not be significant; but for others, it could mean that the anticipated rural idyll is really a communications nightmare – and at such times, litigious former clients are liable to visit their nightmares on surveyors who they believe should have sounded suitable warnings. Difficulties regarding the availability of cable or satellite television services are rather less likely to leave a surveyor exposed in the same way, but many owners would still consider it a significant matter and if the surveyor is aware that properties in a particular locality experience such problems then it would be as well to report them.

Individual properties may have additional services such as a burglar alarm system or air conditioning units. The surveyor need only report the presence

454 Automatic electric gates are increasingly popular and the surveyor should recommend checks into their installation and servicing to ensure they pose no risk to users.

of these services, along with a recommendation for the client to make sure the installation is in working condition and to check the nature of any servicing arrangements. Particular attention should be paid to automatic electric gates (**454**). These are often found at the entrance to private estates or blocks of flats but are increasingly encountered at individual homes. A detailed assessment is beyond the scope of any normal survey, but following the tragic deaths of two children in separate incidents involving automatic gates in 2010, clients should be recommended to check that automatic gates were installed by members of the Automatic Entrance Systems Installers Federation (AESIF) and that the gates are inspected and serviced annually. Any plant and equipment to facilities, such as swimming pools, require specialist inspection and testing and should be excluded from the inspection. Ideally, they should have been identified during the initial conversation with the client and specifically excluded by the written terms and conditions.

If the surveyor is inspecting a flat, the common services are not inspected but a general comment should be made on any that are specific to that flat. This will include such matters as entry phone systems, lighting and heating in common areas, passenger and goods lifts, CCTV, refuse bins, chutes and stores. The presence of common facilities such as a swimming pool, fitness room or sauna should be reported with a general comment. They should not be inspected in detail but the client should be advised to check the extent of rights and responsibilities relating to such facilities.

25

Heating systems

Water heating

The design of any hot water supply system needs to provide, so far as is practicable, hot water at the locations, in the quantities and at the temperatures required by the user at the least overall cost taking into account installation, maintenance and fuel costs. There are many different systems, some deriving hot water from the space heating installation and others which are independent. As with the other components of the services, the surveyor's responsibility is to visually inspect and describe what is present, making relevant observations, but not to carry out any tests. If undertaking a Survey Level Three or Building Survey then, where appropriate, the surveyor will ask the owner or occupier to switch on the heating appliances or system – but only to check the basic operation and not to test efficiency or safety.

An early method of combining heating and hot water provision was to use direct hot water systems, usually gravity fed and supplied from solid fuel back boilers backed up by an electric immersion heater in a hot water cylinder. These are usually old, typically dating from the 1960s or 1970s, and are now out of date. If encountered, the client should be advised that the need for complete replacement should be anticipated.

The more common installations are indirect systems. Until relatively recently, vented systems were invariably installed. These are low-pressure systems which utilise a header tank, often in the roof space, to keep the system topped up and to provide an expansion capability. The requirements for the header tank, regarding insulation, covering and overflow provision, are similar to those for the larger cold water storage tanks. Unvented systems have become increasingly popular over the last few years. These operate at mains pressure and because they are closed systems and rely on an integral pressure relief system which is not visible, they are potentially more dangerous than vented systems. The requirements governing installation are more rigorous and it is particularly important to ensure that a service history is available (**455**).

Both forms of indirect water heating system have hot water cylinders. These should be as close as is practicable to the heating source and be well insulated, either with sprayed factory foam or an insulating jacket. The cylinder should be fitted with a thermostat to regulate the water temperature. Often, an immersion heater will be fitted to provide backup when the heating boiler is switched off (**456**). Vented systems should have an expansion pipe at the top, which is arranged to discharge over the header tank located somewhere above. It should be ensured that this does not discharge over the larger cold water storage tank because it would cause contamination by chemicals in the water from the heating system.

Independent water heating can be provided by gas water heaters. Multi-point installations provide instant hot water to a number of outlets and have balanced flues that discharge externally. These should be assessed in the same way as gas heating boilers (as outlined in the next section on space heating): in the absence of documentation confirming a recent service and safety check, a test must be recommended. Small flueless gas heaters are found much less frequently. Once popular, they are now rarely used because they discharge exhaust gases into the room where they are located; and although they can be maintained in a working condition, it is better that clients are advised to replace them.

Combination or 'combi' boilers have become popular with developers and as replacement installations but are really only suited for relatively small dwellings. These have no hot water storage capacity, which saves space within the property, and they provide instant hot water. Flow rates can be problematic if there are a number of hot water outlets that may be used at the same time or if there are relatively large distances between the boiler and the taps. Once again, these should be assessed and reported in the same way as gas heating boilers.

Hot water can be provided by instantaneous electric heaters, but usually these are relatively expensive to use and have a low flow rate so they are usually confined to locations such as cloakrooms. In locations without a mains gas supply or in electrically heated homes, electric storage water heating systems can be provided. These are usually vented although unvented systems are available. Off-peak systems are typically cheaper to run but require a larger water cylinder to provide a reservoir of hot water for use each day.

An increasing number of properties feature some form of solar water heating installation, the most common being from roof-mounted flat plate collectors or evacuated tubes. As described in the section 'Solar generation installations' at the end of Chapter 4, flat plate collectors are sufficiently similar in appearance to electricity-generating solar PV panels to mislead the unwary and careless, but an incorrect identification during a survey would be a very basic error not speaking well of a surveyor's capabilities. Evacuated tubes are generally thought to be the more efficient of the two types, although it is interesting that the results of a field trial published in 2011 by the Energy Saving Trust (*Here Comes the Sun: A Field Trial of Solar Water Heating Systems*) reported no significant difference between them. These installations are connected to the hot water cylinder and tend to produce modest savings on

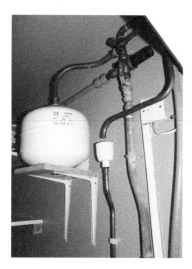

455 *(left)* The pressure vessel for an unvented system with copper, plastic and older galvanised steel pipes. Why were the older pipes not replaced when the newer system was installed, and is this installation being regularly serviced?

456 *(middle)* What is the surveyor to make of this? The features can be identified and reported: the hot water cylinder serves an indirect system; it is insulated and has a thermostat at the bottom and an immersion heater fitted at the top; there is a pump for a power shower (lower right); central heating pumps and switches are visible to the left. As for the quality: it looks rather untidy and seems to have been fitted piecemeal; pipes are in a variety of materials and are not fully supported; the electrical wiring has questionable aspects including a junction box. Tests should be recommended.

457 *(right)* A combined solar thermal and ground source heat pump system being fitted in a new-build bungalow. Surveyors should make a point of becoming more familiar with the different parts of such installations as they become more common.

the costs of hot water provision, but there can be great variations due to such matters as insulation levels and the pattern of use by the householder.

The surveyor cannot be expected to make detailed comments on solar heating installations but it is appropriate to have a basic understanding of the principles and to be on the lookout for obvious defects or leaks, as with any other part of the plumbing (**457**). As with PV panels, the external part of the installation should be inspected as far as possible for evidence of any problems (see Chapter 4) and the client's legal adviser should make appropriate checks (see Chapter 30).

Space heating

The RICS Guidance Note *Surveys of Residential Property* 2013 and the respective Practice Notes for each of the three options under the RICS Home Survey suite all deal with the heating component of the services section of the report with a couple of paragraphs of description and some bullet point lists. These set out the different levels of inspection and identify the relevant parts of the services installations to which they apply. They do not, nor indeed are they expected to, describe how the surveyor should set about inspecting the heating installation at a dwelling or what it is the surveyor should be looking out for, be wary of or be seriously concerned about. The reporting responsibilities are essentially restricted to providing a description of the heating installation and a statement specifying whether or not a specialist test is required. The Practice Notes give no indication of how this judgement should be made.

This is not a criticism of the Guidane and Practice Notes, which achieve their stated objective of defining the scope and extent of the surveyor's responsibilities in this area. As such, the Practice Notes can, as stated at the outset of this book, be expected to provide a benchmark against which residential surveyors and their performance are likely to be measured for the foreseeable future. Nevertheless, it does bring to mind the apocryphal hospital manual for brain surgeons which listed in great depth, over many pages, the administrative and clinical details for admitting and preparing patients for surgery and then for managing their post-operative care and recuperation, while the entire section dealing with the actual surgical procedure itself comprised the words: 'Undertake the brain surgery'. No doubt it was assumed that the surgeon, the expert, had a good grasp of the requirements. Unfortunately, the same cannot be said about surveyors inspecting heating services, not least because there is no expectation that they should have expertise in this area.

This part of the inspection does not primarily consider any of the main structural components, the bread and butter of the survey process; yet any significant defects in parts of the services, especially the heating system, ought perhaps to be reflected in a valuation of the property or the price being paid since, if not dealt with or at least identified prior to the sale, these might have a major impact once the client takes occupation. It is unlikely that failures by the surveyor to report such matters will be readily overlooked by an aggrieved client; and because this is an area where the surveyor does not have expertise, it poses one of the greatest levels of professional risk to the surveyor of any part of the survey process.

How, then, should the surveyor approach this part of the inspection? The first and, indeed, the obvious thing to do is to identify the fuel source for the heating. Not so long ago, this information alone would have been almost enough to deal with the whole of this section because the selection from gas, oil or electricity almost defined what the rest of the system would be like. Sadly, for surveyors at least, such days are a rapidly receding memory and there is now not only a multiplicity of systems available for each of these fuels but also an increasing availability of alternative sustainable fuels, such as solid fuel pellets, and air or ground source heat pumps, which might be encountered (**458**). At the same time, boilers and control systems have become increasingly sophisticated and will continue to do so in response to the demand for progressively more efficient use of heating fuel.

Clients may occasionally enquire about the feasibility of installing 'sustainable' alternatives to conventional systems but advice about the relative merits of different heating systems is beyond the scope of normal survey services. As an aside, however, it is much more energy efficient and environmentally sustainable to improve the insulation of conventional housing than to expend funds on installing any form of capital-intensive system to provide 'free' energy. A highly insulated dwelling, with appropriate measures in place to maintain good-quality air for a healthy and condensation-free environment, will cost relatively little to heat and have a very low carbon footprint; so this 'fabric first' approach should be the one to adopt. Heating a standard unimproved dwelling is very much like trying to inflate a leaking bicycle tyre; no matter how little the energy costs, much of it will just go to waste. A concluding observation on this theme is that the most cost-effective fuel is still mains gas and if that is available then no others need be considered.

Having identified the heating fuel, the surveyor should attempt to determine the basic form of the system. The most fundamental division is between wet central heating systems and other types. Wet systems include the old type of vented, gravity-fed system with a few radiators, often run from a back boiler. These are no longer common and are unlikely to be suitable for retention. More likely to be present are open-vented, fully pumped systems, which have a header tank and a hot water cylinder while sealed, fully pumped systems have a hot water cylinder but no header tank. Combination systems have neither. The final wet system, which is a thermal storage system, is also a sealed system but uses a very large water cylinder as a central reservoir for both the water and heating circuits. In principle, any fuel can be used in any of these systems although some fuels are more closely associated with some systems than others. Systems which do not use water to distribute heat around the system include those using warm air, which is ducted around the dwelling, and independent space heating methods that are separate from the water heating system, such as electric storage heaters or individual gas or electric room heaters.

Once the basic system has been identified, the various components will be seen around the property and should be considered as the inspection progresses. With all wet systems, there should always be a lookout for any leaks that may be occurring, especially at joints, because leakage can lead to rot in adjoining timbers (**459**). Underfloor heating has increased in popularity now that extended runs can be laid with plastic pipework without the risk of leakage from joints, which bedevilled underfloor heating with copper pipework. But laying it properly is a skilled job. Substandard workmanship may not be readily apparent but the surveyor should be alert for anything untoward, including evidence of dampness in unexpected locations that might originate from concealed underfloor pipework.

When it comes to gas, oil or solid fuel heating appliances, there are two important and related matters that should be considered. The first is to think about whether there is adequate provision for combustion ventilation and the other is to check the provision for combustion exhaust. If the appliance is room sealed, it does not rely on ventilation from within the room for combustion but uses exterior ventilation through a balanced flue. There may still, however, be a requirement to provide ventilation around a boiler if it has been enclosed in a small space such as a cupboard in order to maintain its operating temperature at a safe level; this requirement will be determined by the individual boiler manufacturer.

Balanced flues are another area where developing technologies have made the surveyor's life more complicated than it used to be. When there was only one type, the large, rectangular, natural draught balanced flue, it was possible to have a few rules of thumb about the necessary distances the flue should be from nearby windows or gutters; but the introduction of fanned balanced flues and condensing boilers has reduced many of these distances, sometimes more than halving them (**461**). To complicate matters further, it is acceptable for manufacturers' recommendations to overrule measurements given in Approved Document J of the Building Regulations. Matters that are easier for the surveyor to check are the prohibition on flues discharging into enclosed areas, including covered passageways between dwellings (**462**); the need for protective wire terminal guards on any flue which is less than 2m above a pavement, ground level or balcony; and whether the flue is likely to be obstructed by plant growth or whether it has been damaged or vandalised.

The heating system controls should also be examined as the inspection progresses. These are an important part of ensuring the system operates at the most efficient level, and as energy costs continue to increase, clients will usually welcome advice on potential improvements. Systems should have a central programmer to control the operation of the boiler and there should be a thermostat on any hot water cylinder. Ideally, radiator systems should have individual thermostatic radiator valves (TRVs) but often there will only be programmable room thermostats or a central room thermostat, frequently in the hall. These are less flexible than TRVs but at least they are better than nothing. Larger properties, those above 150m^2, are now required to have

458 An integral solar thermal panel on the roof and a ground source heat pump (centre) are intended to provide hot water and central heating for this new bungalow.

459 The blue/green discoloration indicates leakage from the joints at the end of this radiator. If not repaired, such leaks can lead to rot in adjacent timbers.

multiple timing zones so each zone can be controlled independently. If a system does not have the minimum of heating controls then its efficiency will not meet modern standards; this also suggests the system is either of some age – and probably has scope for upgrading – or that it was not installed to the appropriate standards required by current regulations.

460 A modern radiator connected, in a roundabout way, to older galvanised steel pipes. These pipes will be prone to rusting internally, which will reduce their efficiency. Eventually they will need to be replaced, but an external inspection is unlikely to provide an indication of the period before this will become essential. The surveyor can only report that the installation is, at the very least, less satisfactory than would have been the case if the pipes had been renewed. Seeking the advice of an appropriately qualified person should be recommended.

461 An older natural draught balanced flue on the left beside a more modern fanned balanced flue.

462 A potentially lethal situation where the construction of a rear porch completely enclosed the flue serving a wall-mounted gas boiler located in the adjacent kitchen. The circular vent in the porch roof is a wholly inadequate gesture at providing an outlet for exhaust fumes. The installation was immediately shut down by the Gas Board until it could be operated safely.

This very brief summary has tried to highlight a few of the basic areas a surveyor should be considering in respect of the space heating provision which may be present at a property but has, admittedly, imparted relatively little in the way of detail to assist this process. The numerous systems now available and the varying standards that apply to them mean that real expertise is only available from fully qualified practitioners in each of the specific areas. There is still a requirement, however, for surveyors to be able to demonstrate a broad level of insight and understanding about this important subject. This can only be gained from a personal commitment to seeking a degree of knowledge which is beyond the scope of this book, whether from professional seminars, from reputable internet sites or from literature.

This begs the question: at what level should this knowledge be pitched? The ultimate answer to this question is that level which the courts would expect of the reasonable surveyor. At present, this could arguably be defined as that set by the level of knowledge outlined in the book mentioned previously, Phil Parnham's *Assessing Building Services*. Some might feel that this would be setting the knowledge threshold rather too high for the 'averagely competent surveyor' but at least surveyors can be reassured that the threshold is unlikely to be set any higher, certainly for the foreseeable future. In a professional area where reassurance is a rare commodity, every opportunity should be grasped with both hands wherever it is available, and Parnham's excellent book is strongly recommended for those who are serious about bridging their personal knowledge gap in this increasingly complex but crucially important area.

26

Sanitary fittings and drainage

Sanitary fittings

Having considered how water – whether cold, hot or both – has been provided and delivered where required, thought now needs to be given to the fitments where it is used. A surprising range of sanitary fitments will be encountered, some of them conceivably the same age as older dwellings themselves. However, the aim to be 'up to date' – an attitude typical of the 1930s and the improvement process applied to older dwellings carried out since – has meant that many have been replaced by efficient fitments made by reputable manufacturers in accordance with British Standards. Some are not so keen to have 'standard' fitments, so the surveyor may meet up with reproductions of fitments previously considered obsolete along with those of novel design. During the room by room inspection of the dwelling, every sanitary fitment needs to be examined and a note made of the type, the material used and the condition. Yet again, the surveyor's role is to inspect and report but not to test, and this is another area where the option to provide an intermediate Condition Rating 2 is available if less serious defects or concerns are identified.

Many purchasers consider the quality and condition of the sanitary fittings to be an important aspect of the property they are hoping to purchase, so the surveyor must resist any temptation to treat them superficially in comparison with the more substantial structural elements. The first thing to determine is whether they are of a standard and condition which is commensurate with the type and style of the dwelling. Are they of a quality that will last or are they likely to require replacement in the near future? If the fittings are chipped or cracked, they will be insanitary and should be renewed. Have they been properly secured or are they loose or unstable to the touch? Seals around fittings should be examined to ensure there is no risk of leakage (**464**). Persistent water seepage around the edge of a bath and especially showers can lead to serious rot. Ceilings in areas beneath sanitary fittings on upper floors should always be examined closely for signs of telltale staining caused by leakage or dampness. Homeowners may not notice such stains caused by small amounts of dampness, but these are still capable of providing sufficient moisture for rot to become established in the dark, unventilated area in the floor beneath the appliance. Recently renewed decorations in areas below sanitary fittings should be taken as a warning sign that further investigation or enquiries are needed.

Changes in regulations over the last couple of decades, aimed at cutting water consumption, have progressively reduced the size of WC cisterns. The oldest high-level types can hold up to 12 litres of water but the standard size prior to 1989 was 9 litres, reduced between then and 2001 to 7.5 litres with a single flush. Since 2001, all cisterns should be dual flush, using only 6 or 4 litres. Requirements for overflows were also changed in 2001 so that external outlets are no longer required and overflowing water is now permitted to discharge into the toilet bowl.

Baths are difficult to replace so their condition is important, although metal baths can be resurfaced. Heavy cast iron baths have returned to popularity but may run the risk of overloading the floor, especially if placed in the centre of a room (**465**). Modern PVC or acrylic baths are lightweight but may be somewhat flexible, so they must be properly supported throughout their length to avoid pulling away from the seal at the junction with the surrounding walls.

The level of inspection will depend on the terms and conditions of the inspection being undertaken. The RICS Survey Level Three inspection expects the surveyor to observe the 'normal operation of the services in everyday use', which in respect of the sanitary fittings means 'turning on water taps, filling and emptying sinks, baths, bidets and basins, and flushing toilets to observe the performance of the visible pipework' (*Surveys of Residential Property*, third edition, p. 15). Traps and wastes should be checked for signs of leakage, and in particular, connections at the rear of WCs should be checked along with the floor beneath these vulnerable areas.

In properties where additional sanitary fittings have been installed – perhaps a vanity unit or en suite shower room in a bedroom or a toilet and basin in a former understairs cupboard – the client should be advised that these are not original and that their legal adviser should check whether appropriate approvals have been granted. Meanwhile, the surveyor should do as much as possible to see to what standard these works have been carried out. Sometimes all will seem fine but on other occasions there will be serious shortcomings that cannot be permitted to continue. One typical matter to check is whether there is adequate provision of ventilation. Another, bearing in mind the original arrangements, is to consider what the necessary soil and waste runs are likely to be: Are the falls from the newer fittings to original sections likely to be adequate? Are there any unsatisfactory disposal arrangements, such as waste pipes discharging into rainwater downpipes?

More and more houses, and not only those at the upper end of the market, are having expensive fittings such as whirlpool and spa baths, saunas, and wet rooms installed rather than the more basic traditional sanitary fittings. Health risks more usually associated with commercial premises, including legionnaires' disease, can affect some of these installations if they have not been provided with the necessary regular cleaning and maintenance. Electrical fittings are required for some fittings to function properly, and because bathrooms are subject to strict regulations, there should be Part P Building Regulations approval or similar evidence that the installation was carried out and completed by an appropriately qualified person. Although the surveyor will not be expected to provide detailed advice on such installations, the current owner should, where possible, be asked to provide confirmation that they were fitted to appropriate standards and that the necessary cleaning and maintenance have subsequently been undertaken.

By their nature, lacking a conventional shower tray, wet rooms can pose particular risks, especially if installed above a timber floor. Specialist contractors offer installations with warranties but there are no recognised national standards, and DIY kits are available where the standard of conversion is likely

 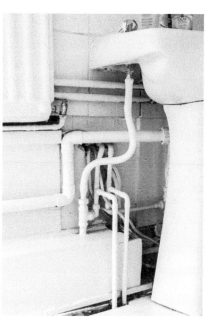

463 A Shanks' 'Combination Levern' close couple syphonic WC suite with matching mahogany cistern enclosure and seat, over 100 years old but still giving good service. Pulling the handle on top of the cistern forward and releasing it operates the mechanism satisfactorily without any need to 'pump'. The plumbing feed has been modernised so the surveyor is probably limited to acknowledging the antiquity and rarity of this fitting and pointing out potential difficulties with maintenance and repair.

464 The surveyor should examine below all sanitary fitments for slow or intermittent leaks and to check that a suitable trap is provided. Most cases will be more straightforward than unravelling this intriguing maze of lead, galvanised steel, copper and plastic pipework, not to mention the unnerving interwoven loose electric cables.

465 This cast iron roll top bath is unusual in that it is square at one end with pillar taps instead of the usual globe design, soap trays and a pop-up waste, but no provision for overflow. At one time, the bath stood on a lead tray – presumably due to the absent overflow – but most of the lead has been removed leaving the floor and, particularly, the ceiling below very vulnerable to careless use. Popular once more, these baths are very heavy and impose a significant loading when fashionably placed in the centre of a room. It is as well for the surveyor encountering this situation to recommend that the bearing capability of the joists should be checked.

466 Judging by the staining on the ground and the splashing on the adjacent wall, these overflow pipes have been leaking for a considerable period. Dripping on the projecting balanced flues has spread the water more widely, and a backfall on the lowest pipe has resulted in water simply running back to the wall rather than falling clear. The number of failures indicates a common fault with the internal fittings, but ignoring this easy and cheap plumbing repair may have led to significant problems elsewhere.

to be subject to less quality control. Joints and seals will be particularly critical above timber floors where a degree of flexibility is almost inevitable and the implications for any leakage are potentially serious. Once again, as much information as possible should be obtained from the owner and the surveyor must make as thorough an inspection as possible in all accessible and surrounding areas. The report should include appropriate recommendations and warnings to the client.

Above ground drainage

Between the outlets of the internal sanitary fittings and the underground drainage system there is the vertical plumbing which connects the two. For many years, this pipework was attached to the exterior of the building and was accessible for visual examination; but now it is usually enclosed within the building envelope and not exposed to view. Historically, dwellings would be fitted with one of two systems: the two-pipe system where soil and waste are piped separately and the ventilated one-pipe system where they are conveyed together but separate ventilating pipes are used to prevent 'siphonage' of water traps in appliances (**467** and **468**). The use of ventilating pipes for both of these systems confusingly doubles the number of pipes used at times. Research into the mechanics of soil and waste water within pipes revealed that separate provision of ventilation was not necessary if the system was properly designed; so since the 1960s, the simple single-stack soil and vent system, in which there are no separate ventilating pipes, has been used for housing purposes.

The seal of water in a trap can only be broken if the pressure changes in the branch pipe leading from the trap are of a size and duration to overcome the head of water in the trap itself. Such pressure changes can be brought about by liquid flow in the main stack – that is to say, by induced siphonage – or by liquid flow in the branch pipe – that is to say, self-siphonage. Induced siphonage depends on the flow load in conjunction with the diameter and height of the stack while self-siphonage depends on the design of the appliance and the length, fall and diameter of the branch pipe.

Both the two-pipe system and the fully ventilated one-pipe system virtually eliminated the risk of seal breakage: the first by interrupting the continuity of the system at hopper head and gully; the second by ensuring, with the help of vent pipes, that the pressure in branch pipes never deviated appreciably from atmospheric pressure. With the single-stack system, the combined soil and waste stack alone is relied on for ventilation and the system needs to be of good design and workmanship to ensure that it functions correctly.

In houses built prior to about the 1950s, the surveyor will invariably find the two-pipe system. This involves the soil stack taking discharges from the WC, or WCs, and the waste pipe, or pipes, taking the waste from the baths and basins. This system ensures that the sewer gases cannot enter the house. The ventilated soil stack and the double break at the hopper head and at the gully to the waste pipe, with its trap, provide all that is necessary for an effective barrier. Even so, the surveyor will observe that the hopper head can become foul and offensive and the same criticism has also been levelled at the gully. For two- or even three-storey dwellings of this period, this system is reasonably satisfactory for the purpose unless major improvements are proposed. If that is the case, it will be necessary for them to be planned on the basis of the one-pipe system in view of the banning of the use of hopper heads for new installations or the renewal of existing installations.

What needs to be looked at critically on the elevations of two- or three-storey dwellings built up to around the 1950s is the condition of all visible pipes carrying either soil, waste or rainwater and whether there are any cracked, damaged or loosely fixed sections; whether there are any signs of leakage at the joints; and the condition of the hopper heads, which is likely to be dependent on the adequacy of size and positioning to cope with the number of pipes discharging into them, hopefully without splashing and staining the adjacent wall. In particular, the surveyor should examine and, wherever possible, feel behind cast iron hopper heads and downpipes for rust and holes since repainting over the years is often neglected due to the difficulties of access. At the base of the waste pipes – which can of course also be carrying rainwater if the systems are not separate – connection needs to be made to a trapped gully, preferably by means of a back or side inlet below the grating but above the level of the water in the seal. Alternatively, generally more often but less satisfactorily, the discharge into the gully is by means of a shoe above the level of the grating, which in the case of waste pipes from sinks, can lead to smelly and greasy conditions compounded by leaves and other debris when the shoe is carelessly fitted and even when the curb to the gully is in reasonably good condition (**471**).

Finally, high-level ventilation to the drains needs consideration. High vents should, in all cases, be taken to a point above the eaves or flat roof and not less than 900mm above the head of any window or other opening into the building within a horizontal distance of 3m from the vent pipe and be finished with a cage or other perforated cover that does not restrict the flow of air. Anything less – and top lengths of vent pipes and wire balloons are very often missing – can leave the possibility of foul air being blown into the dwelling.

Where the two-pipe system was used in two- and three-storey dwellings, there was seldom any need to ventilate fitment traps to combat possible siphonage. Providing there was only one WC serving the upper floor, or floors, and branches were of reasonable length, not more than about 3m for waste fittings, and of a size not less than 32mm for wash-hand basins and 40mm for baths and sinks, siphonage was unlikely to occur. Even if siphonage did occur, it may not amount to more than the smell from a hopper head. It is when additional fitments are installed and little or no consideration given to the effect on the existing fittings that problems can arise. The surveyor should keep this aspect in mind when there are signs of additions to an older plumbing installation. Odours and gurgling when appliances are brought into action can be a clue that there are siphonage implications.

In larger dwellings and blocks of flats built up to the 1920s, the surveyor will often see the tangle of pipes usually associated with a two-pipe installation. The trap to nearly every fitment will have an anti-siphon pipe because of the reduction in the number of hopper heads where they would have been inaccessible for easy cleaning and maintenance. Both soil and waste stacks will be taken up high to roof level as vents along with their associated anti-siphon pipes although, of course, these could be turned in to connect with the respective soil and waste stacks at high level above the topmost fitments. As with smaller dwellings, it can be the addition of extra fitments far in excess of those envisaged when the block was built, such as washing machines and dishwashers, without consideration to their effect on others, that can cause problems.

As for the materials from which pipes, stacks and traps are made, the surveyor will usually encounter plastic on modern dwellings; but on older properties, there will be traditional cast iron, both sand cast and spun, or lead and lead alloy, or copper and copper alloy. Steel tubing is also encountered, but rarely. Asbestos cement pipes were popular for a while early in the second half of the twentieth century. When painted, they can look similar to cast iron pipes; but they emit a dull sound when tapped, compared with the metallic ring of cast iron. Asbestos cement pipes do not require removal if they are in good condition, but if they should need to be removed then the appropriate precautions should be taken and this may add to the cost of such work.

External pipes requiring painting or other protective coating need a free space for access all around the pipe. The clearance with the wall should be 30mm and it is not desirable for pipes to be set in angles or in chases. The walls of houses are often so formed that pipes are tucked in angles where they are exceptionally difficult to inspect, let alone maintain.

Where pipes pass through walls or floors or other parts of the structure, they are best provided with an aperture formed with a sleeve of inert material

467 Typical two-pipe plumbing system on a 1930s two-storey semi-detached dwelling, though unusually positioned on the front elevation. Here the small first-floor bedroom has been converted into a bathroom. The cast iron soil pipe has been correctly extended to roof level as a vent but is of insufficient height to comply with the Building Regulations and the wire balloon is missing. Waste water from bath and washbasin, along with rainwater, are taken to the hopper head.

468 A range of pipes added over the years to a mid nineteenth-century multi-storey dwelling that has been divided into flats. Looking from the right, the first pipe is a 100mm light iron rainwater pipe fixed by clips to the brickwork. Next to it is a 50mm cast iron waste pipe fixed through ears on the sockets to the brickwork, as are all the other cast iron pipes on the elevation. There are four 32mm branch waste pipes in lead and an anti-siphon pipe of the same size, also in lead. Both waste pipe and anti-siphon pipe are taken up to just below eaves level to act as vents, less effectively than they might have been if taken further to above eaves level. Further left, between the pair of windows at ground-floor level, there is a 50mm lead anti-siphon pipe with three branches, which is connected at top-floor level above the highest fitting to a 100mm soil and vent pipe. This is in lead for the top two floors and is satisfactorily carried up as a vent to above eaves level, but lower down is in cast iron with branches in a mixture of lead and cast iron and with visible cleaning eyes. All the lead pipes on the elevation are fixed to the brickwork with cast iron tacks soldered on to the lead.

469 *(left)* A 100mm uPVC combined soil and waste stack on a one-pipe system taking WC and waste water branches and fixed to the brickwork with clips. Although there is a sharp bend at a change of direction to the near horizontal with a liability to blockage, generous cleaning eyes are provided. There are no external signs of ventilation to traps, which is essential to avoid siphonage. Resealing traps may have been used but their presence needs to be checked if they have not already been picked up internally.

470 *(middle)* Gullies on the two-pipe system for breaking the continuity of direct connection between the traps of waste water fittings and underground drains should be arranged with adequate kerbs to prevent undue splashing and, if taking a sink waste, ease of cleaning. This one at the foot of a rainwater pipe needs repair and re-rendering although the shoe to the pipe is set at a suitably low level. The adjacent subfloor air vent is partially blocked by the kerb, but more seriously, any blockage in the gully might overflow through the vent into the subfloor area.

471 *(right)* Ignoring a blocked drainage gully will inevitably lead to unsightly and insanitary consequences. However, it cannot be assumed that the gully alone is at fault: there may be a blockage in the drain beyond. Equally, staining on the wall above, now dry, may indicate an older and more fundamental problem than simply a blocked gully. Further investigation is required to make sure the real problem has been identified.

to protect them. Many existing pipes are not provided with adequate support. The most vulnerable are, of course, horizontal pipes of plastic or lead. Cast iron pipes are normally fixed by means of ears on the sockets, with or without distance pieces. Lead pipes require double or single cast lead tacks, soldered or lead burned to the pipe, and fixed to the structure with galvanised nails or gunmetal screws while sheet lead milled tacks are also used. Asbestos cement pipes are fixed with galvanised mild steel holderbats for building into or screwing to the structure. Copper pipes may be fixed with copper alloy holder-bats or strap clips. Standard fittings are readily available for plastic pipes but the lightness of the pipe runs can provide a temptation to skimp on the provision of adequate support.

Below ground drainage

The underground drains of a dwelling are 'out of sight' and ought, too, to be 'out of mind'. In other words, they should fulfil their function without the owner's attention being drawn to them for many years at a time. There is no reason why drains, provided they have been designed as they should have been to be self-cleansing, should require any examination or maintenance for years. Indeed, the point when when a dwelling changes hands – say, on average, every 7 to 10 years – is probably as suitable a time as any between inspections.

Whether the drains of a dwelling are performing in this desirable manner can only be established through evidence given by owners or tenants who have been in occupation over a period of years and, of course, on the assumption that they are being truthful. Many dwellings built during the 50-year period from 1880 to 1930 do in fact achieve such a performance. This was a period of good design for drains, particularly in relation to laying at adequate self-cleansing gradients, good workmanship and, primarily, being laid at a depth substantially greater than during later periods.

The extent to which the drains will be examined will depend on the level of inspection and report that has been commissioned. For a mortgage valuation inspection, there has never been any requirement that drain inspection chamber's covers will be lifted, but the valuer is still expected to report any obvious indications of problems with the drains that may have a bearing on the valuation. The Survey Level One report – the RICS Condition Report – also states, perhaps more surprisingly, that the surveyor will not lift inspection chamber covers but will make a visual assessment of the system. This is a significant restriction on the extent of the inspection but one which must be complied with as it is enshrined in the terms and conditions of the service. Two points arise. First, the Survey Level One is intended primarily for modern properties where there are unlikely to be significant problems, so the drainage system should have been constructed with modern materials to meet modern standards. If a problem is already suspected when the survey is commissioned then this level of inspection is not appropriate and a more detailed level of inspection should be requested. The second point is that, on many older properties, there are no drain inspection chambers visible or it is not possible to lift those which are present. At such times, whatever level of survey is being conducted, the surveyor's examination of the drains is effectively limited to the same as those at Survey Level One; yet constructive comment is still frequently possible despite these restrictions.

Even without the benefit of lifting drain inspection covers, the surveyor can still comment on their location and condition, or their absence; the assumed lines of the drainage runs can be assessed for signs of unevenness or collapse; or perhaps there may be a risk of root damage from any trees or large shrubs growing on or close to the line of the drain runs. If conservatories, porches or extensions have been built above the line of the drain, checks of the foundation specification should be recommended to ensure that suitable provision was made to prevent damage to the drains. Although a visual inspection provides only limited information, it can provide an indication of whether or not the drains will give relatively trouble-free performance over the years and whether further investigations ought to be carried out.

Knowing or assessing the age of an older dwelling may not, in itself, tell the surveyor much about its drainage system. Whereas the surveyor could be inspecting and reporting on a dwelling built 250 years or so ago, exhibiting most of the characteristics of a building of that time and still basically little altered, this will hardly be so with the drains. The drainage system is unlikely to be even half that age because the remains of any earlier systems will have long since been cleared away and replaced. What enabled the subsequent water carriage system of drainage to be developed was the spread of a piped water supply to individual dwellings from about the middle of the nineteenth century. Dwellings built after the Public Health Act of 1875 generally still have the drainage systems with which they were built. Those from before that time were later improved by the construction of up-to-the-minute drainage systems, embodying the same newish principles.

Thus most dwellings built before 1950 have drainage systems embodying the principles developed in the mid nineteenth century, both in design and character, and are constructed in the same materials that were developed around that time. For example, until around 1950 the only choice of material for drainpipes lay between ceramic, in either fireclay, earthenware or stoneware form, and cast iron. As pipes and fittings in cast iron cost twice as much as those in the ceramic materials, it is not surprising that practically every dwelling built up to 1950 is equipped with ceramic drain pipes. It is only after 1950 that dwellings will be found with alternatives such as concrete, asbestos cement or plastic pipes.

Pitch fibre pipes were also used for a short period – mainly between the 1950s and early 1970s – but these were found to be too insubstantial. Unfortunately, some 50,000 properties still have pipes of this material, which can be difficult to identify as the sections visible within the drain chamber are usually ceramic. When they fail, some insurers may decline a claim on the grounds that the pipes have just worn out due to age and gradual deterioration and are thus excluded from cover – a nasty surprise for the homeowner. Local knowledge of affected areas may assist in identifying suspect properties.

For an inspection of a dwelling's drainage system, it is useful to record the information obtained on a sketch plan showing the position of the dwelling and its relation to the boundaries of the site on which it stands. This is best prepared on lightly printed graph paper, roughly in the correct proportion if at all possible. Reference to the sketch plan will make it easier to compile the paragraphs for inclusion in the report.

For the typical suburban dwelling of the 1930s, and for quite a few from other periods as well, it may be that the surveyor will have observed near the front gate the cover to the last inspection chamber of the dwelling's system before the main drain crosses the boundary to connect with the local authority sewer. With other types of dwelling – for example, the inner-city multi-storey terrace with basement – the last inspection chamber of the system could be fairly close to the front wall and the main drain could receive quite a number of branch connections at this point.

The last chamber in a dwelling's system is often referred to as the 'intercepting chamber' because it has long been thought that foul air from the sewer should be prevented from entering the dwelling's system. This can effectively be achieved by the provision of an intercepting trap positioned at the outfall from the last chamber. The trap has a dual function in that it also deters rats from entering the household system. The interceptor trap needs to be provided with a rodding eye, closed off by a secure stopper, to enable the clearance of a blockage between the chamber and the sewer. The provision of an interceptor trap is a requirement of many local authorities, particularly in urban areas where it is considered that there is an unduly high concentration of offensive gases in the sewers. However, interceptor traps have their disadvantages. They interrupt the smooth flow, often retain foul matter in the water seal and are liable to be the cause of blockages. Possible blockages in the interceptor are the reason the Building Regulations recommend their omission in new drainage schemes unless they are a specific requirement of the local authority. Surveyors should, therefore, not be surprised if there is no interceptor in the last chamber.

Where there is no requirement for an interceptor trap in new systems, either the high-level ventilating pipe at the head of the system or an internal air admittance valve is usually sufficient to introduce a through current of air within the system. Where an interceptor trap is provided in the last chamber of the dwelling's system, it used to be essential to provide for the entry of fresh air to ventilate the whole of the system. This was usually achieved for the average suburban house by the provision of a low fresh air inlet situated behind the front garden fence and connected by a pipe to a high-level point in the chamber. The fresh air inlet draws air in through a light mica flap valve, secured behind a protective metal grille; and when air passes across the top of a ventilating pipe fixed high at the head of the system, this produces a through current of air (**472**). Low-level fresh air inlets are prone to being damaged and should be renewed if necessary; otherwise, there is a risk of foul air escaping. The ventilating system can always be reversed if necessary, with a high-level vent at the front and a low-level one at the rear; although in close urban situations, high-level inlets at both front and rear may be preferable.

For the Level Two and Level Three Survey inspections, the RICS HomeBuyer Report and RICS Building Survey respectively, drain chamber covers should be lifted if this is practicable and safe. First, and self-evidently, covers to inspection chambers need to be readily removable. For this, they should be provided with lifting handles or holes for lifting keys. If these are missing then advice should be given to renew the cover; otherwise, the need to lever up with a long, strong screwdriver or bolster and hammer, to an extent, negates their purpose. On the other hand, difficulty experienced by the surveyor in lifting covers because they are partially covered by earth or for other reasons helpfully provides evidence that there may have been no need to disturb them for a while because no troubles have been experienced.

Inevitably, the surveyor will periodically lift a chamber cover to reveal an unpleasant pool of raw sewage, sometimes only lurking in the bottom of the chamber but occasionally right up to the brim. Warning of the condition within is sometimes given by seepage from around the edge of the cover onto the surrounding area, but usually the discovery is a disagreeable surprise (**473**). In some cases, there may be clues about the cause of the blockage but usually the surveyor can only recommend that a drains specialist be instructed to clear the blockage and investigate the condition of the drains. Until the investigation has been completed, it is not possible to know the extent or likely cost of any repair work which may be necessary. The best scenario is that no work will be needed once the blockage has been cleared, but if any drainage works are required then they will almost inevitably be expensive and the client should be suitably warned. If the dwelling with the blockage is occupied, the surveyor should tell the occupier about the problem so that it can be cleared before it becomes any more of a health risk.

A routine should be established by the surveyor when examining inspection chambers to ensure that essential information is not missed. For example, in the typical intercepting chamber, there is the inlet where the size and material of the main drain can be ascertained. This can be the position where any differential movement between the drainpipe and the side of the chamber may have affected the true alignment of the drain, introducing a visible ridge or sinking in the channel and possibly cracking or breaking the collar of the pipe. Channel sections should be smooth and neatly jointed, and there should be no deposits suggesting either blockage from slow-running sewage or muddy water suggesting infiltration with grit or gravel from the earth outside. The presence of any root growth should be noted. Well-formed, smooth-sloping benching should be provided on either side of the central channel to enable a person to stand and do what is necessary to clear a blockage. This needs to be in good condition and not likely to crumble away and cause a possible blockage, as does any rendering on the walls of a chamber; if not rendered, then the pointing to the brickwork needs to be kept in good condition. Step irons, soundly built in, need to be provided for any chambers where rodding cannot be done from ground level.

Access to the last chamber of the dwelling's system will provide the surveyor with an indication, by following the line of the inlet, of where the penultimate chamber might be. Having found the next chamber and lifted the cover, it is a good idea – provided there are no children or elderly people in occupation and it is safe to do so – to continue the procedure and track down all chambers on the site, lifting the covers of each and leaving them off temporarily. This will enable the surveyor to check that the main drain traverses a straight line between chambers and will enable, progressively, the allocation of each branch drain connection to a particular sanitary fitment or group of fitments, if necessary via a trapped gulley. Hopefully all can be accounted for; but if not, it is possible that connections are made to the main drain without access being obtainable at a chamber. Along with the possibility of the main drain taking a convoluted route between chambers, these represent a risk of substantial problems should blockages occur.

At each chamber where there are branch connections, the direction of the inlet should indicate the purpose when related to the sanitary fitments in the dwelling or in relation to the gullies externally. If the indication is not very clear then the surveyor may need to do what is necessary to establish what function each branch fulfils. It will be found that some branches practically double back on themselves to join the main drain in the direction of flow. For these cases, it is important that the correct angle bends are used to enable a smooth and free flow to be obtained. If this is not being achieved, there may be signs of heavy fouling to the benching and even to the sides of the chamber that can lead to insanitary conditions, particularly if the branch is to the base of a soil pipe where the force of the flush from an upper-floor WC fitment, compounded by the drop, leads to a rapid discharge into the chamber.

Waste water fitments, baths, shower trays, wash-hand basins, bidets and sinks, in contrast to soil fitments, are taken to discharge over trapped gullies and branch drains so these tend to flow at a gentler rate. Signs of grit or gravel deposits, fouling by grease or misuse from the deposit of building materials can provide evidence that all may not be well and that partial blockages may be causing a backup of liquid to the next chamber above (**474**). Pacing the length of underground drains can usefully be combined with an examination of the ground above. Suspicions about the condition of the drains might well be reinforced by signs of subsidence or by the presence of trees or shrubs nearby. This could be so particularly if the subsoil is of shrinkable clay and the tree or trees are of a variety likely to cause damage (as discussed in Chapter 8) or, again, there is what appears to be the excessive growth of vegetation nearby. Of course, if the drains are running near a part of the dwelling where movement has taken place then straightforward leakage of the drains washing away the support to the foundations could well be the cause.

One of the objectives in lifting all the covers to inspection chambers is to identify the purpose of each branch drain with a visible connection to the main drain while also identifying those branches which are connected without the benefit of access since they may prove to be troublesome in the future. In comparatively rare cases, there may also be a connection to the main drain which has no identifiable source. If the branch looks dusty and disused then it could, of course, be from a fitment now removed – an outside WC, for example – but on occasion, it may appear to emanate from adjoining premises. The attention of buyers and their legal advisers needs to be drawn to these points together with any information that the seller may volunteer on the purpose of the branch and any information on liability for maintenance and repair. Of course, there is yet another possibility. It could be that the dwelling's main drain enters the last chamber within the boundary and, instead of continuing on and connecting with the local authority sewer, it is itself a branch drain connecting to another underground drain within the dwelling's boundaries. This is particularly likely to be the case if the dwelling is one of an estate. The surveyor needs to establish, if possible, the status of this 'drain'; but more likely, a study of the title deeds will be necessary by the client's legal adviser as the position can be complicated by various factors, including the location of the dwelling.

Where shared drains are suspected, a recommendation for legal advice should always be made. The situation where drain runs from one property pass over the land of one or more other properties before entering the public sewer used to pose potentially serious difficulties if ever any works were required because the rights and responsibilities of the respective owners were not always easy to define. In 2011 this situation was greatly simplified because shared drains within the boundaries of a property were defined as 'lateral drains' and the local water company or sewerage undertaker was given the responsibility of maintenance and repair. This has clarified the matter in most cases but responsibility for some drains was not transferred, including privately owned septic tanks or cesspits, or privately owned sewage treatment works or pipes connected to them. These facilities are considered in the following section.

Increasing awareness of environmental issues has brought certain matters to the fore that perhaps were previously not considered by the surveyor but which should now be taken into account. The misconnection of foul water

472 (*left*) An unusual treatment for the fresh air inlet to an interceptor chamber, of which there is no sign externally. Despite the probable original intention of avoiding damage – which would have been better achieved by taking the vent up to roof level – the brass grille and mica flap valve are missing and need to be replaced. Drain air could, at times, be circulating around the entrance door, welcoming visitors. Even with grille and valve renewed, the pipe buried in the brickwork will eventually corrode and leak foul air.

473 (*upper right*) A comprehensively blocked drain chamber with little prospect of determining the cause of the blockage. Apart from seeing which of the other accessible chambers are similarly affected, there is little to do except for recommending that the blockage should be cleared and the condition of the drains checked by a specialist. If the property is not vacant, the occupier should be told about the problem so it can be dealt with urgently.

474 (*lower right*) A partial blockage may be causing less obvious problems but the cause still needs to be investigated and remedied. The cement benching within this chamber is cracked and sections have broken away. One of the inlet pipes seems to be damaged and there are signs of fouling on the benching opposite the other inlet. Are other chambers in a similar condition?

or waste water – also referred to as 'grey' water – to surface water drainage can lead to pollution, and water companies and the Environment Agency are becoming much more active in tracing such pollution. Fines can be imposed where serious incidents occur or remedial action is not taken after faults have been pointed out to offenders. This problem frequently occurs when DIY plumbing connects en suite facilities, dishwashers or washing machines to surface water drainage with the risk of polluting watercourses. Of less significance for pollution, but still considered a misconnection, is where surface water drainage is connected to a foul sewer in a system that treats the two types of water separately; this is because the burden on sewage treatment facilities is unnecessarily increased.

A further area that should be noted by surveyors is where paved hardstandings have been created by homeowners. This is typically where parking spaces have been created in front garden areas, but it also applies to any area of more than $5m^2$ that is paved with impervious paving where water is likely to drain onto a public highway. Creating or replacing such an area now requires planning permission unless suitable arrangements can be made to safely retain the run-off within the boundaries. If a recently paved area is noted where run-off is likely to drain towards the public highway, a query should be raised in the report for the legal adviser to check whether consent was obtained.

Private drainage systems

Until this point, it has been assumed that disposal of sewage will be to a drainage authority sewer. It is now necessary to consider those circumstances where this is not the case and where sewage has either to be stored for collection in a cesspool or taken to a small domestic treatment plant, with waste and surface water taken to a sump or to soakaways.

The Environment Agency is progressively tightening the regulations that apply to small sewage discharges (SSDs) from private septic tanks and package sewage treatment plants, and this is being backed up by wide-ranging enforcement measures and sanctions. Although the Environment Agency's stated intention is to respond with proportionality to the seriousness of any breach of the regulations, the ultimate sanctions available can be very severe and punitive. Measures recently introduced or under consideration include permits for use and registration or exemption from registration of SSDs. The regulations are continually being amended – more than once a year at times – while deliberations about the most suitable way of addressing the issue remain unresolved; so it is important that vendors should be asked what measures, if any, they have taken to ensure that their installation complies with the regulations which are current at the time of the inspection and that clients should be advised to obtain the most up-to-date guidance from the Environment Agency.

There are essentially two options available for those without access to mains sewers. The first and most basic is a cesspool – a watertight tank which is used to store untreated sewage until the tank is emptied by a specialist company, typically every three to five weeks. The cost of emptying will depend on a number of factors but could approach £10,000 a year, so it is a significant factor for most owners. The alternative is some form of private treatment facility. Traditionally, these were provided by septic tanks, two linked chambers where biological processes break down the raw sewage and the treated effluent flows out usually through some form of drainage field. A more modern alternative to a septic tank is a packaged treatment plant – specialist installations which use

475 The unsavoury mixture of liquid matter leaking from this septic tank onto the surrounding land is not only unhygienic but a potential cause of pollution to nearby watercourses.

differing methods to treat the effluent to a higher standard than a septic tank, usually well enough to permit discharge into a watercourse as well as to the ground. Both septic tanks and packaged treatment plants require maintenance and annual clearance.

There are limits to what the surveyor will be able to tell from an inspection of any private drainage facility, but enquiries should be made about the type of facility, arrangements for and regularity of maintenance, emptying, and location and direction of any discharge. The answers to these questions often reveal a surprising, not to say worrying, level of ignorance amongst owners about their private sewage facilities. Equally, assurances should not be taken at face value. The inspection of the facility should include a consideration of its proximity to any buildings, watercourses or water sources, such as wells or boreholes. The ease of access for emptying and maintenance should also be noted. Any indications of waterlogged ground, unduly lush plant growth that might suggest leakage and, worst of all, any obvious signs of raw sewage are warning signals (**475**).

This is a specialist area and the surveyor should have no qualms about recommending that a specialist should be requested to inspect and advise on the installation, especially in view of the dynamic state of the regulatory environment. The consequences for the client of a catastrophic failure of the facility make it essential that it is thoroughly checked before a purchase is completed. Finally, the client's legal adviser should check the legal aspects of the arrangement, including whether there are, or ought to be, any legally defined rights or responsibilities associated with adjoining owners, such as any necessary rights of drainage across adjoining land.

PART 5

The surroundings

27

Garages, conservatories and outbuildings

Garages

An integral garage should be inspected and considered as part of the main building, but special thought should be given to the likelihood of fire spread from an integral garage to the living accommodation. The risk of fire from garages is not only due to the probable presence of flammable liquids, such as fuel in a car or lawnmower or stored paints, thinners and varnishes. Because cars are nowadays so much less prone to rusting and there is no longer the same need for covered garaging, owners are more likely to use garages for storage and other activities such as hobbies, which may also pose a greater risk of fire than normal domestic uses.

Garages will vary from small asbestos cement structures at the rear of a terraced or semi-detached house built in the 1920s and 1930s to a large, imposing, detached structure of brick and tiles, perhaps with a flat or a playroom above. Inspection of the garage will, of course, follow, in the same way, the rules for the inspection of its parent structure and does not require any exhaustive description (**476**). It is worth mentioning, however, that any defects in materials – for example, facing bricks or roofing tiles – which the surveyor has found in the main structure may well be repeated on the garage and outbuildings. The surveyor will often find many examples of defects that are likely to be caused from neglect. Outbuildings are never maintained with the degree of care given to the dwelling house itself, and it is commonplace to find advanced attacks of beetle in the timbers that have never been treated or troubles from dampness caused by blocked gullies and fractured gutters that have never received attention.

The consequences of skipping lightly over outbuildings when carrying out an inspection on a comparatively modest three-bedroom detached house were brought home to the surveyor in *Allen* v *Ellis & Co.* [1990] 1 EGLR 170, concerning the separate garage used as a utility room. The surveyor described it as 'brick built in 9″ brickwork and in a satisfactory condition'. The judge found that the garage should have been described as of 'breeze block construction with an asbestos sheet roof which was brittle and fragile, likely to split and crack, scantily supported, much repaired and at the end of its useful life'. The judge concluded that 'if the roof had been accurately described . . . the plaintiff would never have been in peril of suffering the injuries which he did in fact suffer' when he fell through the roof on investigating a leak. In this case, damages were awarded not only on the basis of diminution in value on account of the wrongly described garage but also in respect of the injuries suffered by the plaintiff. Fortunately for the surveyor, the plaintiff's injuries were not as severe as they might have been. It matters not that every surveyor knows, or should know, that it is unsafe to walk on any asbestos roof, new or old. The plaintiff in this case was not told.

The surveyor should also look for faults of design that can cause trouble due to insufficient attention to detail when the garage was built. Even though every care may have been devoted to designing the dwelling house, the garage might not have received the same degree of attention. Alternatively, it might have been erected after the main house was built and such matters as awkward roof gutters that are not adequately drained or provision for gullies might have been overlooked. For smaller garages, lack of drainage is no particular disadvantage.

Garages, especially integral ones, are often converted for domestic use. When this has been done, the surveyor must be particularly vigilant (**478**). Were all necessary Planning and Building Regulations consents sought and obtained? Even when they were, does the conversion comply in all respects with the regulations applicable at the time of conversion? Common failings

476 Serious cracking like this requires investigation even when in a garage, and the questions are the same as if it was affecting the main building: What caused it? Is movement continuing? What remedial work is required? What is the effect on the value of the property?

477 Although not visible in this photograph, there is serious cracking to the concrete blockwork side wall of this garage, so investigation is required to determine not only the cause but whether remedial work is economically viable for such a poor-quality structure. Is the garage unsafe or can it still be used if it is not repaired? What are the implications for the valuation?

include inadequate insulation to external walls, poor or missing damp-proofing to floors, and potentially dangerous means of concealing electrical and gas meters, and even boilers, in an effort to hide the origins of the newly converted living space.

One thing that the surveyor should look out for, especially on developments carried out in the period between about 1960 and 1980, is allocated parking spaces and garaging in dedicated areas some distance from the site of the dwelling. These parking and garage blocks are often subject to poor levels of maintenance, if not high levels of vandalism. If the specific garage or space can be located then it should be inspected and reported in the normal way. Garage blocks were often built at this time with flat roofs, so these should obviously be checked for signs of deterioration. At times, however, the specific unit or space cannot be identified and it will be necessary to make a general comment on the area as a whole with a recommendation to check the condition of the specific unit when that has been located (**479**).

Additionally, when separate parking allocations are encountered, recommendations should be made for the client's legal adviser to ensure that there is proper title to a specifically allocated unit. The full extent of rights and responsibilities should be carefully checked. It is not unheard of for titles to be completely separate and for a garage or parking spaces to have been sold off at an earlier time. Equally, management and maintenance liabilities of access aprons and areas can be either undefined or potentially subject to onerous responsibilities.

478 The integral garage serving the property in the middle of these three has been converted to residential accommodation. Viewed from the outside, it fits in well; but what was the standard of the conversion and were all appropriate consents applied for and complied with?

479 Which of these garages or parking spaces has been allocated to the property being inspected? The number of the allocation is often not the same as the number of the house. What are the rights and responsibilities regarding the access? The answers are likely to have a bearing on the valuation.

Conservatories

As the replacement double glazing market matured towards the end of the twentieth century, the suppliers diversified and started to offer double-glazed conservatories. At the same time, gardens were becoming the focus of popular television makeover programmes. Increasingly, therefore, the standard of construction of conservatories improved from being modest, basic, timber-framed, single-glazed structures suitable only for keeping tender plants and occasional summer use to substantial, year-round rooms, fully double-glazed and heated, often fully integrated with the main accommodation and providing significant additional living space (**480**).

Unfortunately, some surveyors have not kept up with this change. They retain a perception of the conservatory as a simple addition that has little value and can be dismissed for reporting purposes as an 'outbuilding' (**481**). Consequently, their critical faculties are not fully applied to this part of the property. The homeowner who considers the conservatory as an extra living room will be disappointed if the structure does not meet their expectations, whether due to some intrinsic flaw in design or deterioration as a result of substandard materials. When defects arise, a modern conservatory is not often easy to repair; but it will be expensive to replace, and if a surveyor has not reported matters that should have been noticed during the inspection then an expensive claim is likely to result.

A further complication arises when considering how the Building Regulations relate to conservatories. To be considered an exempt structure under Part L of the Building Regulations, a conservatory must be thermally separated from the main dwelling and the heating system for the dwelling must not extend into the conservatory. This means any form of heating cannot come from the central heating installation and there must be a means of closing off the conservatory area from the main dwelling. If the surveyor finds a permanent opening from the living accommodation or if there is a central heating radiator in the conservatory then the 'conservatory' structure needs to meet the full requirements for residential accommodation; for all intents and purposes, it is then regarded by the Regulations as part of the house (**482**). It therefore needs to meet all requirements for habitable accommodation and the surveyor needs to inspect and report on it as such.

Even if it can safely be approached as an ancillary structure, the conservatory should still be subjected to a proper appraisal. How has the structure been designed? Will the roof suffer outward thrust because of inadequate restraint? Is there a visible damp-proof course externally and is the absence of dampness in the floor suggestive of an effective damp-proof membrane? How is rainwater dealt with, especially at the junction with the main building? Are there properly formed flashings? Are there any valley gutters which will be prone to blockage? Are there maintenance implications if the presence of the conservatory will prevent access to upper levels of the property above it? What is the relationship of the conservatory with the main building and the

480 Since uPVC-framed double-glazed conservatories became popular, owners' expectations have been transformed; they now regard the conservatory as an additional room that they often expect to be able to use all year round. Unfortunately, many surveyors have not kept pace with this change in perception and in many cases conservatories are still assessed and reported as if they are of little value, a serious error both practically and financially.

481 Basic conservatories like this used to be seen all the time and they could reasonably be dismissed with a line or two in the report because owners recognised their limitations and did not expect much of them. This is no longer sufficient.

482 Although this addition looks like a conservatory, there is a permanent opening between it and the main building; so for Building Regulation purposes, the addition should be treated as any other residential extension and all requirements for the construction of additional living accommodation must be met.

483 On superficial examination, this timber-framed conservatory appears to be in a reasonable condition; but a more critical appraisal will quickly reveal that it has been poorly constructed. Drainage around the base is very poorly arranged, joints have opened, timbers are rotting and there are a number of other defects which mean it will not be very long before it needs to be replaced because repairs will not be practical or economical.

surrounding grounds, especially if it is a later addition where these matters have frequently not been properly thought through? These are only some of the questions the surveyor should be asking. A full appraisal may not take very long if carried out in a structured manner and the content of the final report may not be more than if line or two if all is well, but it should reflect a careful critical assessment and the site notes should clearly demonstrate that all of these matters have been properly considered.

Other ancillary structures

During the initial reconnaissance of the surroundings to the property, the surveyor will have noted any outbuildings. In some cases, there may be no more than a timber garden shed; but in larger properties, there may be a whole range of substantial structures that contribute significantly to the value of the property. These cannot simply be ignored. They must be assessed and reported in a consistent way which relates directly to their importance to the property.

This importance, itself, needs to be considered in different ways. A modest but suitably fitted out detached timber-framed structure may create a home office which, while not particularly expensive to provide, increases the value and marketability of the property above similar neighbouring dwellings (**485**). Conversely, a large, old, stone-built store – perhaps pre-dating a modern residential development but retained and incorporated into the garden of a house on that development – might add relatively little in value but when assessed for insurance purposes, could conceivably exceed the reinstatement cost of the newer house.

As far as possible, the extent of any outbuildings and ancillary structures should be determined during the initial conversation with the client. The agreement reached about the proposed level of inspection and reporting of these additional areas should then be encapsulated in the terms and conditions document sent to the client for signature. Despite one's best efforts, however, it will sometimes be the case that, on arriving on site, there will be unexpected structures to consider. In most cases, these can be excluded from the main part of the report but it may occasionally be necessary to seek the client's further instructions before proceeding. In all cases, however, it is important that a note is made of all structures, however temporary or dilapidated, so that there can be no doubt after the event that they were seen and taken into consideration on the day of the inspection.

It has been the practice of many surveyors to simply list outbuildings and dismiss them without further consideration. In some instances, such as small, timber garden sheds, this may be appropriate; but even in these cases, a cursory glance should be given to ensure that the structure is not so clearly unsound that it might pose a danger to homeowners or their children. Brief details of any structure of a more permanent nature should be recorded in the site notes and, as an absolute minimum, given a brief mental appraisal.

484 These external stores are rather unprepossessing, but they are permanent structures so they should be assessed and reported as such. This may not take very long on site or take up much space in the report, but they should not be overlooked.

485 This timber-framed structure looks like a simple summer house but is actually a purpose-built home office in the rear garden of a 1930s house. It is fully insulated, double-glazed and serviced with electricity, telephones and broadband so it will add something to the value and saleability of the main dwelling. As it is not a permanent structure, the surveyor need not devote much time to it. But it would be remiss to ignore it and a brief comment on its apparent suitability for a continuation of its current use might be appropriate.

486 An outbuilding such as this could take a considerable time to inspect properly, and depending on its condition, reporting on it could also be time-consuming. Hopefully, the surveyor would have been aware of it while discussing instructions with the client so that the agreed fee and the terms and conditions either reflected the level of detail required by the client or excluded the outbuilding from consideration completely. If caught unawares, the surveyor must make a judgement based on the level of inspection being conducted and then make sure the report accurately reflects what was and what was not undertaken on site so that the client can be in no doubt.

The surveyor should continue to bear in mind the relationship of this outbuilding to the property as a whole, not only physically and in respect of relative value but also regarding any potential liability for future maintenance and repair. As soon as any outbuilding or ancillary structure impacts in any of these areas, the surveyor must reflect this in the extent of the inspection and the detail of the report.

This is not to say that every outbuilding should be subjected to the same detailed inspection and report as the main property. Usually the surveyor's critical appraisal will enable a conclusion to be reached that can be accurately reported in a sentence of description and one or two brief sentences summarising the overall condition. The Level One and Level Two services certainly require no more; and in most circumstances, this will probably also be appropriate in a Level Three Building Survey. If questions should later be asked, it is an optimistic surveyor who expects to be able to rely as a defence on the proposition that any outbuilding not mentioned in the report should automatically fall outside the scope of consideration. Former clients facing hefty but foreseeable maintenance costs on outbuildings will be unlikely to share this view.

Any leisure facilities within outbuildings, such as swimming pools, gyms and tennis courts, are specifically excluded from all three Survey Levels, but they still need to be noted by the surveyor and mentioned in the report as confirmation that they have been seen and not just missed. They do, however, need to be included in any calculation for reinstatement purposes, so appropriate measurements and descriptive notes of the construction will need to be taken when an insurance calculation is required. Some purchasers may not realise that swimming pools are expensive items to maintain and, in some circumstances, can detract from rather than add to value; so appropriate warnings should be given even when the facility itself is not being reported on in detail.

If an outbuilding is particularly large or an important feature of the property – for example, a large workshop or a purpose-built home office – the client may wish the surveyor to report in detail on its condition (**486**). In these circumstances, the surveyor may be taking on as much liability for this part of the property as with the main building. It should be ensured that the fee reflects the amount of time required to do justice to the outbuilding, both on site and when reporting. If the outbuilding is large or complex or if its condition is less than ideal, it is not unheard of for the time involved on site to double. If this was not factored into the initial planning, the pressure to complete the job within the original fee and time estimate can easily lead to shortcuts and errors. The surveyor's liability is not limited by such miscalculations. The importance of the preliminary conversation with the client, the establishment of the client's needs, the management of the client's legitimate expectations and a properly related fee quotation, all enshrined in the terms and conditions, bear repeating once more.

28

Gardens and boundaries

Gardens

When conducting Level One or Level Two services, surveyors are expected to carry out a visual inspection of the grounds during a general walk around. This should, when necessary and appropriate, also be conducted from adjoining public property. The list of matters to take into consideration includes such external features as retaining walls, gardens, drives, paths, terraces, patios, steps, hardstandings, dropped kerbs, gates, trees, boundary walls, fences, non-permanent outbuildings, and rights of way. This amounts, in effect, to anything that might be costly to resolve and could potentially affect the client's purchase decision (**487**) or where some feature presents a hazard that should be remedied once the client takes occupation (**488**). For Level Three services, the surveyor's review of the same features is expected to be 'thorough' and the surveyor should inspect and report in greater detail on all matters associated with the grounds, including being prepared to 'follow the trail' of suspected problems to a greater extent than for Level One and Level Two surveys. Examples include assessing retaining walls in danger of collapsing, deeply sunken paths or driveways, and dilapidated boundary walls or fences as well as considering the legal and insurance implications.

Despite this intimidating list of items, it is usually sufficient in most cases to describe the condition of the garden in general terms. More attention is desirable where trees are present. It may well be necessary for individual trees to be described separately and their condition noted if unstable or with loose elements likely to cause injury, even apart from cases where spreading roots may exert a pressure on the house or outbuildings under review. The chief danger with trees is that they absorb the moisture from the subsoil. In the case of a survey in an area of shrinkable clay soil, the surveyor should take great pains to gauge the probable effects not only of trees on the site but also of any that are visible nearby. The surveyor should also be able to recognise the early growth of tree species likely to establish themselves without any assistance – such as ash and sycamore – which are frequently found self-sown in extremely unsuitable locations close to houses and which should be removed before they become established (**489**).

Enough has been said about trees in the section of that title in Chapter 8 (on foundations) not to need repetition here. Sufficient has also been said (in the section 'Ground movement' in the same chapter) about other aspects which may be seen in the garden, such as sloping ground, areas which have sunk perhaps because of leaking drains, cavities below ground, mining subsidence or settlement of sites filled with unsuitable material. The boundary between long since consolidated unfilled land and comparatively recently filled land can sometimes provide spectacular effects, but all indications of movement need recording and including in the report with conclusions drawn and, if necessary, a recommendation given to obtain further advice. Of particular concern are the effects of such movement on steps and pathways, which may have rendered them dangerous.

Any visible evidence of wildlife in the garden should be included in the report on the basis that it might be of concern to a buyer. The persistence of moles in digging up a lawn can be considered by some as a nuisance; and badgers, while not an endangered species, are protected by law. The antipathy between dogs and badgers is well known and disturbance of setts and blocking of trails is more than just frowned upon, so buyers with dogs as pets will wish to know of any indications of their presence. Foxes are to be expected in country areas and urban foxes, now commonly found in many towns and cities, are difficult to get rid of. A family of them can ruin a garden by fouling and digging, and the noise made at night has to be heard to be believed. If noted during the inspection, evidence of their presence should be included in the report because some buyers will be dissatisfied if not told, even though it might be debatable whether a buyer would succeed in an action at law against a surveyor for not mentioning this.

At most domestic properties, the surveyor will see most of the garden simply by walking around to check the location and condition of the boundaries. Occasionally, a garden will be so overgrown that parts are effectively inaccessible; and in larger gardens, there will come a point when some areas will remain unseen. In such cases, the areas not inspected should be reported along with the reason for the omission and any warnings that may be appropriate about what might be found when these areas are inspected.

One particular risk, especially where gardens or adjoining land have been neglected or where gardens are bounded by a watercourse, is that Japanese Knotweed may be present. This invasive non-native plant has achieved an unenviable reputation that had led some mortgage lenders to decline mortgage applications where it was found to be present or even in the proximity of a property (**490**). A more measured response thankfully followed the publication in 2012 of the RICS Information Paper *Japanese Knotweed and Residential Property*. Japanese Knotweed is not easy or quick to eradicate, but with persistence and a professional approach by a specialist contractor, it can be done. It only becomes a serious problem if it is neglected − a feature shared, incidentally, with most defects which affect properties. Mortgage lending where Japanese Knotweed is present should no longer be a problem if surveyors and valuers follow the risk assessment protocols outlined in the Information Paper and contractors meet the standards established by the Invasive Weed Control Group, a specialist subdivision of the Property Care Association. These standards include providing the guarantees or bonds required by lenders. Effective chemical treatment will take several seasons, so inspectors will sometimes encounter properties that are subject to a Japanese Knotweed management plan prepared by a specialist contractor. The treatment is deemed successful and complete when there has been no regrowth for two successive years, and the surveyor can assess the progress of the treatment against the plan.

Chemical treatment is the remedial measure of choice for most domestic circumstances, but if any building works are required above where Japanese Knotweed is actually growing then complete removal will be necessary. The growth below ground, the rhizomes, can be as deep as 3m and spread 7m outwards from the visible growth, so substantial amounts of waste material may require disposal. It only takes a piece of a plant the size of a fingertip for new growth to become established, so the clear-out has to be thorough and site hygiene scrupulously observed to avoid any contamination. Even worse, once the waste is taken off the site, it becomes designated as contaminated waste, which can only be taken to a licensed waste facility. The costs of excavation can, therefore, be significant. This was the reason for the former concerns – before the protocols for chemical treatment were established – amongst lenders about the possible impact on value. If building works are carried out without proper regard to the presence of Japanese Knotweed then the plant will inevitably reappear. In exceptional circumstances, where the plant has been built over, it may even grow up internally between building components – perhaps around a floor slab or at the junction between an original structure and a conservatory or an extension. Incomplete clearance of existing growth is the usual source of the horror stories loved by the press about Japanese Knotweed 'invading' homes.

There is now an expectation that inspectors will recognise Japanese Knotweed at all times of the year, including during the winter when only dead 'canes' are visible; so it is just as important for the surveyor to be able to identify Japanese Knotweed as it is for any of the common types of tree. The RICS Information Paper provides some illustrations to assist identification, and there are many more available from simple searches on the internet. One difficulty for the inspector is that the risk assessment protocol outlined in the RICS paper asks whether Japanese Knotweed is present within 7m of the building or the site boundaries, so there is an implied but clear requirement to look beyond the boundaries of the property being inspected. The question which follows, and which the surveyor must also answer when confronted by a garden which is overgrown, is just how far it is necessary for the surveyor to go in the face of such difficulties as impenetrable undergrowth or high boundary fences or high hedges. There is obviously no simple answer and the

487 Whatever the level of inspection, deterioration such as this should always be reported. On a steeply sloping site, remedial work will inevitably be expensive. It might be a problem big enough to deter some potential purchasers.

488 It is not unusual for paths to be uneven but a ridge like this, due to tree root growth, just inside the entrance on an otherwise level path could cause anyone to stumble and perhaps fall. The surveyor has to decide what is reasonable and what is not acceptable for any given situation. This should be reported as an unacceptable risk.

490 Japanese Knotweed in the early summer, with its characteristic zigzag stems and spade-shaped leaves with a flattened base. It is important for surveyors and valuers to be able to recognise Japanese Knotweed at all times of the year, including when only the dead canes are visible in the winter. It is equally important that appropriate advice should be given to clients, reflecting the most recent guidance from RICS, so that remedial treatment is properly managed to meet the all-important lender requirements.

489 The plant on the right is a buddleia – a common self-sown straggling shrub that can grow to 3m or even 4m in height if ignored. It is not ideal in this location but can be treated as a garden shrub as long as it is regularly cut back. The plant on the left, however, is a self-sown ash – a forest tree that has no place so close to a building and should be removed immediately.

measure can only be the same one which the courts would ultimately use if the surveyor's judgement on the day should be seriously questioned: the measure is that of the 'reasonably competent' surveyor. The surveyor should apply the test of reasonableness and then record that decision in the site notes, ideally with photographs to be able to demonstrate, if later asked, why it was reasonable not to have inspected a particular area or not to have noticed Japanese Knotweed growing on adjoining land.

Boundaries

When considering whether to pursue the purchase of a dwelling, it is usually the internal layout and the condition that are paramount in a buyer's mind. The boundaries do not feature prominently and, indeed, may have received no more than a glance; yet when in occupation, they can take on a degree of importance not only as a matter of convenience and comfort but of expense as well. If the dwelling is of any size then the surveyor, having completed the inspection of both the interior and the exterior along with the services, might well consider moving outwards to inspect the boundaries before dealing with immediate concerns around the dwelling. This will make for a bit of variety and may even allow a little time for reflective thought on what has been seen so far; and, of course, it offers a longer-distance view of the dwelling itself, which can be useful for considering the relationships of the various parts.

The case of *Bolton* v *Puley* 1983 serves to illustrate the aspect of expense in relation to boundaries. It concerned the survey of a house in Somerset purchased in 1977 for £27,000. This had a 200-year-old random stone boundary wall, 55m long. Over 17m of this length served as a retaining wall for a patio some 3m above the adjoining roadway. There was a partial collapse of the wall in 1980 and engineers were called in. They established the rate of movement in the wall, which by then was leaning 155mm out of plumb; and the judge considered half this amount would have been visible in 1977. He decided that the surveyor should have mentioned this along with the bad condition at two points and should have advised a partial repair, which would have cost £5,900 at that time. This would have provided the purchaser with a better wall and, accordingly, he reduced the damages to £4,425 plus £500 for distress, vexation and worry. The trial took 12 days and, in addition, costs must have been considerable, the whole of which would have had to be met by the losing defendant.

In hilly areas, the boundary will often comprise a retaining wall holding back the soil of the garden. Indeed in such areas, retaining walls may form a feature of terracing in the garden itself. Collapse of such walls can be a very serious matter, particularly if they adjoin a public thoroughfare but equally so if they are the cause of death or injury to one of the occupants or their family.

The surveyor does well to err on the side of caution when reporting on retaining walls. Old walls rely on the mass weight of stone or brick to prevent the soil exercising an overturning motion. The foundations, if any, and the thickness of the wall at its base are just the sort of elements which cannot be seen. The surveyor must, therefore, make a careful visual inspection along the full length for cracks and fissures and use a plumb line to check whether the wall is upright or not. A fractured and leaning wall must be a candidate for rebuilding, and even a severely leaning wall without a fracture must be considered ready for rebuilding because the appearance of a fracture would probably coincide with collapse. The surveyor must clearly recommend taking further advice from an engineer in the case of suspect retaining walls (**491**).

It is not only retaining walls that are candidates for possible collapse. Prior to 1900, brick walls were most commonly provided to form boundaries some five or six feet high and the local brick of the district was invariably used, even if not always wisely. It is a tribute to many of the local bricks produced that considerable numbers of these boundary walls, if they have been carefully maintained, are in as good condition today as when they were built. When neglected, however, they can prove troublesome. Insufficient foundations, the lack of vertical movement joints and the absence of damp-proof courses together with open joints and long, unstiffened lengths of brickwork, can cause leans, bulges, fractures and settlements to a spectacular degree. The surveyor must carefully determine whether a boundary wall in such a condition can be saved or whether it is essential to undertake rebuilding on the grounds of safety, perhaps reusing the old bricks (**492**).

The government's Planning Portal website describes garden and boundary walls as among the most common forms of masonry to suffer collapse and one of the commonest causes of deaths by falling masonry. Should such walls be neglected, it is possible that an insurance claim could be rejected. Besides general deterioration and ageing, the Planning Portal lists four possible ways in which a wall might be affected later in life. The first is by an increase in wind loading or driving rain following the removal of a nearby structure; the second is the felling of mature trees close by or the planting of new trees adjacent to the wall; the third is by changes leading to a greater risk of damage by traffic; and the fourth is by alterations to the wall itself, particularly the removal of part, perhaps for a gateway, or an addition to the height.

For inspecting garden or boundary walls, the Planning Portal lists very much the same matters for consideration as those which have already been dealt with in relation to the dwelling's external walls; but these assume even greater importance when a wall is entirely free-standing and exposed to the elements on both sides. These include crumbling of the surface and cracks in the brickwork; defective pointing; adjacent tree action, both by the roots and from lateral pressure; penetration of vegetation such as ivy; loose copings; and impact damage from vehicles. It also suggests further investigation if a wall of half a brick, one brick, or one and a half bricks leans out of plumb by more than 30mm, 70mm or 100mm respectively. The Planning Portal website includes a table prepared by BRE for safe maximum heights for brick and blockwork walls, on level ground, when situated in four UK zones. As adapted,

Table 6 Maximum heights for free-standing walls

UK Zones	Maximum height for wall type and thickness (mm)					
	Brick			Block		
	½ (100mm)	1 (215mm)	1½ (325mm)	100mm	200mm	300mm
Zone 1	525	1,450	2,400	450	1,050	2,000
Zone 2	450	1,300	2,175	400	925	1,825
Zone 3	400	1,175	2,000	350	850	1,650
Zone 4	375	1,075	1,825	325	775	1,525
Zone 1:	South-east England, south and east of lines drawn southwards from Birmingham to Bournemouth and eastwards from Birmingham to Ipswich					
Zone 2:	East Anglia, eastern, central and western counties of England into mid Wales but excluding west Cornwall; the south of Northern Ireland					
Zone 3:	Cumbria; south-western counties and eastern counties of Scotland up to Aberdeen; west Wales; west Cornwall; central Northern Ireland, including Belfast					
Zone 4:	Scotland north of Aberdeen and the western counties of Scotland; the north of Northern Ireland					

Note: In very sheltered conditions and where piers are used, taller walls may be acceptable.

Source: based on: https://www.planningportal.gov.uk/permission/commonprojects/fenceswallsgates/.

these heights are set out in **Table 6**. Very much in the interests of safety, all these aspects need to be checked by the surveyor.

Although there is no law requiring owners to enclose their land, most choose to do so to keep casual interlopers at bay and avoid arguments that they or their goods might inadvertently be encroaching on a neighbour's land. Some will even be happy to bear the cost of erecting walls and fences for this purpose on all sides and, within the boundary, proudly proclaim their ownership of the constructed features and be prepared to accept the entire cost of maintenance, invoking the Access to Neighbouring Land Act 1992 if opposition is encountered to entering a neighbour's land to carry out the necessary work. Most, however, will wish to limit their expenditure to those boundaries where they have legal responsibility. Soundly built walls or fences of sturdy character and constructed in the correct positions according to measurements on deeds represent the ideal for any owner. A surveyor will usually be pleasantly surprised to find boundaries in this condition because, all too often, it is necessary to report that at least some are ill defined or that the ownership of walls and fences is unclear. Unfortunately disputes between neighbours regarding boundaries can become distinctly unpleasant with feelings, tempers and, ultimately, legal costs frequently out of all proportion to the physical size or financial value of the matter in dispute, whether involving the precise position of boundaries or the ownership and repair of walls and fences.

The surveyor must inspect and report on what can be seen of the type and condition of boundary walls and fences, on what has been said by the vendor and on whether there is any evidence on site to suggest confirmation of what the surveyor has been told about ownership. Piers to a brick wall, horizontal arris rails on a close-boarded fence or concrete posts for a chain link fence ought to be located on the side of the boundary where responsibility for maintenance lies. However, occupiers are not always in possession of the correct information and walls and fences get put up wrongly, not least because some homeowners misguidedly prefer to have what they consider the more attractive 'outer' face of a fence or wall facing inwards towards their own garden. In the absence of the surveyor being handed a copy of the deeds before the inspection to enable a check to be made on measurement and the incidence of the traditional 'T' mark on plans to denote ownership, it may be necessary to advise reference to the legal adviser for investigation. Of course, brick walls between gardens may be built on the line of junction and ownership deemed to be shared by the owners on either side. In the absence of firm information on the deeds, the walls may have to be treated as shared, under the Party Wall etc. Act 1996, if alterations or repairs are required.

One or two, sometimes overlooked, points need to be borne in mind about brick boundary walls. In view of their cost, most of those encountered will tend to have been built before about the 1920s and without a damp-proof course. Where abutting the dwelling, they may bridge the dwelling's damp-proof course and be the source of damp penetration, requiring the insertion of a vertical damp-proof membrane at the point of contact. Another aspect concerns the detailing at the top of the wall. While brick on edge and tile creasing may be an ineffective barrier to downward penetrating damp, it is usually more secure than other forms of capping such as clayware, stone or concrete sections, which are often loose and can be easily dislodged. The presence of a large, heavy gate without proper strengthening of the wall by piers can loosen sections by frequent use; and even if piers are provided, impact from vehicles can dislodge and render them unsafe. Failure to provide a brick relieving arch above roots spreading out at the base of a growing tree can lead to severe fractures and a danger of collapse.

For timber fences, a note on the type, condition and approximate age should be given along with an idea of likely durability. For example, on close-boarded fences, where arris rails are bowed, split or discoloured green from moss or dampness, the strength is bound to be affected and the need for renewal may not be too far off. The most common defect with timber fences of this type, however, is that the wooden posts have become rotted at their bases. The surveyor should rock the fence slightly to see if it is stable as it will have to withstand gale force winds and driving rain. The surveyor may consider that the provision of concrete spurs is sufficient to extend the life of the fence for a number of years ahead or it may be that more radical repair work is required. Very often, an apparently safe-looking close-boarded fence may be on the point of collapse (**494**). Fence quality can vary considerably, from the good-quality contractor-erected close feather edge boarded with arris rails and 100mm posts set in concrete – by no means cheap to replace in long lengths

– to the flimsy woven panel or waney edge lapped panel type nailed to 50mm uprights set in short 'metposts' and put up by the DIY handyman.

Brick walls and timber fencing are generally considered suitable enclosures for gardens to provide privacy, but their height has been limited by the Planning Acts over the last 60 or so years to 2m so that they do not interfere unduly with the light to gardens and neighbouring dwellings. This is not so, however, with natural vegetation, which can grow to inordinate heights and become a nuisance to neighbours. Without cooperation on cutting back or reducing height, sadly lacking in many cases, nothing could be done about this until Part 8 of the Anti-social Behaviour Act 2003 came into effect. This enables those complaining of high hedges and neighbours refusing to reduce their height to refer the dispute to the local authority who, if necessary, will issue a formal notice to have the hedge reduced in height by its owner or, on failure to comply, will enter, do the work and recover the cost. The spur to the passing of this Part of the Act was the proliferation of Leyland cypress evergreen trees planted by owners to screen their garden from neighbours. While an isolated mature specimen of this evergreen in a corner or on a large lawn is not unattractive, its main characteristic is its unusually rapid growth, and when planted as a hedge it is a different story. If not regularly pruned, it soon becomes too high for cutting back from a pair of steps and can eventually deprive a neighbour's garden of all sunlight.

Railings are another form of enclosure that will be found – apart from brick or stone walls, timber fencing or hedging – particularly along the frontage of dwellings built in the eighteenth and nineteenth centuries to protect persons, owners and visitors from falling down into the basement areas, so much a feature of the multi-storey terraced dwellings of the time. For their purpose, they need to be strong, securely fixed and with rails set close enough together to prevent small children from slipping through. The surveyor needs to check on these aspects, applying a little pressure to see that they are secure (**495**). If they are not then they will need either repair or replacement to ensure that accidents do not happen. Rust at the base and elsewhere is the principal cause of disrepair.

As to the materials used for railings through the ages, these are wrought and cast iron and mild steel. Both wrought and cast iron were used for railings from about 1710 up to around 1830 – cast iron for the stouter, heavier support sections and wrought iron for the subsidiary and decorative members. From 1830, with the Industrial Revolution in full swing, cast iron reigned supreme, and the maker's stamp will still be found on many sections of castings in the style of the times. Mild steel became available in the late nineteenth century and took over from cast iron almost entirely from the 1920s onwards. The sections are very much thinner than those in cast iron and tend to look distinctly inferior. Cast and wrought iron are less affected by corrosion than mild steel but all need coating with a paint system and for this to be renewed at regular intervals.

Many of these railings were deliberately topped with impressively sharp spikes to discourage anybody who might be tempted to climb over them. These

492 Differential movement as dramatic as this is unusual, but remedial work is necessary and will be expensive. Who will be responsible for the work and liable for the cost?

491 A leaning retaining wall is potentially dangerous and can cost many thousands of pounds to repair. When it is on a boundary, there may be uncertainty about the responsibility for repair. This can have a significant bearing on the value of a property so careful inspection is always required and it should be ensured that the report makes clear the full implications and likely liability associated with ownership.

days, the risks to personal safety are taken much more seriously and the deterrent effect of a spiked barrier must be balanced against the possibility that the railings might somehow cause an injury, however unlikely the circumstances or however reckless the recipient. The measure of reasonableness needs to be applied but it is certainly worth reflecting for a moment or two on the proximity of steps or staircases or the circumstances in which someone might fall against the spikes.

When framing recommendations for work of modifications to boundary walls or railings, care must be exercised if the property stands in a conservation area or if it is a listed building because appropriate consents may then be needed prior to any works being carried out.

The proliferation of car ownership during the post-war years led to the use of dedicated parking and garaging areas on developments of smaller dwellings, and these were often some distance from their related properties (as mentioned in the previous chapter; see **479** in Chapter 27). This trend peaked between about 1960 and 1980, but even with older properties, there may be areas of garden or garaging on areas of land separate from the main site – perhaps on the far side of a rear lane or with a communal footpath

494 (*above*) Timber fencing is always likely to be vulnerable to damage from high winds, but briefly checking the firmness of some individual fence posts while walking the grounds will help to gain an indication of their general condition so that the client can be put on warning if their condition seems to be unduly poor.

493 (*left*) This dwarf boundary wall has fallen away from its original position by the building, either because of an inadequate foundation or more serious general subsidence in this area. Displacement of the downpipe beyond suggests that the latter may be the cause, in which case the drains may be at risk and further investigation is required to ensure more serious damage does not follow.

495 Metal railings are intended to deter intruders, but are they dangerously spiked in a way which is hazardous to the general public and are they firmly in place or liable to collapse if pressure is exerted on them? If they are in a conservation area, the owner may not have a free hand in dealing with them.

passing immediately at the rear of a terrace with the garden beyond. When this is encountered, the legal adviser should determine the full extent of rights and responsibilities and any unusual aspects should be reviewed in case there is an impact on value or future saleability.

Any shared means of access should be reported so that all rights and responsibilities can be determined by the legal adviser. Shared access driveways or access spurs have become increasingly popular with developers, but they are a perennial source of difficulty between neighbouring users so purchasers should only proceed with full knowledge of the legal position. Where there are not any obviously shared access arrangements, the surveyor should ensure that access to the property is direct from public land that actually borders the property. If there is no direct access from public land then there is the possibility that the owner of any intermediate strip of land may be able to prevent access or impose a cost for the right to pass over the intermediate land. This is more likely to be an issue outside urban areas because boundaries may be less formally defined; but even in urban localities, there may be uncertainties with older properties. The issue of access to private properties across common or private land has been the source of much legal argument. The decision by the House of Lords in the case *Bakewell Management Ltd* v *Brandwood* 2004 to reverse earlier onerous decisions improved the situation somewhat for owners and occupiers of such properties, but uncertainties still remain and the long-term consequences of this decision remain to be seen. For the inspecting surveyor, the important matter is to identify any uncertainty about access rights so that these can be properly investigated and resolved before the client has made a commitment to purchase.

PART 6

Risk and legal matters

29

Risks

Risks

Safety in the home is an increasing concern to the public. Many more people die from accidents in the home each year than die on the roads – some 4,000 according to the Royal Society for the Prevention of Accidents. The media are not slow to report the relatively rare fatalities from gas explosions and carbon monoxide poisonings; but much more common incidents that rarely make the front pages – such as simple falls down stairs or against non-safety glass – can be fatal, and even non-fatal injuries can be life-changing.

The surveyor is doubtless mindful of the prevailing mindset in these risk-averse days, when the totemic cry of 'health and safety' seems to have spread a blanket of cold water and fear over many activities previously considered wholesome and innocuous. The Health and Safety Executive itself has a far more open attitude than it is often given credit for, and its work in improving standards and reducing serious accidents – not least in the construction industry – is generally to be praised. Nevertheless, there is a widespread perception among the public and the media that any mishap, however small, must be somebody's fault and that compensatory cash should therefore be extracted from that somebody, or their insurer, through litigation if necessary. This belief has been unhelpfully promoted by some firms of solicitors which have based their businesses on the premise.

As a consequence, the surveyor needs to strike a balance in terms of warning the client of any serious or unusual hazards that a property will present to an occupier without listing every one of the minor potential risks which any homeowner might encounter on a daily basis. Some hazards, such as a trip hazard due to an uneven path or paving, may be readily capable of remedy. Others – for example, an uneven, steeply winding staircase in a period property (see **434** in Chapter 20) – might well pose a serious potential risk to children, the elderly or the plain unwary, but at the same time they provide much of the character that makes such properties attractive to purchasers and hence contribute to their value. To remove such historical features is not only cultural vandalism of the highest order but may also be illegal if the property is listed.

The term 'risks' in this context is more widely drawn than simply 'health and safety'. Indeed, it is significant that the RICS Guidance and Practice Notes refer only to 'risks' and they specifically make the point that the surveyor is not undertaking a formal health and safety assessment. However, reflecting public attitude and recognising that the surveyor has an important role to play, the latest edition of the RICS Guidance Note *Surveys of Residential Property* includes a specific section dealing with the reporting of risks, unlike any preceding edition.

It should also be appreciated that the term 'risks' relates not only to the obvious issues of personal risk but extends also to risks that might physically affect the property and those which might adversely affect its occupation or value. Many of these may not be building defects – the bread and butter of the surveyor's work – but they reflect the need for the surveyor to have a wider awareness of matters that could potentially have a significant impact on the occupier of a building but which might not be readily apparent to a potential purchaser when viewing the property prior to purchase. The surveyor should describe any particular hazard in the relevant section within the main body of the report, but it may also be helpful for the client to have a summary of the main matters of concern, cross-referenced to the parts of the report where there is more detail. The RICS Home Survey format has a dedicated section on risks and helpfully separates these into four subsections: those relating to the building itself; those within the grounds; those directly affecting people; and a catch-all 'Other risks' section. It is helpful to use these distinctions in considering the type of risks which an inspector should note and report to the client.

Risks to the building

Risks to the building will usually be dealt with in the relevant part of the report but sometimes implications will extend beyond their immediate location. Where dampness is present, for example, there will be a risk to adjacent timbers; or it might be that structural inadequacies with the roof may affect the supporting walls or, conversely, structural movement in the walls may affect the floors or the roof. There may be features relating to specific forms of construction that pose particular risks which are not associated with more traditional forms: the risk of mundic in the south-west is one example and uncertainties associated with unconventional forms of construction would be another. Any inappropriate uses of accommodation within the building, such as bedrooms in damp basements, inadequate light or ventilation to rooms or substandard roof conversions, can also be reported in this section.

There will be occasions when some aspect of a property's location may present an unusual risk which should be reported. For instance, where a property is located below a cliff, there may be a risk of falling rocks causing

496 These pine trees and the gravelly subsoil in this locality are not usually associated with damage to drains or foundations; but if they should be blown down in severe storm conditions and fall towards the house, they are near enough to hit it and even falling branches might land on and damage the garage. It is a risk worth reporting.

damage; or there may be large trees nearby which could fall against and damage the property if they should be blown down during high winds (**496**). Usually the risk will also relate to the grounds – for example, when there is a risk of flooding.

Risks to the grounds

Risks associated with the grounds will include matters which may have been identified during the preliminary desktop research, such as contaminated land, flood risk or mining. If there is a significant risk that flooding may affect the property then it may be appropriate to recommend that the client obtains advice on flood prevention measures, such as barriers to doorways and air vents or fitting anti-flooding non-return valves to sanitary fittings. A further step is to consider making the property 'flood resilient' so that, after being inundated by a flood, it can be restored to habitable use much more quickly than is the case with unprepared dwellings.

Where properties are located in areas affected by radon gas, the risks of radon, the means of assessment and the potential implications if radon is found to be present should be reported. This is becoming increasingly important as the risk to occupants may rise as dwellings become more airtight to increase energy efficiency levels.

Some risks associated with the grounds are more likely to be noted during the inspection. If Japanese Knotweed is found then it should obviously be reported; but if none was noted on or immediately around the site during the inspection yet it is known to be present in the general locality of the property then it may be appropriate to mention it in this section of the report so that the client can keep a lookout for it in the future and deal with it promptly should any growth appear.

Risks to people

When reporting risks to people, the surveyor should concentrate on potentially serious or unusual matters and not list every minor hazard which is a part of normal occupation of the home. The risk must be clearly identifiable and not too remote. Typical matters include unsafe services; the risk of carbon monoxide poisoning; items which may fall, such as loose slates or dangerous cornices; missing or unsafe balustrades (**497** and **498**); inadequate provision of emergency escapes; the actual or possible presence of asbestos in textured finishes, cement boards or insulation; the risk of lead in paint; and the absence of safety glass (**499**). If any of these matters are identified during the inspection, they should be reported in the appropriate section of the report but they can also be included in a section of the report summarising risks. It may be thought that some would be obvious, but at times, it is better to sound a warning to make sure that the client is aware of the matter (**500** and **501**).

Other risks will routinely be noted by surveyors but may not have occurred to some prospective purchasers, such as the proximity of power lines (**502**) and the risks associated with electromagnetic fields (**503**). It is unlikely that automatic gates, for example, will be generally recognised as a potential hazard, so it is worthwhile bringing them to the client's attention and reinforcing the recommendation for checks made elsewhere in the report.

At times, the decision to report a particular risk to people will be a matter of judgement based on the circumstances. The paths and driveways of many properties are uneven and may present trip hazards. A difference in height of 25mm is considered to be an unacceptable trip hazard on public streets and, while there are clearly differences between a public highway and a garden path, there comes a point where a degree of unevenness or an unexpected change in levels may become potentially dangerous and should be reported to the client, especially when in areas used more particularly by visitors such as the postman (as illustrated in **488** in Chapter 28).

Another hazard that warrants a mention is any unprotected area of water, including even modestly sized garden ponds. It is a sad fact that every year about five small children drown in garden ponds. Adults are less likely to suffer fatal encounters, but the distractions of a social event such as a summer barbeque or the subtle camouflage of wet paving or a light covering of snow can result in the unwary experiencing an embarrassing, unpleasant or limb-breaking

498 Balcony balustrades should meet the same standards as for staircases, so the gaps on this balcony are too large and a particular danger to any children who may use it. The condition of the supporting timber decking should also be considered but can simply be reported in the section dealing with external joinery, unless it has clearly deteriorated to a dangerous condition.

497 Comments about parallel rail balustrades from the 1960s should be included in the section of the report covering internal joinery; but they pose such a hazard that further mention should certainly be made if there is a separate section in the report dealing with risks.

499 Glazed panels, such as these beside a staircase, were a popular feature in 1960s dwellings but safety glass was rarely used. The risks associated with breakage and falling should be pointed out in the report unless the surveyor can see markings on the glass to show the original panes have been replaced with safety glass.

500 Arrangements like this are so hazardous that it is difficult to know where to start. Presumably this is only for the short term, but the significant dangers still need to be reported.

501 This is not an open doorway from a kitchen into the garden but a single, fixed, double-glazed window. The client should be advised to fix sighting strips to the glass to stop the unwary trying to walk through it, especially when the lighting conditions make it impossible to make out.

502 Most people will recognise when a property is close to an electricity pylon and be able to make a judgement about whether they are prepared to live there, but the surveyor should still point out that proximity may have an impact on the property's value and future saleability.

503 Some people will not want to live beside an electricity substation but, unlike a pylon, not everyone will recognise one even when it is in close proximity to a dwelling. The surveyor should be on the lookout for such features and point them out so that the client can make an informed decision about their purchase.

soaking. A word of warning in the surveyor's report may help to prevent a tragedy and, even if ignored, means that the surveyor can sleep with an easy conscience while aggrieved nursers of broken limbs look elsewhere when seeking redress for their injuries.

Other risks or hazards

This final grouping covers matters which the surveyor wishes to report but which don't fall within the previous, more clearly defined categories. Many of these will be of a more general nature and relate to the surrounding area. Clients moving from outside a locality may have little awareness of vexatious matters which are common knowledge to local residents such as flight paths, factories which are sources of intrusive noise or unpleasant smells, or perhaps proposals for major infrastructure changes, bypasses or housing developments. Clients will particularly welcome observations about such matters when the surveyor's local knowledge can add real value to the service being provided.

On the other hand, surveyors must ensure that they do possess the appropriate degree of local knowledge when accepting instructions. One surveyor, asked to comment on possible noise from aircraft, said that the dwelling was 'unlikely to suffer greatly from noise, although some planes would inevitably cross the area' – probably a fair enough comment if that was his view after a single daytime inspection. The buyer, after moving in, considered the noise excessive, sued (*Farley* v *Skinner* [2001] UKHL 49) and was able to show that if the surveyor had asked the aviation authorities then he would have been told that 'aircraft stack over the area for two periods a day'; the complainant secured an award of £10,000 in damages. The Court of Appeal reversed this decision saying there was no evidence of physical discomfort or inconvenience and the nature of any harm was not compensatable. The determined complainant took the case to the House of Lords where it was held that the surveyor's accepted obligation to report on noise was an important and exceptional one and became a contractual guarantee overriding the obligation of reasonable care, the purchaser requiring an expectation of peace and pleasure; the original award of damages was reinstated.

30

Matters for legal advisers

Legal matters

During the course of an inspection, it is almost inevitable that matters will arise which have possible legal implications. The surveyor is not directly responsible for resolving such queries but should report them to the client so they can be investigated and clarified by the appointed legal adviser. The surveyor is in the privileged position of seeing and examining the subject property and, in many cases, is able to ask questions of the vendor face-to-face, whereas in almost every other instance, no other adviser appointed to assist during a transaction will have those advantages. The surveyor, therefore, has a responsibility to act as the eyes and ears of the client's legal adviser and must be particularly alert to matters which may not be apparent to someone who has not seen the property.

The majority of matters will involve routine checks such as ensuring that planning consent was granted for an extension and that works were signed off as compliant with the appropriate regulations on completion. Having inspected the property, the surveyor may have little doubt that everything will prove to be in order, but it is still necessary for the legal adviser to make sure the documentation is present and complete so the client can be fully reassured that everything is as it should be. Without a comment in the surveyor's report, the legal adviser may not be aware that a property has been extended and relevant enquiries may not be made.

Arguably of greater significance are the cases where the surveyor sees something which suggests that works have not been carried out in compliance with the appropriate regulations. The implications of this are twofold. First, there are circumstances when the regulatory authority may seek to remedy the breach. This is likely to have associated legal and cost implications. Second, if works are non-compliant then there is an automatic question regarding the overall standard of whatever has been carried out: does the non-compliance involve a relatively minor detail or is it a reflection of shoddy or grossly substandard work? After that, there come questions about how the matter should be dealt with. Is there an impact on the value of the property and the price being paid? Should the vendor deal with the issue prior to completion or should the property be purchased as it stands, leaving the purchaser to remedy the matter?

Some related legal matters which come to the surveyor's attention will be regarding the extent of rights which the property owner may have or responsibilities to which the owner may be subject. Sometimes a helpful vendor will make a comment which seems to answer a query posed by the surveyor, perhaps about a right of access; then it is only a case of reporting the statement and asking that the legal adviser seeks verification. There is, it has to be admitted, some merit in treating vendors' assurances with the proverbial pinch of salt until independent confirmation can be obtained, and there will be occasions when the surveyor has the very distinct impression that a vendor is not being wholly straight about something. It is understandable that a vendor might seek to gloss over some awkwardness – perhaps a difficult neighbour or an adverse local planning issue – but it is always disconcerting when a vendor makes a statement in blatant and direct contradiction to something which the surveyor categorically knows to be the truth. At such times, the professional alarm bells ring loudly, and at the very least, the surveyor will mentally question every statement the vendor has made.

The range of possible legal issues is the same for all levels of inspection but the amount of explanation given in the report will vary depending on the level of service being provided by the surveyor. A significant proportion of the matters that will need to be checked fall into two main subject areas: regulatory matters and guarantees. After that comes a broad range of issues, the most important of which relate to the tenure of the property. The frequency with which the remainder are likely to crop up will partly depend on the types of property routinely being inspected and the locations where the surveyor practises.

Regulation

This category of legal matters reflects legislation and the legal framework which may affect the property and any work which may have been carried out or may be proposed. Typical examples include:

- details of conservation areas or whether the property is a listed building, and the extent to which either of these designations may relate to works which have been carried out or which the surveyor has recommended as necessary;
- matters relating to Planning and Building Regulation consents for alterations or extensions;
- whether any trees at or around the property are subject to Tree Preservation Orders;

- whether a mining report should be recommended in certain parts of the country;
- checking whether a remediation certificate exists on sites which may previously have been contaminated – this does not apply only to new-build sites although it is more likely to apply to them;
- if any works have been carried out under any of the various 'competent persons' schemes, the relevant certification should be checked – for example, FENSA for glazing or HETAS and Gas Safe for solid fuel and gas installations respectively;
- in some cases, it will be appropriate to check the likely future use of adjoining land or there may be proposals for significant public or private developments in the vicinity which might have an impact on the property and its future value and saleability.

Guarantees

The surveyor is not responsible for confirming whether any guarantees or warranties relating to the property are valid or whether they can be transferred to the new owner. However, if made aware of their existence or when recognising works of the type which usually benefit from guarantees, the surveyor should recommend that the client's legal adviser makes the necessary checks and, if possible, effects the transfer on completion of the purchase. The guarantees that are available include the following:

- With new and nearly new buildings, there is usually some form of certification, most often from NHBC (National House Building Council) but there are alternative providers. There are always limitations to these, and the client's legal adviser should scrutinise the terms carefully as the cover they provide is usually much less far-reaching than many people assume.
- For buildings of nontraditional or unconventional construction built since 2013, accreditation may be provided by the Buildoffsite Property Assurance Scheme (BOPAS), described in 'Accreditation and warranties' in Chapter 15. Insurance under this scheme lasts 10 or 12 years but construction details will remain available from the BOPAS website (www.bopas.org) for the lifetime of the building.
- Any major work to the structure of the property – such as underpinning, extension or conversion work, construction of a conservatory, cavity wall tie replacement, damp and timber treatment, installation of replacement glazing or cavity wall insulation – is likely to come with some form of guarantee from the installer or contractor that carried out the work. If work such as an extension was professionally supervised, the firm or person responsible may have provided certification that it was being carried out and was completed to a satisfactory standard.

- The installation or repair of the services is likely to have some form of guarantee.
- In some circumstances, it may be worthwhile for the new owner to take over the existing building insurance policy held by the previous owner, especially where a claim has been made or investigation has been carried out or is in hand in relation to structural movement. This can help avoid the problem of trying to establish the exact date on which an insurable event occurred, or could first be noticed as having occurred, and of having to determine which of two insurers might be liable in the event of a claim.

Other matters

The tenure of the property is clearly of primary importance. Even with freehold properties where matters are usually relatively straightforward, there can be complications. The difficulties of having an allocated garage or parking space in a location remote from the site of the main property are touched on in Chapter 27 and illustrated in **479**. Matters such as flying or submerged freeholds ought to be, but sometimes are not, reflected in the details of the title.

If the property is leasehold, the surveyor is entitled to make reasonable assumptions about the lease terms and assume that the lease will be checked by the legal adviser. The RICS Home Survey service has a very good two-page annex entitled *Leasehold Properties Advice* which sets out the assumptions being made by the surveyor under the terms of the service; it lists a range of questions which the client should raise with their legal adviser about leasehold properties. Those not offering RICS Home Survey products would do well to consider including something similar within their own report format. During the inspection, the surveyor may have noticed things like multiple occupation or tenancies that might affect the tenure but would not be obvious to someone simply inspecting the legal documentation, but which should be clarified before a purchase is completed.

Another important consideration, usually taken for granted but of great significance for some properties, is to determine rights and responsibilities for access. These may be shared with adjoining owners – increasingly common on modern housing developments with shared driveways – or may be across privately owned land (**504**). It is unlikely that the surveyor will be able to determine these facts, but where they are in question they should be clearly raised as matters requiring legal guidance (as described at the end of Chapter 28). Conversely, some owners may believe that they have the benefit of access rights but the surveyor may have doubts about the legitimacy of such claims, and the legal adviser should be asked to clarify the position (**505**). Sometimes the situation will be completely unclear and the surveyor will simply be unable to make any assumptions; he or she will have to report the difficulty and leave the resolution to the legal adviser (**506**).

504 Access to the first-floor flats in the background, above retail premises, is along a private shared access road leading to this service and parking area. The legal adviser should be asked to determine the access rights of the flat owners and the extent of their potential liabilities. If these are not clearly defined, the flats might be effectively unsaleable.

505 The timber panels in this hedge are gates from the rear gardens of the properties in the background to playing fields in the foreground. Is this an informal arrangement with no right of access, does the owner have a right to free access, or is it a licensed access dependent on payment of a licence fee? The surveyor may not know the answers but the client's legal adviser may not even realise the questions need to be asked if the surveyor doesn't mention the arrangement and ask for clarification.

506 There is no clue who owns or has right of access along this footpath which runs between two properties. In the absence of any information, the surveyor can only mention its presence and seek clarification from the legal adviser, not least in order to determine whether the owner of the footpath or each adjacent property has the responsibility of maintaining the respective boundaries.

Other matters that the inspector should or may need to report as requiring legal advice include:

- arrangements relating to any sustainable energy installation, such as solar PV panels, where there may be a benefit under the Feed-in Tariff scheme, but also where there may be long-term liabilities associated with the service provider that installed the panels, typically a 'roof lease' arrangement;
- any arrangements made under the Green Deal or any similar scheme where a property becomes subject to a long-term repayment arrangement following receipt of loan finance to pay for energy-related improvements;
- ownership and responsibility for maintaining property boundaries and any evidence of boundary problems, such as poorly defined boundaries, matters relating to party walls, rights of light or works in progress on adjoining land;
- the availability and status of service connections, including easements, servitudes or wayleaves, shared or private drainage and any apparent need for rights of discharge from private drainage facilities;
- private or unadopted roads, footpaths or sewers;
- hardstandings where surface water drains on to a public highway;
- requirements for parking permits;
- when protected species, such as bats or badgers are found to be present at the property.

Appendix

Further information

Images and resources from both this edition and the first can be viewed on the accompanying website at: www.routledge.com/cw/santo.

One of the difficulties with listing possible sources of further information at the conclusion of a book such as this, which has taken a broad view of the inspection process, is that all of the areas that have been considered have their own dedicated specialists, many of whom have committed their thoughts and opinions to print. The following section, therefore, only identifies some primary sources of information along with some individual publications that have been of special assistance during the preparation of this second edition or which the author commends as being of particular merit in a specific field. Whenever the practitioner needs to undertake an in-depth investigation into some form of construction or defect, therefore, it is inevitable that specialist sources not listed here will need to be consulted. It is hoped, however, that the following information may provide a pointer towards the appropriate direction, if only during the initial stages of an enquiry.

RICS

The Royal Institution of Chartered Surveyors produces a range of professional guidance and standards products for members, some of which are mandatory, while others recommend best practice or give information. There are currently five tiers of these documents, which are entitled, respectively, International Standards, Professional Statements, Codes of Practice, Guidance Notes and Information Papers. The distinctions between the tiers are important but not always fully understood so, before listing some papers of particular relevance to residential surveyors, a brief word of explanation is worthwhile.

International Standards are international, high-level, principle-based standards developed in collaboration with other relevant bodies and which are mandatory for members.

RICS Professional Statements (formerly known as **Practice Statements**) provide members with mandatory requirements or a rule that a member or firm is expected to adhere to. This term encompasses Practice Statements, Red Book Professional Standards, Global Valuation Practice Statements, Regulatory Rules, RICS Rules of Conduct and government Codes of Practice. Professional statements are relevant to professional competence in that each surveyor should be up to date and should have informed him or herself of the relevant mandatory content within a reasonable time of their publication (the 'effective date' specified on the document).

RICS Codes of Practice are approved by RICS, and endorsed by another professional body or stakeholder. They provide users with recommendations for accepted good practice as followed by conscientious practitioners. Each document states whether the codes described in the paper are mandatory or recommended good practice.

RICS Guidance Notes provide users with recommendations or approaches for accepted good practice as followed by competent and conscientious practitioners. They are not mandatory but are intended to provide advice to RICS members on aspects of their work. Where procedures are recommended for specific professional tasks, these are intended to represent 'best practice' – i.e. procedures that, in the opinion of RICS, meet a high standard of professional competence.

RICS Information Papers contain practice-based information that provides users with the latest technical information, knowledge or common findings from regulatory reviews. Information Papers may recommend or advise on professional procedure to be followed by members. They are relevant to professional competence to the extent that members should be up to date and have knowledge of Information Papers within a reasonable time of their coming into effect.

RICS cautions members that when an allegation of professional negligence is made against a surveyor, a court or tribunal may take account of any relevant Guidance Notes or Information Papers published by RICS in deciding whether or not the member has acted with reasonable competence. RICS has the strategic objective over the next few years of separating papers with mandatory content that is capable of regulation, such as Professional Statements, from those which are primarily intended to provide information. In the longer term, therefore, many papers currently published as Guidance Notes or Information Papers are likely to be refocussed, updated, and if appropriate recategorised so that it is clearer whether or not they reflect mandatory requirements.

The following RICS papers, and indeed all other such RICS papers, are available to members as free downloads from the RICS website (www.rics.org).

Building Survey, RICS Practice Note, 1st edition (2012).
Condition Report, RICS Practice Note, 1st edition (2011).
Description of the RICS Condition Report Service – RICS public information leaflet.
RICS Home Surveys Information Sheet – RICS public information leaflet.
HomeBuyer Report, England, Wales, Northern Ireland, Channel Islands and Isle of Man, RICS Practice Note, 4th edition (2010).
HomeBuyer Report, Scotland, RICS Practice Note, 1st edition (2010).
Flat Roof Coverings, RICS Information Paper, 2nd edition (2011).
The 'Mundic' Problem, RICS Guidance Note, 3rd edition (2016).
RICS Valuation – Professional Standards (the Red Book), January 2015 edition.
Surveying Safely, RICS Guidance Note, 1st edition (2011).
Surveys of residential property, RICS Guidance Note, 3rd edition (2013).
Valuation of historic buildings, RICS Information Paper, 1st edition (2014).

There is also a subscription service at isurv (www.isurv.com), an online information portal from RICS, which contains a wealth of practical information, guidance and insight on surveying issues from RICS and industry experts, including an illustrated database of building defects.

Books

Bravery, A. F., Berry, R. W., Carey, J. K. and Cooper, D. E. (2003) *Recognising Wood Rot and Insect Damage in Buildings*, 3rd edition, BRE Press.
Murdock, John R. and Murrells, Paul G. (1996) *Law of Surveys and Valuations*, Taylor & Francis.
Murdoch, John (2002) *Negligence in Valuations and Surveys*, RICS Books.
Parnham, Phil (2012) *Assessing Building Services*, RICS Books.
Parnham, Phil (2011) *A Surveyor's Guide to RICS Home Surveys*, RICS Books.
Practical Building Conservation Series (2012–2015, ten volumes), English Heritage.
Ross, Keith (2005) *Modern Methods of Construction: A Surveyor's Guide*, BRE Trust.
Swindells, David J. and Hutchings, Malcolm (1993) *A Checklist for the Structural Survey of Period Timber Framed Buildings*, A Building Conservation Group Publication, RICS Books.

Technical guidelines and leaflets

Anthrax and Historic Plaster, Managing Minor Risks in Historic Building Refurbishment (1999), English Heritage.
Bats and Buildings (2012), Bat Conservation Trust.
Bats in Roofs – A guide for Surveyors (2002), English Nature, available from: www.arborecology.co.uk/resources/batsinroofs.pdf.

Bat Mitigation Guidelines (2004), Mitchell-Jones, A. J., English Nature.

Choosing between Survey Types (2011), Independent Surveyors and Valuers Association.

Cotswold Stone Slate Roofing – Technical Guidance for Owners and Occupiers (2004), Tewkesbury Borough Council.

Domestic Building Services Compliance Guide (2013 edition – for use in England), HM Government.

The Dorset Model, Thatched Buildings – New Properties and Extensions (2009), Dorset Building Control Technical Committee.

Driving at Work: Managing Work-related Road Safety (2003), INDG382, Health and Safety Executive.

Environment Agency Regulatory Position Statement 116 v4.0 (2014), Registration and permit requirements for small sewage discharges in England, Environment Agency.

Fire Safety for Thatch, Thatch Advice Centre.

The Guide to Copper in Architecture (2006), European Copper in Architecture Campaign.

Guide to Installation of Renewable Energy Systems on Roofs of Residential Buildings (2011), NF30, NHBC Foundation.

Living with Bats – A Guide for Roost Owners (2013), Bat Conservation Trust.

Managing Change in the Historic Environment – Doorways (2010), Historic Scotland.

Natural Stone Masonry in Modern Scottish Construction – A Guide for Designers and Constructors (2008), Building Standards Division, Directorate for the Built Environment, Scottish Government.

NHBC Standards 2014, NHBC.

The Removal of External Asbestos Insulating Board (AIB) Soffits (2013), Asbestos Liaison Group.

Screeds with Underfloor Heating: Guidance for Defect-free Interface (2003), Co-Construct.

Surveying Historic Timber-Framed Buildings, Oxley Conservation.

Timber Shake Installation/Fixing Specification (2011), John Brash & Co. Limited.

Typical Details for Historic Buildings and Conservation Areas (2010), Essex County Council Historic Buildings and Conservation.

UK Guide to Good Practice in Fully Supported Metal Roofing and Cladding, 2nd Edition, Federation of Traditional Metal Roofing Contractors.

Window of Opportunity – The Environmental and Economic Benefits of Specifying Timber Window Frames (2005), WWF-UK.

Working Alone: Health and Safety Guidance on the Risks of Lone Working (2013), INDG73(rev3), Health and Safety Executive.

Research and Information Papers

A Survey of Low and Zero Carbon Technologies in New Housing (2012), NF42, NHBC Foundation.

Cellulose-based Building Materials: Use, Performance and Risk (2013), NF55, NHBC Foundation.

Here Comes the Sun: A Field Trial of Solar Water Heating Systems (2011), Energy Saving Trust.

Non-traditional Housing in the UK – A Brief Review (2002), Council of Mortgage Lenders.

Risk Assessment of Structural Impacts on Buildings of Solar Hot Water Collectors and Photovoltaic Tiles and Panels – Final Report (2010), Building Standards Division, Directorate for the Built Environment, Scottish Government.

United Kingdom Housing Energy Fact File (2013), Department of Energy and Climate Change (DECC).

Planning

The Planning Portal is the UK government's online Planning and Building Regulations resource for England and Wales: www.planningportal.gov.uk/wps/portal/portalhome/.

The Planning System in Scotland: www.scotland.gov.uk/Topics/Built-Environment/planning.

The Planning Portal for Northern Ireland: www.planningni.gov.uk/index.htm.

Building Regulations

Downloads of the most recent versions of the Approved Documents that support the 14 technical 'Parts' of the Building Regulations for England are available free from this website: www.planningportal.gov.uk/building regulations/approveddocuments/.

The 14 Approved Documents are:

Approved Document A (Structural safety)
Approved Document B (Fire safety)
Approved Document C (Resistance to Contaminants and Moisture)
Approved Document D (Toxic Substances)
Approved Document E (Resistance to Sound)
Approved Document F (Ventilation)
Approved Document G (Sanitation, Hot Water Safety and Water Efficiency)
Approved Document H (Drainage and Waste Disposal)
Approved Document J (Heat Producing Appliances)
Approved Document K (Protection from Falling)
Approved Document L (Conservation of Fuel and Power)
Approved Document M (Access to and Use of Buildings)
Approved Document N (Glazing Safety) From 6 April 2013 – only relevant to Wales
Approved Document P (Electrical Safety)

The equivalent site for Scotland is at: www.scotland.gov.uk/Topics/Built-Environment/Building/Building-standards.

For Northern Ireland: www.dfpni.gov.uk/index/buildings-energy-efficiency-buildings/building-regulations.htm.

For Wales: wales.gov.uk/topics/planning/buildingregs/publications/?lang=en.

Websites

The difficulty with providing lists of websites is that many do not stand the test of time. The majority of the following organisations, however, are well established; and even if some individual website addresses fall by the wayside, it is reasonable to expect that the most of the remainder will be accessible for the foreseeable future.

Asbestos Information Centre
    www.aic.org.uk/FAQAC.html
Association of British Insurers
    www.abi.org.uk/
Automatic Entrance Systems Installers Federation (AESIF)
    www.aesif.org.uk/Default.aspx
Bat Conservation Trust
    www.bats.org.uk/
BOPAS (Buildoffsite Property Assurance Scheme)
    www.bopas.org
The Building Societies Association
    www.bsa.org.uk
Clay Roof Tile Council
    www.clayroof.co.uk/default.htm
Copper Development Association
    www.copperalliance.org.uk/
The Council of Mortgage Lenders
    www.cml.org.uk
The Electrical Contractors' Association (ECA)
    www.eca.co.uk
The Electrical Safety Council
    www.esc.org.uk
The Environment Agency
    www.environment-agency.gov.uk
The Federation of Traditional Metal Roofing Contractors
    www.ftmrc.co.uk
Gas Safe Register
    www.gassaferegister.co.uk
Health and Safety Executive (HSE)
    www.hse.gov.uk

HETAS
  www.hetas.org.uk
Historic England
  https://historicengland.org.uk
Historic Scotland
  www.historic-scotland.gov.uk
Lead Sheet Association
  www.leadsheet.co.uk/
Microgeneration Certification Scheme (MCS)
  www.microgenerationcertificationscheme.org
Natural Slate Standards
  www.buildingconservation.com/articles/slate/slate.htm
  www.cembrit.co.uk/BS_EN_12326–12004–22281.aspx
NHBC (National House Building Council)
  www.nhbc.co.uk
NHBC Foundation
  www.nhbcfoundation.org
Oil Firing Technical Association Limited (OFTEC)
  www.oftec.org
PlantTracker – a site established by the Environment Agency, the Nature
  Locator team at the University of Bristol and the Centre for Ecology and
  Hydrology to help combat the spread of the UK's most problematic
  invasive, non-native plant species, including Japanese Knotweed.
  http://planttracker.naturelocator.org/
Private Water Supplies
  www.privatewatersupplies.gov.uk
Property Care Association (PCA)
  www.property-care.org
Solid Fuel Association
  www.solidfuel.co.uk
Thatch Advice Centre
  www.thatchadvicecentre.co.uk
Thatched roofs, the 'Dorset Model'
  https://www.dorsetforyou.com/building-control/help/technical-
  committee/thatched-roof
TRADA (Timber Research and Development Association)
  www.trada.co.uk
United Kingdom Liquid Petroleum Gas (UKLPG)
  www.uklpg.org
Water Regulations Advisory Scheme
  www.wras.co.uk

Last, but by no means least, an immensely rich and often underutilised source of information is available through Google Images, the Google product that searches for photographs and illustrations on the web. If in doubt about the identity of a fungus, plant or tree, or if wanting to see the consequences of – or possible solutions to – a building defect such as roof spread or cavity wall tie failure, queries using this search engine will usually come up with numerous, often helpful and sometimes graphic images related to the subject in question. Naturally, the standard cautions about the reliability of web-sourced information must be heeded, but the professional fascination of seeing so many variations on the theme of, say, dry rot outbreaks is such that a frequent risk is simply that of being distracted for far too long from returning to the matter which actually started the search in the first place.

Index

Note: Page numbers followed by 'f' refer to figures, followed by 'p' refer to photographs and followed by 't' refer to tables.

T - #0629 - 071024 - C566 - 246/189/29 - PB - 9780080971315 - Matt Lamination